GERMAN JET ENGINE
AND
GAS TURBINE DEVELOPMENT
1930–1945

GERMAN JET ENGINE
AND
GAS TURBINE DEVELOPMENT
1930–1945

Antony L. Kay

Airlife
England

First published in 2002 by
Airlife Publishing, an imprint of
The Crowood Press Ltd
Ramsbury, Marlborough
Wiltshire SN8 2HR

www.crowood.com

This impression 2012

© Antony L. Kay 2002

All rights reserved. No part of this publication may be reproduced or transmitted in any form or by any means, electronic or mechanical, including photocopy, recording, or any information storage and retrieval system, without permission in writing from the publishers.

British Library Cataloguing-in-Publication Data
A catalogue record for this book is available from the British Library.

ISBN 978 1 84037 294 6

This book contains rare photographs and the Publisher has made every endeavour to reproduce them to the highest quality. Some, however, have been technically impossible to reproduce to the standard that we normally demand, but have been included because of their rarity and interest value.

Typeset by Rowland Photosetting Limited, Bury St Edmunds, Suffolk

Printed and bound in India by Replika Press Ltd

Acknowledgements

The author wishes to thank most sincerely the following institutes, companies and individuals without the help of whom this history would have been the poorer. Many of the following have since died, been disbanded or absorbed.

In England:

National Lending Library for Science and Technology
Mr M. Cox and the National Gas Turbine Establishment
Armour School, Royal Armoured Corps Centre
Ministry of Technology
Cranfield Institute of Technology
The Stationery Office
Imperial War Museum
National Reference Library of Science and Invention
Sir W.G. Armstrong Whitworth & Co. (Engineers) Ltd
Flight International (Anne Tilbury)
Phillip Jarrett
Eddie J. Creek
Bill Gunston

In the USA:

Dr Hans-Joachim Pabst von Ohain
Department of the Army, Washington
Smithsonian Institution
Dipl.-Ing. Peter G. Kappus
Fairchild Hiller (Republic Aviation Division)
Ford Motor Company
Wright-Patterson Air Force Base
Library of Congress
Richard T. Eger (Luftwaffe Archives Group)
Dr Bruno W. Bruckmann
Dipl.-Ing. F.J.A. Huber
Dr-Ing. Anselm Franz

In Germany:

MAN-Turbo Gmbh
Fritz Hahn
Prof. Dr-Ing. Otto Lutz
Brown Boveri & Cie, AG
Bundesarchiv, Koblenz
Bundesamt fur Wehrtechnik und Beschaffung
Budesverband der Deutschen Luft-und Raumfahrt-industrie e.v.
Daimler-Benz AG
Deutsches Museum, Munich
Ernst Heinkel AG
Siemens AG
Deutsche Versuchsanstalt fur Luft-und Raumfahrt e.v.

In France:

SNECMA, Paris
AGARD, Paris
Dr-Ing. H. Oestrich

Regrettably, some helpers may have been inadvertently omitted from this list, but I must mention Ms Cecilie M. Kay who so willingly typed up the author's very cursive script into a decent manuscript. Special thanks also go to Richard T. Eger for so often finding new material requested by the author during the recent revision of the original manuscript.

Preface

This book was begun as a project some thirty years ago, and then, for reasons which shall remain another story, it was shelved. Thus, the reader will notice in the acknowledgements that many of the people and organisations who gave assistance no longer exist. When the project was recently revived, thanks in no small part to the kind help and interest of Bill Gunston, I had the singular experience of seeing just how much working methods had changed, thanks to the introduction of the computer, superior copying machines, plastics and all manner of new developments.

Power units of the jet engine and gas turbine types have played an enormous part in transforming the world, largely for the good. They have provided the most efficient means of power in many spheres and transformed the world of transport and travel. The world is now much more available to the everyday person, and the 'world village' in commerce has arrived. Efficient and enormous power is now available in many fields.

The inspiration for this book was that, although the jet age was largely launched in the industrial and scientific workplaces of Great Britain and Germany, only the British story (mainly centred around Sir Frank Whittle and Sir Stanley Hooker) had been told in much detail. Therefore, although I am justifiably proud of the British achievements, I felt that the complementary and fascinating German story was well worth the telling, and so here it is. The approximate time period 1930–1945 has been chosen as covering the time when serious work began in Germany until it more or less ceased with the German collapse in May 1945 at the end of the Second World War in Europe. Some space is also given to subsequent world-wide exploitation of the German work. Of course, without straying too far from the title's subject, one must not forget the independent gas turbine work in other countries such as the Soviet Union (A.M. Lyul'ka, from 1936) and Hungary (G. Jendrassik, from 1932).

In piecing together this historo-technical mosaic, original sources were almost exclusively used. Where possible, contemporary German documents were studied, but to a larger extent and almost as immediate, Allied Forces reports were also used. These intelligence reports were made by teams which followed extremely closely on the heels of the Allied armies as they captured and advanced across German-held territory and on into Berlin itself. There was an urgency to capture equipment and documents before the Germans could destroy them because it was known that there were many technological prizes to be won, not least the much-vaunted Wunder Waffen (wonder weapons).

A little later, more thorough examinations and interrogations were made and more reports written up. Later still, in the 1950s, symposia were held and more reports made. All were grist to the mill. In examining drawings from the period 1930–45, it was found that they varied enormously in quality, and some are now virtually unreadable. Often, the clearest part left on a document was the red stamp warning: Geheime Kommandosache! (Top Secret). However, within a reasonable timescale, I have made every effort to work even the most damaged drawings up to a reasonable standard of clarity, and a large number have been redrawn. On the drawing (and document) front, another difference was found between now and thirty years ago: then the information was plentiful and freely given, but now much has disappeared – often because of relocation. Where there are gaps, I will always be most interested if any reader can fill them.

My first acquaintance with one of the subjects of this work was at the sharp end of the V1 flying bombs which crossed south London, where we lived, on their way to central London. Many of these Doodlebugs, as we called them, fell short and exploded in the strip known as Buzz Bomb Alley. The nearest explosion to me was in the next street, where the houses were flattened. One always listened attentively to the distinctive sound of the bomb's pulsejet engine because, when the sound stopped, the bomb dived straight down. That was in 1944.

Finally, I hope the reader has as much enjoyment reading or delving into this history as I have had in piecing it together.

Antony L. Kay
Selsey, West Sussex.

Contents

Foreword		8
Section 1	Introduction	9
Section 2	Gas Turbines for Aviation	15
Section 3	Gas Turbines for Land Traction	156
Section 4	Marine Gas Turbine Units	174
Section 5	Gas Turbines for Industry	192
Section 6	Gas Turbine Research and Development	209
Section 7	The Jet Helicopter	231
Section 8	Pulsejets for Aviation	238
Section 9	Ramjets for Aviation	263
Chronology of German Jet Engine and Gas Turbine Development		289
Sources and Bibliography		291
Appendix	Preserved examples of engines in the world	293
Index		295

Foreword

Dr G. M. Lewis, CBE MA FEng FRAeS
(Formerly Technical Director Rolls Royce plc)

The first flight of a turbojet-powered aircraft took place in Germany in August 1939, and six years later over 6,000 jet engines had been produced. Behind these events lies a story of immense engineering design and development effort, encouraged and supported but sometimes frustrated by the interplay of government and industrial politics. This thoroughly researched book details the evolution of jet propulsion in Germany from the early concepts through to the end of World War II and includes a fresh insight into the parallel work on gas turbines for tanks, ships and industrial purposes, jet helicopters and the pulsejet engine for the V1 missile. The chapter on ramjets provides a link with the world-leading advances being made in supersonic flight.

The contemporary activity in the UK, spanning a similar period of time, is so far not as comprehensively documented except for the pioneering work of the late Sir Frank Whittle. There is much similarity in the engineering approach to the many technical challenges in the two countries but the contrast is in the level of government support and in the acceptance of new ideas by the established aero engine companies. On the 50th anniversary of the first flight with the jet engine designed by Dr Hans von Ohain, I recall him remarking that if Whittle had enjoyed the level of support he had received in the early days, the RAF could have had a jet fighter in service some three years earlier than the Luftwaffe.

My own perspective dates back to the end of 1945 with the Bristol Aeroplane Company when my task was the preliminary design of the Olympus jet engine, later developments of which power the Concorde and many naval vessels. The rate of progress was such that the requirement was for over four times the thrust and 40 per cent better fuel efficiency compared with production engines then current. The available design data was sparse but of excellent quality, in the form of reports from the Royal Aircraft Establishment and Power Jets. As the design progressed, volumes of translated German research tests became available from a bureau of the US Navy, and these extensive and systematic test data were found to validate and refine our design rules against a very much wider data base. From this time onwards there was little direct influence on British engine developments from the German information. This was in contrast with the contribution made by German engineers to post-war jet engine development in Russia, France and the USA.

This valuable book provides an insight into the management and process of engineering development in a period of continuous innovation, intense time pressure and in a climate of escalating industrial disruption. They failed by a narrow margin to turn the tide of the air war against the Allied forces.

Section 1

Introduction

The early pioneers (1877–1934) — Government supporting organisations (1933–1945) — Methods of government procurement — Notes on data and terminology

By the time the Second World War had ended in 1945, the vanquished Germans had achieved many firsts in the field of jet engine and gas turbine technology. They had flown the world's first turbojet- and pulsejet-powered aircraft and got the first jet fighters, bombers, and flying bombs into production and operational service. In addition, they had flown the world's first jet helicopter and ramjet missiles, had installed gas turbines in industry and were actively engaged in projects for marine and land vehicle gas turbine units. It was, in fact, an age of new technical developments. To mention but a few contemporary developments, the V2 (A4) rocket had flown to the edge of space; radar, infra-red and other electronic techniques had been developed; a range of synthetic products was available; and the atomic bomb was a reality.

All these developments were largely caused by the war, accelerated in the last decade before 1945, but the total time span from when a scientific discovery is made until it is turned into something which affects people's lives has very often been about fifty years. This observation has been made frequently and can be roughly applied to the general families of jet engines and gas turbines, the German development of which this book describes in detail. We look not only at the origins and development of many engines and schemes, but, where appropriate, at their production, service life and subsequent post-war fate. The people and policies involved are not neglected either.

The types of engine covered are all of the air-breathing type, rocket types (which carry their own oxygen to support combustion) being excluded except where they are specifically connected with the jet engine or gas turbine. These latter terms, given in the book's title, are very broad indeed, but are taken to mean an engine which burns fuel supported by oxygen drawn from the atmosphere to provide hot gases which perform work by a thrust on the engine and/or give mechanical power via a turbine wheel. Such engines today, if not yet reigning supreme, are performing ever greater amounts of work in all kinds of fields where previously other types of engine, such as the reciprocating piston (using petrol, oil or steam), held sway. Most obvious, of course, is the turbojet engine which dominates the powering of aircraft today.

Given this important revolution, the reader may well wonder why this history should be devoted to the German side instead of giving a global picture. In fact, there are already a number of books which ably review the global scene, although, even then, the main concentration has to be on British and German work, at least up to the late 1940s. Other books deal only with the British side, but none, until now, has attempted to give a comprehensive (to say 'definitive' would be arrogant) picture of the German work alone.

Notwithstanding this, the German story is a fascinating one because of the range of work tackled and because, for most of the time, this work was performed in a country isolated by war and under the heel of the peculiar Nazi regime. For Germany, the Second World War began on 1 September 1939 and ended on 7 May 1945. During this period the bulk of the country's jet engine and gas turbine work was carried out, almost exclusively for the purpose of maintaining or increasing its military power. As the war situation moved from successful into critical and then desperate periods, so greater efforts were made, until the slim chance of a technical and military breakthrough resulting in time from advanced or obscure projects was preferable to the more certain chance of total defeat.

If, then, war fostered the development of invention, curiosity (rather than necessity) was the mother of invention in the first place. To find out who was first curious in the fields of jet engines and gas turbines, we have to go back to the beginning of, and even before, the twentieth century, when pioneers and inventors were dreaming up schemes and sometimes performing experiments. Their ideas covered a vast field and involved all types and combinations of steam and gas turbine and jet engine. All sorts of applications were in mind, but the big problem for the first serious engineer developers was to choose a promising scheme to work on. The first pioneers were scattered all over the world, but it will, perhaps, be illuminating to look at some of the more interesting ideas which formed the background to later work and could have had an early influence on German workers in particular. For this purpose, a few patents taken out in Germany before 1933 will be mentioned.

The early pioneers (1877-1934)

In the following, each bracketed number is that of the relevant patent (Patentschrift) of the Reichspatentamt, but some patents were also filed abroad, while some foreign patents were also filed in Germany.

The earliest of all German patents on gas turbines (2023) is that of Joseph Wertheim of Frankfurt, covering an *Atmospheric Gas Engine* in which a mixture of gas and air is exploded in a turbine: it was granted in 1877. In the following year, a patent (3944) was granted to Tecklenburg of Darmstadt for a *Gas Motor* which incorporated an explosion turbine operating on gasoline, this being the first patent covering an open combustion chamber. The earliest German patent for a constant-volume explosion turbine is that of Christian Broeker of Mannheim, this patent (53322) being granted in 1890 and specifying a closed combustion chamber in which a gas and air mixture exploded. More interesting is the patent (89297) granted in 1896 to August Rohrbach of Erfurt for his *Turbine driven by expanding air*, which was of the constant-pressure type of gas turbine.

Although such ideas continued to appear and were patented, it was not until 1908 that the first German gas turbine was built. In that year, Hans Holzwarth (whose studies went back to at least 1905) had a constant-volume type of gas turbine unit built. This complicated type never became a serious competitor of the more practical constant-pressure gas turbine, although Holzwarth (and a few others) pursued his ideas with considerable vigour and some of his units were installed in industrial works. In the meantime, innumerable other ideas were being patented.

In 1925 a patent (409,743) was granted to Wilhelm Pape of the Bergmann electrical company, Berlin, for his version of a combustion turbine with an explosion chamber in the circuit. Karl F. Leich of Hamburg took out various patents for combustion turbines also, one example (427,985) being granted in 1926. A curious *Turbine with revolving combustion chambers arranged on a wheel crown* was patented (476,033) by Oscar Hart and Joseph Hetterick of Munich in 1929. Somewhat similar in principle was the patent (454,003) granted to Karl Enders of Dresden in 1930 for his *Combustion turbine with combustion chambers surrounding the shaft*.

The close affiliation at that time between steam turbines and proposed gas or combustion turbines is often noted, and some designs merely specified a 'working fluid'. An example of this is the design for a *Counter-current steam or gas turbine with chiefly axial admission of fluid* patented (502,555) in 1930 by a prolific Karl Roeder of Hannover. Other designs strove to combine the piston engine with a turbine, such as in the patent (545,768) of Georg Dreher of Berlin, whereby a *Piston engine and turbine work together in such a way that a part of the piston engine combustion gases impinge upon the turbine*, granted in 1930. On a more practical level, many combination units were designed and worked on whereby a turbine was used to recover energy from waste gases, such work paying good dividends eventually in the field of turbo-superchargers for piston engines. In this field, many German companies, such as BMW, Rheinmetall, Junkers, and Argus, were active.

Outside Germany, inventors and designers were also busy with ideas for combustion turbines and reaction or jet engines of all kinds, and many patents were granted in Germany by about 1933. From Rumania, Hermann Oberth (later famous for his pioneering rocketry work) was granted a patent (429,462) for a *Combustion turbine with auxiliary liquid* in 1926. From England, Lilian Farrow filed a patent (519,018) in 1928 for a *Combustion turbine with a blower at one end of the casing and a turbine rotor at the other end*. In the same year, Robert E. Lasley of the USA filed a design for a *Gas-steam turbine* which was granted patent (560,273) in 1932. Very active in turbine design at this time was Alfred Lysholm of Sweden, and much of his work was specifically aimed at aircraft propulsion, one such design being filed in 1933 and granted patent (672,114) in 1939. Frederick Ljungstrom also worked in this field, supported Lysholm and filed patents in Germany.

Remarkably, the design of the *world's first turboprop engine*, a 100 hp unit, was begun in 1932 by György Jendrassik of the Ganz wagon works in Budapest, Hungary. This workmanlike engine ran in 1937 and was followed by the Cs-1 1,000 hp turboprop which ran in August 1940. Of considerable promise, the Cs-1 had a 15-stage compressor and an 11-stage turbine, but by 1941 it had been buried for good by the war.

Much work was performed in France, some of the earliest being by Victor Karavodine on explosion gas generators for turbines. René Leduc (later noted for his ramjet work) filed a design for a gas turbine in Germany in 1934, and this was granted a patent (665,954) in 1938, this being the same year when Joseph Loeffler of Czechoslovakia was granted a patent (664,169) for his *Explosion turbine*. Earlier, in 1934, Marcel de Laderriere of France was granted a patent (605,003) for his *Combustion turbine with two coaxial shafts*. French practical work of particular note which did not go unstudied in Germany was that of René Armengand and C. Lemale, who developed a gas turbine between 1900 and 1906, although this gave but a small amount of useful power and was only about three per cent efficient. Work which later provided useful data in the designing of gas turbines was performed in France by Lucie and Auguste Rateau on turbo-superchargers for piston engines.

The foregoing examples formed but a small amount of the data and ideas which were available, through careful patent file searches and other study, not only to German engineers but engineers in general by 1934. To enumerate more would be to labour the point. The breakthrough finally came in Germany when Hans von Ohain's work led to the first true flight of a turbojet-powered aircraft (the Heinkel He 178) on 27 August 1939. Von Ohain's study of the gas turbine for the jet propulsion of aircraft was begun in the early 1930s but, curiously enough, he did not begin with a comprehensive picture of the work and study of previous pioneers. For example, when in 1934 he made a search through the patent records, it seems possible that he did not come across Whittle's patent of 1930, Guillaume's axial turbojet patent of 1923 or the Swedish Milo turbojet patents. The pioneering work in England of Frank Whittle on jet propulsion is widely known, but considerable credit must also go to René Anxionnaz of the Société Rateau. Remarkably advanced was Anxionnaz's French patent 864,397 which was filed in December 1939 for a bypass turbojet engine.

Both Whittle and Anxionnaz were contemporaries of von Ohain, but the latter was the first to demonstrate (albeit secretly from the world) the turbojet principle, and he thereby opened up the field for jet propulsion and enhanced the possibilities for the gas turbine. Some of the names so far mentioned will appear again on subsequent pages, but most of the early inventors were destined to sink into obscurity, as is the regrettable fate of most inventors. Their efforts, usually born of curiosity, were frustrated for many reasons, including impracticality, prematureness, lack of support, distrust of new ideas by others and lack of necessity or application. The introduction of a new machine requires 10 per cent invention and 90 per cent innovation, the latter being provided largely by hard, practical work and financial support. Most of the support for German jet engine and gas turbine development eventually came from government organisations. Therefore, to avoid confusing the uninitiated reader later on, a glance at the structure of these organisations will now be made.

Government supporting organisations (1933–1945)

The following glance at German government organisations, as far as they relate to this history, concentrates on the aeronautical field, since it was from here, and chiefly at a later date, that developments in other fields emanated. Following the First World War, government support for aeronautical developments came from the innocuous-sounding Verkehrsministerium (Communications Ministry), which, in 1931, handed out the first jet engine research contract to Paul Schmidt for pulsejet work. When Adolf Hitler and his Nazi party came to power, Hermann Goering was appointed Reichskommissär for aviation, and soon after, on 29 April 1933, the Reichsluftfahrtministerium (RLM), or German Aviation Ministry, was established (in place of the Verkehrsministerium), with Goering as its head. The RLM went through many changes even before 1939, and at the beginning there were few men in it with any great technical ability. As a first step towards rectifying this deficiency, an engineer corps was set up under Dipl.-Ing. Guenther Bock, but his corps was not a success, few were keen to join it and, from 1942, it was gradually disbanded. In 1935, the Hoehere Luftwaffenschule was established at Gatow, near Berlin, to train officers in all aspects of maintaining the Luftwaffe, and by this time, the RLM was worked into some sort of shape whereby it could control all aspects of German research, development and production relating to air power. In simplified form, the RLM structure from 1935 was:

1935–7

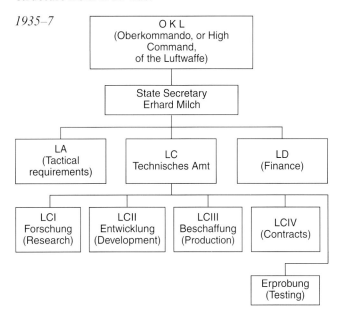

The most important feature of this organisation was the establishment of the Technisches Amt (Technical Office), Section LCII.2 of which began supporting Helmut Walter's development work on ATO rockets for aircraft in 1937. At the same time, research section LCI continued supporting not only Schmidt's pulsejet research work but also a ramjet that Walter was working on.

In 1936, Ernst Udet was ordered by Goering to take charge of the Technisches Amt, and Udet supplanted Bock as chief engineer. To cater for the needs of an expanding Luftwaffe (air force), a reorganisation took place in 1937 which increased the horizontal link between research, development, etc., by linking their various sub-sections (airframes, engines, guns, etc.). This led to the formation of seven professional technical divisions (Fachabteilungen LC7 to LC13) which each dealt with its own development and production affairs. Simplified, the new structure was as illustrated in the 1937–8 family tree overleaf.

This proved to be an interim organisation which only lasted a year and was replaced in 1938 by the so-called Generalluftzeugmeister, or GL, organisation, which was to remain the fundamental organisation for some years. From the 1938–41 chart overleaf it will be noted that the formation of the GL gave seven staff divisions (Stabs Abteilungen-GL1 to GL7) and also three directorates, one of which was the Technisches Amt (GL/C).

1937–8

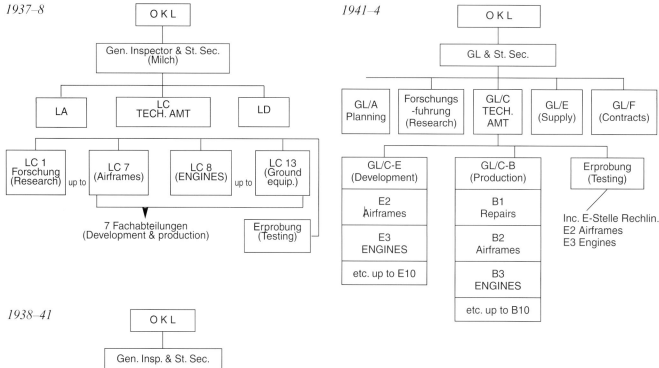

1938–41

1941–4

Development, production and supply matters were handled vertically throughout sections GL/C2 to C8 of the Technisches Amt (section GL/C1 handling research), as well as by the GL/E directorate. Already, the link between the research section and the various development departments was less close than before.

During 1941, the bad turn in the war for Germany brought about the formation under Albert Speer of the Ministerium für Bewaffung und Munition (Speer), or Ministry for Armaments and Ammunition – later designated R.f.Ruk. This ministry took over the main responsibility for military supplies and reduced the RLM's responsibility in this field by about two-thirds, but the subsequent reorganisation of the RLM was more radical than might be supposed. The principal changes included the merging of the Generalluftzeugmeister with the Secretary of State (both posts being held by General Erhard Milch), the inclusion of the Stabs Abteilungen under one directorate of projects (GL/A Planungamt) and the placing of the Forschungsfuhrung (research) on the same horizontal level as the GL/C Technisches Amt.

The final reorganisation of the RLM occurred on 1 August 1944, due to all departments connected with production being transferred to the Speer Ministry, and the GL organisation was then known as the Chef der Technischen Luft Rüstung, or Chef TLR (Chief for Technical Air Armament). Thus, the final pattern of the RLM structure, so far as engines were concerned, is shown below. By this time, the now incompetent Goering had little influence on matters, and he was even replaced, on 25 April 1945, by von Greim as C.-in-C. of the Luftwaffe. In addition, after the July 1944 plot to kill Hitler, the SS Hauptamt unofficially took over control of the RLM, and a certain proportion of all officers, engineers and scientists were forced to join the SS, even if this only meant paying lip service.

The reader should realise that the foregoing charts show only a small part of the RLM structure, and that similar, or more complex, structures existed for the other services. The ministries supporting research, development and procurement for the other services were the Heereswaffenamt, or HWA (Army Ordnance Board), for the Army and Kriegsmarine Technischesamt, or KTA (Naval Technical Office), for the Navy. All the

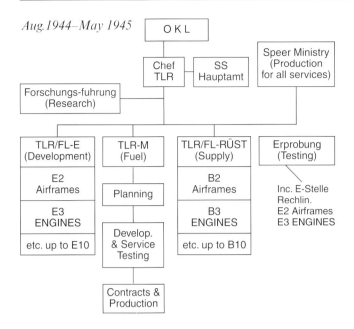

Aug.1944–May 1945

services came under the command of the Oberkommando der Wehrmacht, or OKW (High Command of the Armed Forces), which was created on 4 February 1938 by Hitler and, naturally, commanded by him. A fourth service, the SS, increasingly encroached into all spheres as its sinister power increased under the machinations of Heinrich Himmler and his henchmen.

Methods of government procurement

The general methods by which the government procured and supported the development of equipment (e.g. engines) for service use are briefly examined here, again in relation to aeronautical matters.

Institution of a development policy for the RLM came from the head of its technical department (Technisches Amt, GL and then, by 1944, the Chef TLR), and this policy was carried out by the various sections concerned. In the case of engines, the development section was LCII-2 in 1935, LC8 in 1937, GL/C3 in 1938, GL/C-E3 in 1941 and TLR/FL-E3 in 1944. The development section drew up its specifications (usually dictated by tactical requirements passed down from the OKL) and approached the chosen companies within industry with a view to issuing a development contract. To a large extent, a company was oriented towards one of the armed services. Following an approved design, the company would put in hand a number of engine prototypes (V-series) which would be tested by both the company and at the RLM's test establishment (Erprobungstelle). If successful, pre-production engines (O-series) would be built and tested at the official establishments and in service trials. Once the engine reached the production stage (initially, A-series), it came within the jurisdiction of the appropriate production department.

There were seven official testing establishments (under Oberst Ing. Petersen) the oldest of which was E-Stelle Rechlin (commanded by Major Behrens), which carried the responsibility for aircraft and engines. If Rechlin recommended acceptance to the appropriate development section, production was invariably initiated. The seven divisions at Rechlin covered aspects similar to those of the similarly numbered sections within the Technisches Amt. E3 Rechlin covered the testing of engines (except rocket motors) and, as an example, its official type test for turbojet engines was approximately as follows:

Fighter condition:

(i)   5 minutes at full throttle
(ii) 5 minutes at 80 per cent maximum rpm
(iii) 15 minutes at full throttle
(iv) 10 minutes at 80 per cent maximum rpm
Repeat cycle for 100 hours running 3 minutes at 110 per cent thrust during every five hours. Cold start to idle in one minute. Idle to full throttle in six seconds.

Bomber condition:

(i) 5 minutes at full throttle
(ii) 30 minutes at 80 per cent maximum rpm
Repeat cycle for 100 hours running three minutes at 110 per cent thrust during every five hours. Cold start to idle in one minute. Idle to full throttle in six seconds.

Production engines were subject to the same schedule but for a one-hour running period only. Apart from these static tests by E3 Rechlin, there were also flight tests performed by E2 Rechlin.

Research problems, either basic or which arose out of development or testing work, were put to an appropriate research institution or solved by the company concerned. Contracts for basic research work usually originated from the Forschungsfuhrung, although, from 1941, this department had an increasingly tenuous connection with the RLM since it was thought that research should exist as a separate entity. By the latter half of 1944, the SS Hauptamt was working alongside the Speer Ministry (though certainly not in any amicable sense) and controlled all politically unreliable scientists and engineers. By this time, the SS, Army and Navy were pursuing their own programmes to produce gas-turbine-powered vehicles and boats.

To conclude, we can say that German development or research programmes fell into one of three categories:

(i)   In a backwater, sporadically financed by military organisations or drawing on general funds of the company or institution concerned.
(ii) Were seen as worthwhile and were actively supported by a military organisation, hoping that full priority would be given later.
(iii) Were given full priority by having the interest of Hitler. This could be a two-edged sword, however,

because Hitler would sometimes affect the work programme or its outcome by making 'intuitive' decisions without regard to military, industrial or scientific reality, while, at the same time, ignoring the suggestions of his advisers and experts.

German jet engine and gas turbine developments in many cases (e.g. the turbojet and pulsejet) passed through all three categories, which was quite a normal process unless Hitler interfered. Each of the following eight sections of this book is devoted to a particular type of engine or field of application, and is broken down into companies or institutions in, as far as possible, chronological order. Each section has a conclusion (which glances at the post-war fate of the work), but Section 2 has a conclusion after each company because of the length of that section.

Notes on data and terminology

Because of the diverse sources of information, strictly comparative data cannot be given throughout, and, indeed, there are many gaps yet to be filled. In addition, discrepancies were often found between different sources so that, where official German figures could not be found, a decision sometimes had to be made on the basis of collation, calculation and deduction. All data are given in metric terms with British equivalents in parenthesis. The following items of data may need explanation:

Thrust:	Given in kp (kilopounds), which is simply thrust in kilograms
Speed:	(e.g. 9,500 rpm). Given at full load or maximum speed
Weight:	Sometimes dry, sometimes not, but impossible always to qualify
Pressure ratio:	Given at full rpm (or thrust) and static
Airflow:	Air mass flow
Specific fuel consumption:	One figure denotes kg/kp.h or lb/lb.h and is given for static thrust and sea level conditions
Specific Weight:	Weight per unit thrust given for static, sea level conditions

Unless stated otherwise, the above qualifications apply.

On the question of terminology, the Germans gradually evolved a useful system of abbreviations which were used by both companies and official bodies. The most common abbreviations follow:

GT	Gas turbine
TL	Turbinenluftstrahl (turbojet)
TLR	TL combined with rocket (Rakete) booster
TLS	TL with afterburning or reheat
PTL	Propellerturbinenluftstrahl (turboprop)
ML	Motorenluftstrahl (piston-driven ducted fan)
MLS	ML with afterburning or re-heat
MTL	Ducted fan with combined piston engine and turbine drive
L or S	Lorin-Triebwerk or Staustrahltriebwerk (subsonic ramjet or athodyd)
Tr	Trommsdorf-Triebwerk (Trommsdorf drive or supersonic ramjet)
IL	Intermittierendes-Luftstrahl (pulsejet or aeroresonator)

Thus, for example, jet propulsion by turbojet is referred to as TL-Antrieb and a turbojet engine as a TL-Triebwerk or TL-Gerät. Other, less common, abbreviations (e.g. ZTL, PTLR, LR, RL) will be found in the text.

Section 2

Gas Turbines for Aviation

Ernst Heinkel AG — Junkers Flugzeug-und Motorenwerke AG — Bayerische Motoren Werke AG (BMW) — Daimler-Benz AG — Dr-Ing. h.c. F. Porsche KG

By far the greatest German effort in the gas turbine and jet engine field went towards the development of gas turbines to power aircraft. In this form the gas turbine was used in turbojet engines and was planned to be used in turboprop and other hybrid engines. Such engines are entirely rotary and are capable of higher speeds and powers than the reciprocating piston engine. A mixture of fuel and air is burnt in a combustion chamber (or chambers) to provide a hot gas stream. The acceleration of the airflow through the engine provides a propulsive thrust, but some of the energy in the hot gases is extracted by the turbine, which drives a compressor forcing compressed air into the engine. If, in addition, airscrews are to be driven to provide a turboprop engine, more power must be extracted by the turbine, leaving less energy for purely jet thrust. There are, of course, many arrangements and designs for even the main components, such as the compressor, combustion system and turbine.

To develop the turbojet engine, which preceded the turboprop and other types of engine, required not only a certain level of technical development and ability but also considerable resources to pay for skilled teams and special equipment. Following her defeat in the First World War, Germany's aviation industry suffered a decline, but by 1926 ways were being found around the restricting Treaty of Versailles and the industry was again picking up. By 1933, when Hitler's Nazi party came to power and brought with it general expansion, the industry was among the best in the world and included the aero-engine companies of Argus, BMW, Bramo, Daimler-Benz, Hirth, Junkers and Walter. These companies were, in general, still working hard to produce reliable piston engines of high power, but a few engines in the 800 hp class were ready for production (some being based on foreign designs) or were under development. For such development work, government support was available. Few at that time had any interest in jet engines (although, as we shall see in another section, a research contract was given to Paul Schmidt for his pulsejet work), one reason being that piston engines could give an adequate performance into the foreseeable future. A few engineers and scientists probably realised that the turbojet engine would need many years of research and development before it became usable, but only the government could initiate a long-term programme in anticipation of future, high-performance aircraft.

The fact that Germany had an extensive turbojet programme at all was due to the early efforts of a few men within the RLM. In August 1937, Helmut Schelp was sent to research section LC1 of the Technisches Amt, where he was given charge of the pulsejet and ramjet projects of Schmidt and Walter that were already being financed. However, Schelp's enthusiasm was for gas turbines, and this did a great deal initially to foster and finally form the RLM's jet engine programme. By the end of the war in 1945, Schelp was one of the most knowledgeable men in Germany concerning the overall picture of the programme, although his technical knowledge was general rather than particular. He received a master's degree in engineering at Stevens University at Hoboken, USA, in 1936. In that year he returned to Germany and was selected to follow a new advanced course in aeronautical engineering at the DVL research institution, Berlin, this course covering all aspects of aeronautical engineering plus pilot training. The idea of this course was to provide a much-needed nucleus of skilled men destined for positions of authority. With the successful completion of his training, Schelp received his civilian technical degree of Diploma Ingenieur and, for entry into government service, received the rank of Baumeister. (Eventually he rose to the rank of Oberstabs Ingenieur, which was roughly equivalent to a British Lt.-Colonel.)

During the independent study part of his course at the DVL, Schelp theorised on the maximum speed possible for an aircraft, and concluded that, owing to compressibility effects, this would be Mach 0.82, or well below the speed of sound. The next step was to consider types of engine which might power an aircraft up to this speed. Advocates of airscrews for high-speed flight, including Prof. Quick of the DVL, maintained that a maximum efficiency of 71 per cent could be expected for an airscrew at Mach 0.82. In addition, it appeared that the large amount of power needed for the airscrew would result in an unacceptably heavy piston engine, and although Schelp considered piston-driven ducted fans, these too were discarded for weight reasons. Undoubtedly, Schelp was familiar with the work and studies of all the early pioneers mentioned already, and more besides, and he made a careful study of all types of jet propulsion systems which might be suitable for high-speed flight. By the time he transferred to the Technisches Amt of the RLM, he had firmly decided that power units incorporating the gas turbine were the type to pursue.

His enthusiasm was not shared by the DVL or by the LC1 research section in the Technisches Amt, but by September 1938 Schelp had transferred from this section and moved to the engine development section LC8 (which soon became GL/C3). This section was headed by Wolfram Eisenlohr, who at first gave no priority to jet engines, but Schelp soon found an ally in Hans A. Mauch who headed the rocket development section of GL/C3. Mauch's interest sprang from what he had seen of the early Heinkel turbojet work by von Ohain and discussions he had held with the Junkers airframe company. (It is interesting that both these companies were in the airframe and not the engine field.) Nevertheless, official support for turbojet development was denied to these companies precisely because Mauch considered that such radical development should be entrusted to established engine companies, and this was quite understandable. Accordingly, he and Schelp consulted the companies of BMW, Bramo, Daimler-Benz and the Junkers engine division on the question of developing the turbojet and kindred units, but they were greeted with little enthusiasm because of the considered immense technical problems involved. Also, the industry was anxious to concentrate on the full development of its piston engines, and this made sense. Eventually, as we shall see, all the main aero-engine companies, plus some others, worked on turbojet engines and in accordance with the RLM programme.

One of the strongest technical arguments against the turbojet, that axial compressors of suitable efficiency did not exist, was largely overcome by promising research conducted by Ludwig Prandtl, W. Encke and A. Betz at the AVA research institution at Göttingen. The axial compressor, as opposed to the centrifugal compressor, almost entirely dominated German gas turbine design, and was chosen in the beginning by Schelp for turbojets because of the smaller frontal area and less drag it offered. Another design feature which met with official favour, but less universal acceptance, was the annular combustion chamber. On the other hand, another factor which affected the character of German turbojets, most especially their turbines, had more inescapable logic behind it: this was the increasing shortage of imported special metals with which to make heat-resisting steels. A glance at the following table showing the monthly depletion of these special metals, issued by the Speer Ministry in November 1943, makes the situation clear:

	Chromium	Silicon	Molybdenum	Nickel	Manganese
Monthly consumption (tonnes)	3,751	7,000	69.5	750	15,500
Monthly imports (tonnes)*	–	4,200	15.5	190	8,100
Monthly deficit made up from stocks (tonnes)	3,751	2,800	54.0	560	7,400
Month's reserve of stocks (tonnes)	5.6	6.4	7.9	10.7	18.9

*Because of the war situation, imports from the Balkans, Turkey, Nikopol (USSR), Finland and northern Norway were not included, although important shipments of chromium (Turkey), nickel (Finland) and manganese (Nikopol) were still forthcoming at that time.

In fact, equipment such as turbojets had only a very small share of the scarce metals, while the armaments industry, which particularly needed chromium, had the major share. How the German turbojet succeeded in the face of this restriction will be recounted later, but in the beginning at least, development was not hampered. By the end of 1939, Mauch had set in motion a programme of jet engine study and development. Although this programme was opposed by Eisenlohr, who favoured a long-term project, he was overruled by Ernst Udet, who was head of the Technisches Amt. However, by 1941, Eisenlohr too was convinced of the desirability of turbojet development. In the meantime, the programme had gone ahead primarily with Junkers working on the 109–004 engine, destined to attain production status speedily, and BMW working on a more-advanced substitute, the 109–003 engine. (German air-breathing jet engines were officially prefixed by '109' with a second number, between 1 and 499, relating the engine to the company concerned.)

The two turbojet engines mentioned above, which were the only two to reach production status, fell within the official Class I. This class is shown with the others in the following table laid down by the Technisches Amt as a broad basis for aviation gas turbine development:

Class	Approx. thrust (kp) for turbojet	Pressure ratio	Turbine stages	
			Turbo-Jet (TL)	Turbo-prop (PTL)
I	Up to 1,000	3.5/1	1	–
II	1,300 to 1,700	5.0/1	2	3
III	2,500 to 3,000	6.0/1	2	3
IV	3,500 to 4,000	7.0/1	3	5

The plan was to develop turboprop engines from the larger turbojet engines with extra turbine stage/s added for the airscrew drive. It was also intended that the same basic engine would be developed into a ducted fan engine to economically cater for aircraft ranges too long for the turbojet and too short for the turboprop, but this part of the programme received scant attention. Schelp's department, with the key men being Walter Brisken and Emil Waldmann, carried out a lot of studies on turboprop engines after their extensive investigations into turbojets. Much of their study centred around turboprops with heat exchangers for speeds up to 600 km/h (373 mph) at a 10,000 m (32,800 ft) altitude. Although the industry took up the development of such engines, the time span (16 years) of Schelp's carefully worked-out development schedule extended beyond the end of the war and could not be covered.

Since a chief purpose of the turbojet was to attain higher speeds, the aerodynamic side could not be neglected, and indeed posed greater problems in later years. Therefore, while the official jet engine programme was being formulated at the end of 1938, a corresponding programme for high-speed airframes was also being formulated by Hans M. Antz of the Technische Amt's airframe development section LC7 (soon redesignated GL/C2). As far as possible, the two programmes were related, one influencing the other. Although Heinkel was already privately working on experimental jet and rocket aircraft, Antz made his first industrial contact with the Messerschmitt company. Nevertheless, this proved to be a fruitful contact, since, despite technical and political difficulties, the Messerschmitt Me 262 twin-jet fighter went into service use before the war ended. On a minor scale, the Arado Ar 234 jet bomber and the Heinkel He 162 jet fighter also went into service, though the latter type saw no action. Many other jet aircraft were projected, designed or under development.

Before leaving the RLM's jet engine programme, mention must be made of a setback and its sequel. The setback began when the hard-pressed Ernst Udet committed suicide on 17 November 1941. Udet, with his enthusiasm for aviation and radical developments, coupled with his influential position within the RLM, had done much to overcome opposition to jet engines.

With his demise, decisions on development and production were taken by Erhard Milch, who favoured impressing Hitler with large production figures for conventional equipment, while giving little thought to jet aircraft. Hitler was duly impressed, at least at first, and also gave little thought to jet aircraft. Now the Me 262 was ordered into production on 5 June 1943, but much delay was caused through Allied bombing and lack of priority over obtaining the necessary manpower and facilities. In November 1943, when the worsening war situation was silencing the last opponents of jet aircraft anyway, a competent individual was again in a position of authority. This individual was Oberst Ing. Siegfried Knemeyer, who moved from his position as a technical (electronics) adviser to Goering to become head of the airframe and engine development section (GL/C-E, later known as TLR/FL-E) within the RLM. In this position he became responsible not only for the development of all aircraft and armaments for the Luftwaffe but also for air-launched guided missiles.

Knemeyer was qualified in the subjects of electronics, aerodynamics and navigation, he had worked in industry, been trained as an operational pilot and had flown on high-altitude reconnaissance missions. He urged that a speed superiority of at least 150 km/h (93 mph) over enemy aircraft was essential in view of Germany's critical air defence position, and he proposed the abandonment of most conventional aircraft in order to concentrate on jet fighters and bombers. While Goering agreed with Knemeyer, Hitler was swayed by Goebbels and Bormann to continue with conventional aircraft production, which suited the production chiefs very well. Production of piston-aircraft therefore soared to record heights, although there were not enough pilots or fuel and training programmes were being drastically abbreviated. Against such opposition, Knemeyer's policies only gained ground slowly, and it was not until after September 1944, when the first operational Me 262 fighter unit had been formed, that the real importance of jet aircraft dawned on Hitler. Finally, in March 1945, the jet programme moved to a position of absolute priority under the ruthless Obergruppenführer (SS General) Hans Kammler, but by then all advantage had been lost and the war had only a matter of weeks to run.

Ernst Heinkel AG

HeS 2 (TL) — HeS 3 (TL) — HeS 6 (TL) — HeS 8 or 109–001 (TL) — HeS 8 developments (TL and ZTL) — Expansion of the Heinkel turbojet programme — HeS 30 or 109–006 (TL) — The constant-volume HeS 40 (TL) — The Mueller ducted-fan engines (ML and MTL) — HeS 50 (ML and MLS) — HeS 60 (MTL) — Focus on the HeS 011 or 109–011 (TL) at Zuffenhausen — 109–011 V1 to V5 series — 109–011 V6 to V25 series — 109–011 A-0 — Planned application of the 109–011 A — The Tuttlingen engine — 109–021 (PTL) — Conclusion

The Ernst Heinkel AG, a well-established airframe company, but with no experience of engine development, was the first German company to begin practical development of the gas turbine for the purpose of jet-propelling an aircraft. The man who inspired the development was Dr Hans-Joachim Pabst von Ohain, who was born in Dessau on 14 December 1911. His father prospered in Berlin as a light bulb distributor. As a student of applied physics and aerodynamics at the University of Göttingen, von Ohain was also an enthusiastic glider pilot, an activity much encouraged by the government then.

It was at the university, by 1935, that von Ohain began to crystallise his ideas on jet propulsion. He patented a simple jet propulsion system for aircraft which comprised a two-stage compressor (axial fan followed by a centrifugal compressor) and an inward-flow radial turbine. An annular combustion chamber with eight or twelve copper petrol burners was envisaged. This scheme was along the same broad lines as Frank Whittle's patent of 1930, but it is not known if von Ohain was aware, at that time, of Whittle's patent. Sooner or later, however, von Ohain and his later co-workers kept track of patent applications both inside and outside Germany, and they became very familiar with the similar work of Whittle and also of A. Lysholm (sponsored by Milo Aktiebolaget of Stockholm).

While at Göttingen, von Ohain was in the habit of taking his little sports car to the garage of Bartels und Becker, where he found an automotive and railway engineer, Max Hahn, to be unusually skilful and reliable. Therefore, when von Ohain had designed a model to demonstrate the turbojet principle, he went to the same garage for Max Hahn to build the model, the cost being somewhat below 1,000 DM and paid for out of von Ohain's own pocket. In this exercise, we see the efforts of a first-class theoretician and scientist combined with those of a first-class practical craftsman. As he later reminisced, 'I was a physicist who really didn't know what nuts and bolts were'.

Taking the demonstration model engine (the 'garage model') to the backyard of the Physical Institute, University of Göttingen, von Ohain attempted to obtain satisfactory running, but experienced trouble with the combustion system. Attempting to run the engine on petrol, it was never self-sustaining since most of the combustion took place on the exhaust side of the turbine. Flames up to three metres long emerged from the exhaust nozzle, and the electric motor, keeping the engine running at about 8,000 rpm, overheated.

Since von Ohain had exhausted his personal funds with these extra-curricular activities, a sponsor was now needed in order to continue jet propulsion experiments. At this point, his professor (who was Department Head of the First Physical Institute), Robert W. Pohl, came to the rescue by writing to Ernst Heinkel in February 1936 and vouching for the apparent soundness of von Ohain's work. Heinkel was always seeking advances in aviation and was already interested in rocket-powered aircraft projects. He therefore arranged a meeting with his technicians and von Ohain, who took along his notes and stated frankly that he had built his 'garage engine' but that it would not work despite showing promise. The Heinkel engineers looked at his work and pronounced it not good but interesting. Largely because of Ernst Heinkel's enthusiasm for new propulsion systems, the outcome of this meeting was the engagement of von Ohain and also (at the latter's instigation) Max Hahn in April 1936. These men, spurred by an over-optimistic estimate of the time and cost required, then continued the development of a demonstration engine, using as a starting point further work on the 'garage engine'. However, as von Ohain later recalled: 'When I first came to Heinkel the engineers considered me a crazy boy, a physicist who really did not appreciate the problems of materials, materials treatment, castings and machine builder's designs. These shortcomings bothered me a lot and I worked very hard to become a fully-fledged engineer. After two years or two and a half years Heinkel's engineers couldn't tell me anything new about design.'

HeS 2 (TL)

Von Ohain and Hahn took the garage engine to Heinkels and used a bigger electric motor to run it cold at about 1,000 rpm, and then with a little petrol. It was found that the airflow between the compressor and combustor was very messy and there was even a pressure drop caused by a back-flow into the compressor. These tests took about two months. It was then decided to start afresh and design a new demonstration engine using hydrogen as fuel.

Fig 2.1 Hans von Ohain's first demonstration turbojet model (the 'garage engine').

These photographs and captions were supplied by Hans von Ohain

This was the first turbojet engine built by Max Hahn in the automobile garage Bartels & Becker, Göttingen, in 1935. It was designed and financed by Hans von Ohain. Total cost somewhat below 1,000 Marks.

Max Hahn with the model engine.

Rotor: radial-outflow compressor and radial-inflow turbine (in order to show rotor blades, outer circular shroud removed).

The work of developing the HeS 2 (Heinkel-Strahltriebwerk 2, or Heinkel Jet Engine 2), as the demonstration engine was known, was carried out in great secrecy in an isolated hut erected on Marienehe airfield, between Rostock and Warnemünde. Because von Ohain, at that time, had no engineering training, he was assigned Dipl.-Ing. Wilhelm Gundermann to work with him. Gundermann had studied aircraft engineering and turbo machinery engineering at the Technisches Hochschule, Berlin, under Prof. Hermann Fottinger. He was assigned a team of six to eight draughtsmen and stress analysts. Calculations were made using, as a starting point, the characteristics of centrifugal pumps, the Francis inward-flow radial water turbine and formulae for steam turbines.

Meanwhile, Max Hahn soon had six to eight fitters and mechanics in the workshop. Unlike the designers, who were chiefly brought in from outside, Hahn's workers were selected from the best in the Heinkel workshops, much to the annoyance of the shop superintendents. Hahn also played a major role in the design work, especially for the combustion system, because he had been very active previously in this field at Göttingen and had filed several patents. In fact, it is said that his patents eventually exceeded those of von Ohain and Gundermann, but some fogging of this issue is caused by the fact that some patents were filed under Hahn's name for the purposes of secrecy alone.

As the photographs (Figure 2.1) show, construction of

the garage engine was simplified by making extensive use of sheet metal, including a sheet-metal compressor and engine casing. Such construction, of course, obviated the need for large castings or major machining operations, and only a few parts had to be purchased from outside companies. It therefore suited the Heinkel airframe facilities to follow a similar but somewhat more sophisticated form of construction for the HeS 2, construction of which was begun in the late summer of 1936. A local shipyard was able to form sheet-metal parts over hardwood formers. Flanges were then riveted to the sheet-metal forms and machined true. A somewhat crude centrifugal compressor was used and the engine was of large diameter, because mainly the annular combustion chamber was positioned outside the compressor diameter. The reason for this was to give a longer burning length. The turbine driving the compressor was of the radial-inflow type and there was a single anti-friction bearing at each end of the rotor assembly. This simple, lightweight engine was fairly cheap to make.

Von Ohain's reasons for favouring the combination of a centrifugal (or radial-outflow) compressor with a radial-inflow turbine were based on the great simplicity, compactness and, above all, the natural matching of turbine and compressor characteristics which this specific configuration promised. The back-to-back compressor and turbine of the garage engine had been given up and separated more, so that better, less abrupt passages between them could be designed with better flow characteristics.

By about March 1937, the HeS 2 was completed, and with almost no problem at all, it ran a few weeks later after it was mounted on a primitive test stand. The difficult problem of combustion had been simplified by using hydrogen gas as fuel, but not without much burning of the metal. It was not until September 1937 that this engine ran on petrol, but the combustors soon clogged up. Under great pressure from Heinkel, von Ohain worked day and night on the problem. Eventually, a solution was found by Hahn, who built a combustor based on his soldering torch. In this, a boiler-type petrol vaporiser produced vapour to burn in the combustors, like propane combustors.

As Figure 2.2 shows, all the engine's airflow passed through the combustion chamber. In addition, because the rotor was made largely from relatively thin and light sheet metal, its moment of inertia was extremely small and therefore the throttle response was outstandingly fast, almost as fast as that for a contemporary piston engine. The result was that the first sustained runs with the HeS 2 gave an impressive demonstration that the turbojet principle was entirely feasible and encouraged Heinkel to back further development along these lines. Accordingly, the development of a flight engine, the HeS 3, was put in hand, together with the design of a suitable test aircraft, the Heinkel He 178.

Little data are known for the HeS 2 engine, but some approximate figures are:

Static thrust	136 kp (300 lb)*
Speed	10,000 rpm
Rotor diameter	0.61 m (2 ft)
Overall diameter	0.97 m (3 ft 2.25 in)
Length	0.90 m (2 ft 11$^{7}/_{16}$ in)

*von Ohain's figure. Other sources say 550 lb

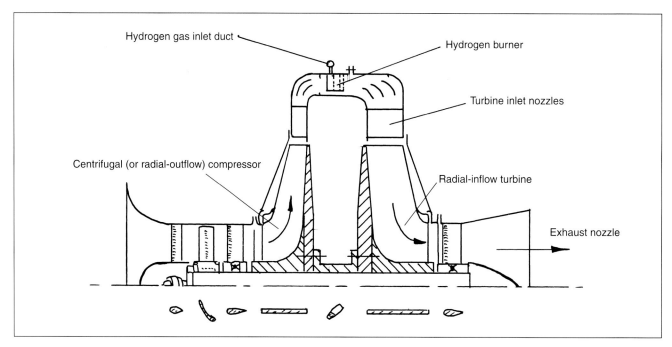

Fig 2.2 Half-section showing the basic layout of the first Heinkel turbojet engine, the HeS 2.

HeS 3 (TL)

In designing this flight engine, the chief aims for von Ohain were a reduced diameter compared with the HeS 2, a liquid-fuel combustion system and a static thrust of about 800 kp (1,764 lb). The new engine had an axial-flow inducer followed by a centrifugal compressor, these being driven by a radial-inflow turbine. With an increased staff in a larger building, work proceeded on the first flight engine, the HeS 3, the first example of which was being bench-run by about March 1938. Unfortunately, the attempt to minimise the dimensions of this engine resulted in an undersize compressor and poor combustion, so that insufficient thrust was developed.

A redesign led to the HeS 3b (Figure 2.3), and two of these engines were showing promise by the summer of 1939. The engine had an eight-bladed axial-flow inducer followed by a 16-bladed centrifugal compressor, these being driven by a 12-bladed radial-inflow turbine. Air from the compressor passed into tangential diffuser slots, after which it was divided partly forward through the reverse-flow, annular combustion chamber and partly rearward to mix with the combustion gases before these entered the turbine section. Curved guide nozzles led into the turbine, and a fixed-area exhaust nozzle was used. The compressor blades were riveted to a disc having a curved boss which led forward to a stub shaft mounted in a ball race. A similar construction was used for the turbine, but the stub shaft for this led rearward to a roller race mounting. A short, flanged tube connected the compressor and turbine discs by bolting. The fuel used with this engine was petrol, which was injected by a series of simple fuel jets spaced around the combustion chamber. Each fuel jet was supported inside the chamber by a small grid of four pipes through which the fuel flowed, the grids serving to pre-heat the fuel and aid mixing by means of turbulence. Pre-heating of the fuel also occurred by using it to cool the rear roller bearing housing (aft of the turbine) before piping it to the fuel jets.

After bench trials, flight tests began in May 1939 with an HeS 3A engine suspended beneath an He 118 (D-OVIE) converted as a flying test bed. (The He 118 was a two-seat, single-engine monoplane. Only a few V-series were built to fulfil a divebomber role, but the production order was won by the Junkers Ju 87.) Secrecy was still preserved, and these flight tests were conducted each day in the dawn hours before the factories opened. The pilot was Flugkapitän Erich Warsitz and the flight engineer was named Kunzel. After a piston-engine take-off, the HeS 3A was switched on and gave, apparently, a bluish exhaust stream because of the petrol fuel used.

Following a series of these air tests, the turbine burned out in the HeS 3A, but by then the He 178 airframe was ready to accept the redesigned HeS 3b. The He 178 was a single-seat, shoulder-wing monoplane with a nose air intake and duct leading to the engine and a rather long exhaust duct of about one third the fuselage length. Little was known of losses in air ducts at that time, and these lengthy ducts were inefficient. Nevertheless, the He 178 flew, following taxiing trials, making a short hop along the runway on 24 August 1939 and, finally, the first true flight on 27 August. These flights made the He 178 the world's first aircraft to fly solely on the power of a turbojet engine. Following this success, Heinkel began to seek backing from the RLM, which had already been made aware of his jet project during 1938. A demonstration of the He 178 before RLM officials on 1 November 1939 did not produce apparent enthusiasm, although, as Hans Mauch later imparted to von Ohain, the He 178 had an enormous impact on the RLM and encouraged the pursuance of jet aircraft development. Nevertheless, Mauch was against the Heinkel airframe works building flight engines, because of their lack of experienced engine designers and engineers. On the other hand, Schelp approved of Heinkel's engine work, and he assumed a more influential position in GL/C3 of the Technisches Amt when Mauch left at the end of 1939. The He 178 was only employed for a short time before the programme moved on, and its maximum speed was only about 600 km/h (373 mph). Data for its HeS 3b engine are as follows:

Thrust	450 kp (992 lb) at 800 km/h (497 mph)
Speed	13,000 rpm
Weight	360 kg (795 lb)
Pressure ratio	2.8 to 1
Specific fuel consumption	2.16 per hour
Airflow	22.5 kg/sec (49.6 lb/sec)
Specific weight	0.8
Diameter	0.93 m (3 ft 0^1 in)
Length	1.48 m (4 ft 10 in)
Frontal area	0.68 sq m (7.3 sq ft)

HeS 6 (TL)

Modification of the HeS 3b produced the HeS 6 engine with an increased thrust, but the weight went up also and the performance was not outstanding. Fuel consumption was, however, much improved over the HeS 3b. Test flights were made beneath an He 111 before the end of 1939, but the development of the HeS 6 was not pursued. Data for this engine were:

Thrust	550 kp (1,213 lb) approx. at 800 km/h (497 mph)
Speed	13,300 rpm
Weight	420 kg (926 lb)
Specific fuel consumption	1.6 per hour approx.
Specific weight	0.73
Frontal area	0.66 sq m (7.1 sq ft)

At about the same time that von Ohain's team turned its attention to the development of a new centrifugal

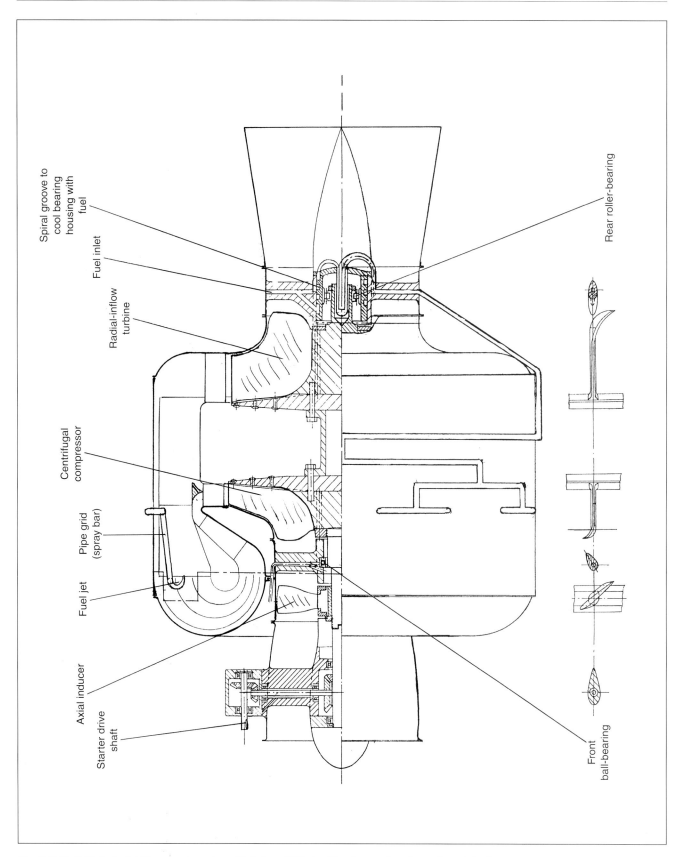

Fig 2.3 HeS 3b turbojet (drawing based on Dipl.-Ing. Wilhelm Gunderman's schematic original)

GAS TURBINES FOR AVIATION

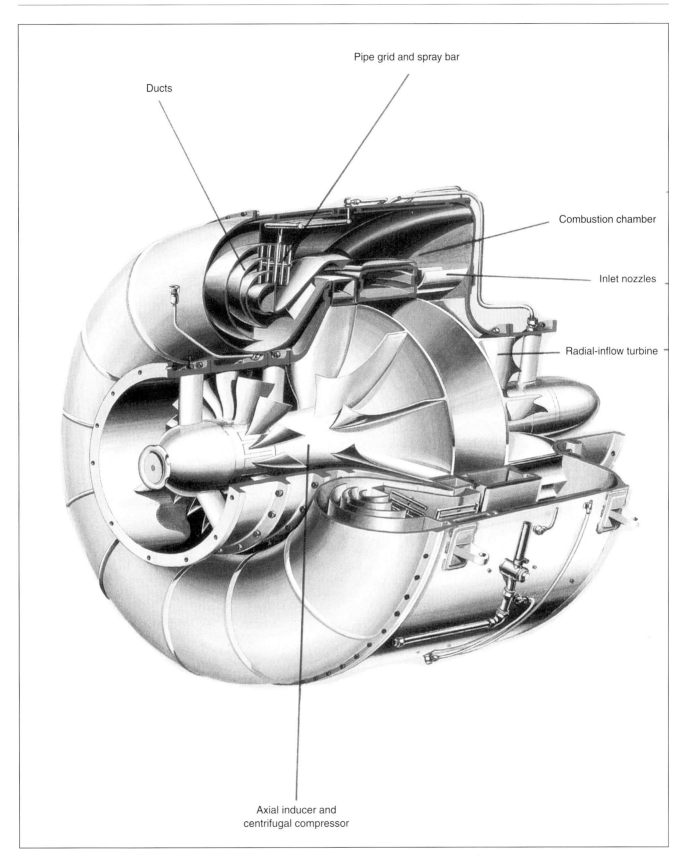

Fig 2.4 HeS 3b turbojet. *(Via Bill Gunston)*

Fig 2.5A Heinkel He 178, the world's first turbojet-powered aircraft.

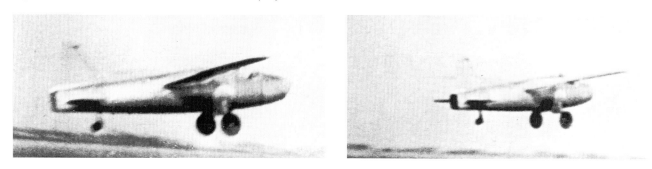

Fig 2.5B, C He 178 taking off.

Fig 2.5D After the first full flight of the world's first turbojet-powered aircraft, the Heinkel He 178 on 27 August 1939, a small celebration was held. From left to right are Erich Warsitz (the pilot), Prof. Ernst Heinkel and the engine's designer Dr Hans von Ohain. The latter two in particular were instrumental in kick-starting Germany's turbojet programme. Von Ohain was never comfortable wearing formal attire.

engine, the HeS 8, Heinkel hired Max A. Mueller and other engineers who had left the Junkers Flugzeugwerke where they had been working on a number of jet propulsion engine projects. These projects, which Mueller brought to Heinkel, included ducted fan engines and one particular axial turbojet engine which became known as the HeS 30. The new engine developments are described below, but of special interest are the centrifugal HeS 8 and the axial HeS 30, which were to be alternative power-plants for the twin-engined Heinkel He 280 fighter being designed.

HeS 8 or 109-001 (TL)

The HeS 8 was the first Heinkel turbojet to receive official financing, and was accordingly given the RLM number of 109-001. The development aimed at producing a centrifugal engine with about 700 kp (1,544 lb) thrust but with a reduced weight and diameter compared with the previous HeS 3/6 series. A reduction in diameter was accomplished mainly by the introduction of a new combustion chamber layout whereby the combustion chamber was behind instead of outside the compressor. Thus, the previous reverse-flow combustion chamber was abandoned for a straight-through design.

A centrifugal compressor was chosen for the HeS 8, not because it was particularly favoured by von Ohain, but because, by this time, the RLM had directed Heinkel to keep von Ohain on advanced development efforts for centrifugal engines. In fact, von Ohain's original plans were to enter into axial turbojet development following his acquaintance with Dr Encke of the AVA (see Section 6) from 1938, and he was familiar with the many advantageous characteristics of axial-flow machinery, considering such to have the greatest promise for future turbojet development. His original choice of centrifugal rotors was motivated by the fact that axial rotors would require large component test stands for properly matching the various axial-flow compressor stages and also for matching turbine to compressor. Such component test stands were not available at the time, and a convincing demonstration of the turbojet principle was first necessary in order to acquire such highly expensive equipment.

The compressor of the HeS 8 consisted of a 14-blade axial-flow inducer in the intake duct followed by a 19-blade centrifugal compressor. The inducer blades were of aerofoil section and were forged from aluminum alloy, whilst the compressor blades were also of aluminum alloy but were retained in a steel hub and riveted to a rear disc. The 14-blade, radial-inflow turbine was of similar construction to the centrifugal compressor, but the blades were in steel. A drum-type rotor with flanged ends connected the turbine and compressor components by bolting into the disc and hub at each end. The revolving system was supported by a bearing housing containing two ball races between the impeller and compressor and by a housing containing a single roller race aft of the turbine.

Air from the compressor passed through two sets of diffuser vanes before entering the annular combustion chamber. Fuel was injected into the combustion chamber through 16 sets of eight nozzles (giving 128 nozzles in all), each nozzle being a small tube. Half the nozzle tubes were of a different length from the others so that fuel could be sprayed onto two annular rings at the combustion chamber entrance. These rings were of steel and had grooved surfaces from which the fuel vaporised, mixed with the air and burned. The HeS 8 fuel system is shown in Figure 2.6.

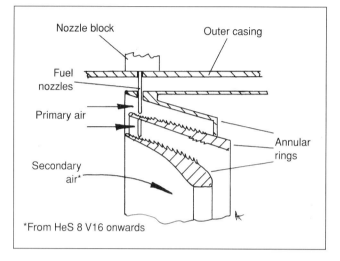

Fig 2.6 HeS Fuel System.

The main casing of the engine consisted of castings with external ribbing for increased rigidity. The exhaust nozzle was of the fixed-area type and all auxiliaries were grouped around the smaller diameter of the air intake section. Drive shafts for the auxiliaries led out through fairings* in the air intake and made their connections through bevel gears.

Whilst it realised the aims of reducing diameter and weight over previous designs, the HeS 8 did not progress so promisingly on the performance side. By September 1940, the airframe of the He 280 V1, first prototype of Heinkel's twin-jet jet fighter, was ready, but the HeS 8 engines were not advanced enough for installation. Work progressed on other He 280 air-frames, whilst gliding flights only had to be performed with the He 280 V1. By March 1941, when this aircraft had made about forty gliding flights, the HeS 8 was producing about 500 kp (1,102 lb) static thrust, and two examples of the engine were installed beneath the wings of the He 280 V1. This aircraft, with the test pilot Fritz Schäfer at the controls, made its first, short, powered flight on 30 March 1941. On this flight, at least, no engine cowlings were fitted

*One shaft was simply in the airstream.

Fig 2.7 An HeS 8, probably pre-V16, and a rotor assembly on display at an exhibition.

because the HeS 8s leaked fuel. One of the most important flights of the He 280 V1 at Marienehe occurred on 5 April 1941 before officials, including Udet, Eisenlohr and Schelp. The success of this demonstration virtually killed off any scepticism that remained concerning the desirability of full official backing for Heinkel's jet programme.

By early 1942, however, performance of the HeS 8 was still disappointing, the thrust being only about 550 kp (1,213 lb), and only 14 V-series engines had been built and tested. In the HeS 8 V15 engine an attempt was made to improve the compressor system by fitting a single-stage axial compressor just after the centrifugal compressor. A design study attributed to the new axial compressor stage 0.5 of the 3.2 to 1 total pressure ratio for the whole compressor system. In the HeS 8 V16 engine (see Figure 2.8), sandwich secondary air mixers were introduced. (Most, if not all, previous engines had used no system of secondary air distribution after the compressor.)

The number of HeS 8 engines built is unknown, although the highest experimental number known is HeS 8 V30. Not until early in 1943 were more engines ready for the He 280, and the He 280 V2 and V3 were flown with HeS 8 engines. By then, the static thrust of each engine had crept up to about 600 kp (1,323 lb), but the development was overtaken by the Junkers 109–004 and work in hand on other Heinkel turbojets. The He 280 V3 continued to fly HeS 8 engines to provide data for other developments, although this aircraft crash-landed on one occasion following a turbine failure. (See Figure 2.10). Other test flights were conducted with an HeS 8A fitted beneath an He 111 (c/n: 5649) converted bomber (see Figure 2.11). The HeS 8 was abandoned by the spring of 1943, even as a research instrument. Typical data for the engine were:

Planned thrust	700 kp (1,544 lb) at 800 km/h (497 mph)
Max. thrust obtained	600 kp (1,323 lb) static
Speed	13,500 rpm
Weight	380 kg (838 lb)
Pressure ratio	2.7 to 1
Specific weight	0.55
Diameter	0.775 m (2 ft 6 in)
Length	1.60 m (5 ft 3 in)
Frontal area	0.48 sq m (5.16 sq ft)

HeS 8 developments (TL and ZTL)

Although full information is lacking, it is believed that the HeS 9 (TL) engine was designed as an outgrowth of the HeS 8 V15 (see Figure 2.21). The indications are that the HeS 9 was to utilise a radial-inflow turbine similar to the HeS 8 but was to have a new compressor set. This was to comprise an axial inducer, a diagonal-flow compressor and, finally, a two-stage axial compressor. This arrangement would fit the HeS 9 into the flow path development sequence leading to the HeS 011 engine (described later), but the only thing known for certain is that 10 HeS 9 engines were ordered by the RLM for development.

Also on order were three HeS 10 or 109–010 (ZTL) experimental ducted-fan engines constructed around the HeS 8. A modified HeS 8 engine was externally faired

GAS TURBINES FOR AVIATION

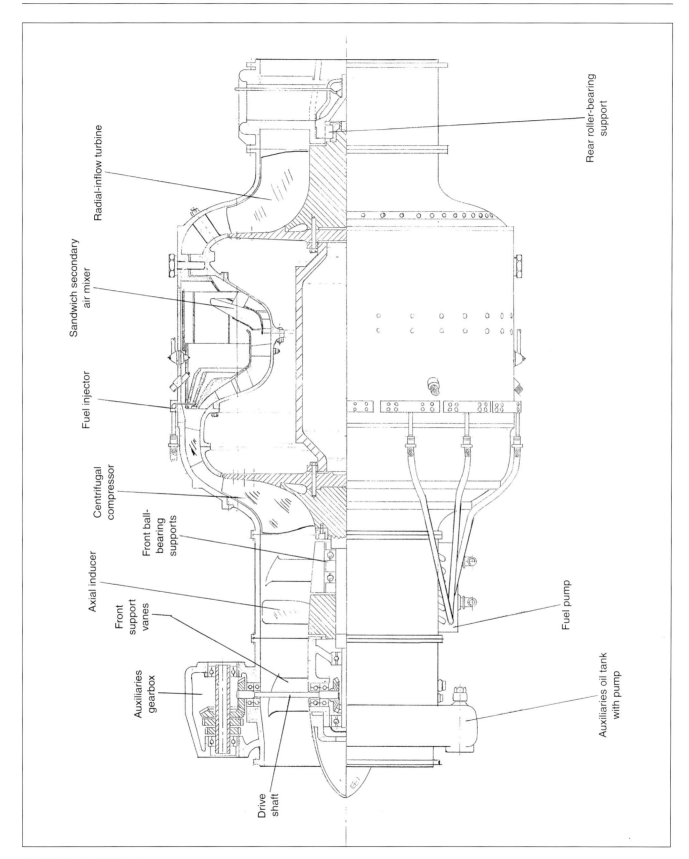

Fig 2.8 HeS 8 V16 turbojet with primary air (to combustion chambers) and secondary air. *(Author)*

RIGHT:
Fig 2.9 The Heinkel He 280 V1 taking off from Marienehe airfield on its maiden flight on 2 April 1941. The HeS 8 turbojet engines are completely uncowled owing to initial engine leaks posing a fire hazard by pooling. This aircraft carried with it the hopes of Heinkel to forge ahead of its rivals in the fighter and jet aircraft field.

Fig 2.10 Heinkel He 280 V3 (GJ+CB).
ABOVE LEFT: With engines uncowled owing to early fuel leaks.
ABOVE: Later, following a landing skid.

LEFT:
Fig 2.11 To speed progress on HeS 8 A development, especially concerning a speed regulator and variable-area exhaust nozzle, many flights were made using this converted Heinkel He 111 bomber as a flying test-bed. Note the streamlined telescopic struts used to lower the test engine into a less-disturbed airflow once airborne. *(R.C. Seeley)*

Fig 2.12 Heinkel HeS 8 A turbojet test engine with inlet fairing (on right) and exhaust duct fitted.

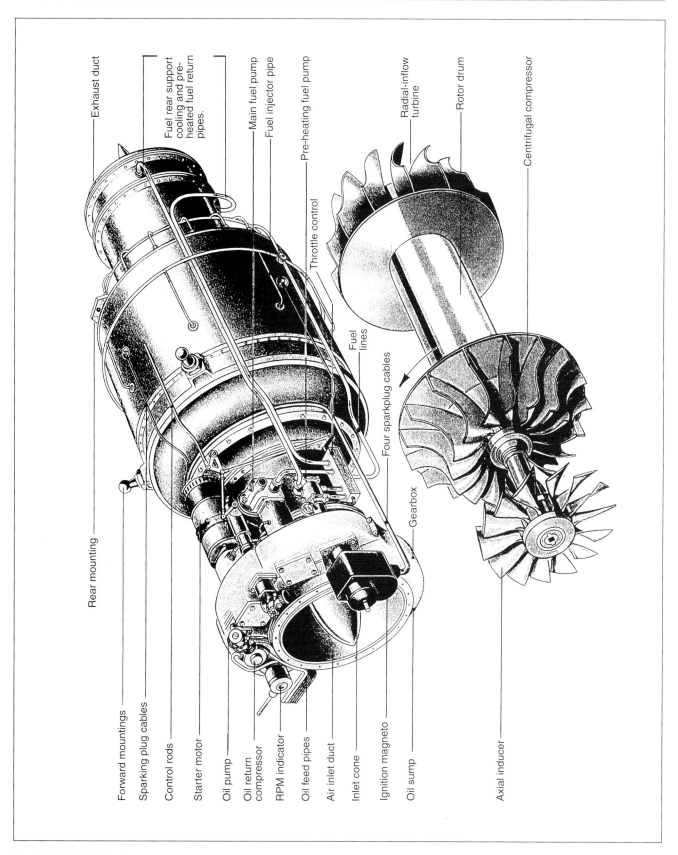

Fig 2.13 HeS 8 A turbojet perspective.

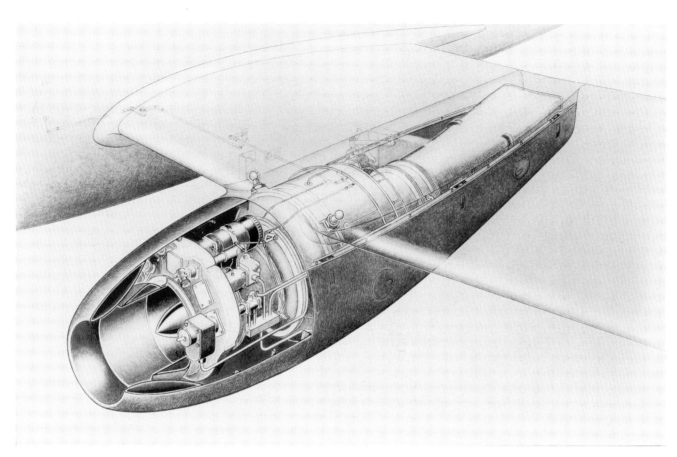

Fig 2.14 Partial section showing the installation of the Heinkel-Hirth HeS 8 A turbojet engine under the port wing of the Heinkel He 280.

into a suitable shape and was supported by struts inside a cowling forming a duct around the whole engine. The fan, near the cowling intake, was driven by an extension of the engine shaft through gearing, and air from the fan, together with the turbojet exhaust gases, emerged together from the cowling exit nozzle. A radial-inflow turbine was used which, in the normal HeS 8 engine, was only sufficient to drive the axial inducer and centrifugal compressor. For the HeS 10, the previous axial inducer was omitted, but it was necessary to provide more turbine power to drive the ducted fan. Thus, a second turbine, of the axial type, was fitted inside a new exhaust nozzle behind the radial-inflow turbine. The second turbine was not connected independently to the ducted fan, which would have been modern practice. From the layout of the HeS 10, shown in Figure 2.15, will be noted the modified intake to the turbojet engine, which also had a combustion chamber on the lines of the early HeS 8 design. With the HeS 10, it was hoped to provide an engine more economical at lower speeds for longer flights. However, the development was not pursued, although at least one test engine may have been built during the early development phase of the HeS 8. Data for the HeS 10, or 109–010, are:

Planned thrust	900 kp (1,985 lb) at 600 km/h (373 mph)
Engine speed	13,500 rpm
Weight	500 kg (1,102 lb)
Specific weight	0.56
Diameter	1.01 m (3 ft 3^i in)
Length	2.86 m (9 ft 4^l in)
Frontal area	0.8 sq m (8.6 sq ft)

Expansion of the Heinkel turbojet programme

By November 1939, Heinkel had about 120 scientists, technicians, and engineers working at Rostock on turbojet engines and aircraft. The number of personnel had been increased in October that year with the arrival of Max Adolf Mueller and about half of his team from Junkers. Whilst von Ohain's team continued development of centrifugal turbojets (the HeS 8 had been started on at that time), Mueller's team continued the previous Junkers developments, the most important of which became known as the HeS 30, 40, 50, and 60. Thus, although the personnel strength had been

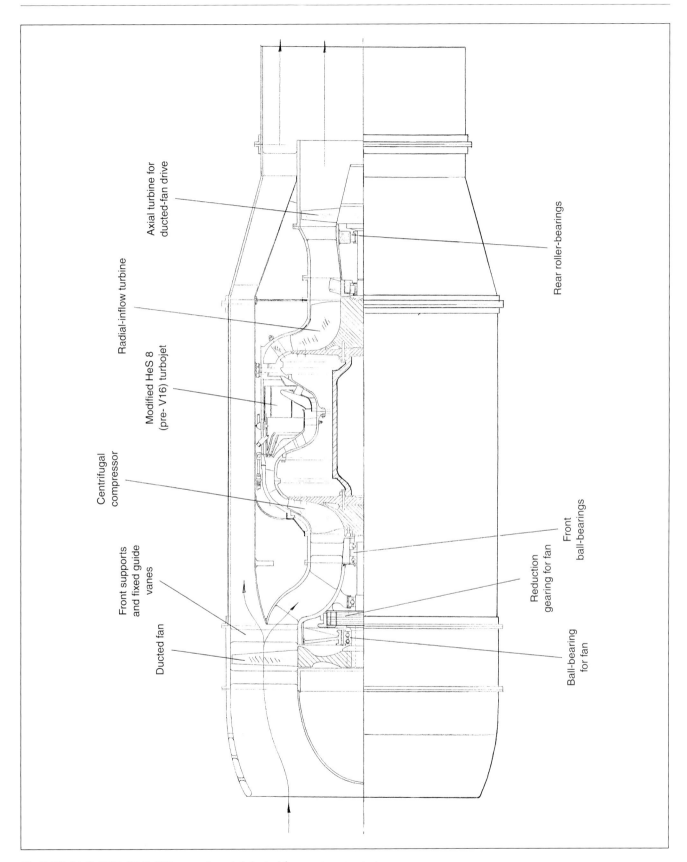

Fig 2.15 HeS (109–010) ZTL experimental ducted-fan engine. *(Author)*

increased, the problem of obtaining the requisite number of skilled men had been made actually worse because of the increase in the number of engine developments. There were, in addition, numerous jet engine projects being studied, largely by Mueller's axial turbojet team. The facilities at Rostock were, however, being expanded, and a start was made on the planning of a new engine development factory and the building of a 14,000 kW compressor test rig to run at 12,000 rpm. (This test rig was not, in fact, completed until after the war.)

In order to acquire more labour and further facilities for the jet engine programme, it was necessary for Heinkel to obtain permits from the RLM, but here there was still some resistance to Heinkel's programme. Schelp objected only to the large scope of the programme, but Wolfram Eisenlohr, who was head of GL/C3 and accordingly allocated labour for aero-engine development, was still not convinced that there was any urgency about turbojet development. However, the day after the flight of the He 280 before officials on 5 April 1941, an impressed Eisenlohr announced in Berlin, 'The kudos which must go to the Heinkel firm for having produced the first jet unit is enormous. This fact justifies their being allowed to go on working on engine development.' Ernst Udet, chief of the GL/C Technisches Amt, was also impressed by the He 280 flight, and he arranged the authority for Heinkel to obtain control through a majority holding in the Aero Bank on 9 April 1941 of the Hirth Motoren GmbH. This company had specialised in the design, development and production of small but efficient piston aero-engines, e.g. HM508 and HM60R, and auxiliaries such as oil pumps. The chief gain to Heinkel in terms of extra labour and facilities was the Hirth factory at Marconi Strasse, Stuttgart/Zuffenhausen, but the smaller plant at Berlin/Grünau (Werk Waltersdorf) was also taken over. This control, which six other firms had also been seeking, allegedly cost Heinkel 50 per cent more than the market price. Complete absorption of the Hirth Motoren GmbH into the Ernst Heinkel AG was achieved by April 1943. Officially, the chief work of the Stuttgart factory was to be the development of a Class II turbojet (later designated 109–011), and it was the agreement to do so which assisted Heinkel to obtain control in 1941. It was not until the end of 1942, however, that the Stuttgart facility had been refitted ready to take on Heinkel work. In the meantime, von Ohain and Mueller continued their developments at Rostock.

HeS 30 or 109–006 (TL)

Although it was abandoned at a fairly early stage in its development, the HeS 30 axial turbojet was the most promising of all the Heinkel engines, and was possibly the only one with any long-term value. The design was begun by Mueller (as one of a number of projects) at the Junkers Flugzeugwerke before 1938, and was a generally very advanced engine of small dimensions for its planned thrust. Of particular interest was the 50 per cent reaction-type blading of the axial compressor, in which the work of air compression was shared between the stator and rotor blades. (This differed from German practice in most other axial compressors, which used impulse-type blading, whereby the stator blades performed almost all the work of compression; such compressors were largely to the designs of Encke and Betz of the AVA.) The HeS 30 compressor, which achieved a pressure ratio of about 3:1 with only five stages, was designed by Rudolph Friedrich, an aerodynamicist at Junkers. Mueller's engine was actually built and put on test by about the end of 1938 but could not then run under its own power. Only by using an external compressed air supply and with a very high fuel consumption could the engine be brought up to about half speed. This engine was, in fact, abandoned by Junkers when that company accepted an official contract to develop a turbojet (later known as the Junkers 109–004) in 1939.

Mueller, however, promised to have his engine running within one year of his arrival at Rostock in October 1939. In fact, the first run was not made until more than two years had passed. It is not known how much the final HeS 30 design differed from the original design at Junkers.

As already stated, it was planned to develop the HeS 30 alongside the HeS 8 as alternative power-plants for the He 280 fighter. For both engines, small frontal area was a prime aim in order to give acceptable wing nacelle mountings and thus avoid troublesome fuselage air ducting, such as with the He 178 aircraft. Of the two engines, the HeS 30 had the smaller diameter.

The layout of the HeS 30 is shown in Figure 2.16, and an artist's impression is given in Figure 2.17. The compressor was of the five-stage axial type and was driven by a single-stage axial turbine. Ten individual combustion chambers were provided, each varying from a circular cross-section at the entry to a heart-shaped cross-section at the exit. Here, a particularly novel feature was introduced in the form of variable turbine-inlet guide vanes. Secondary air was admitted to each combustion chamber by allowing it to pass under five overlapping, progressively larger, tubular sections. The exhaust nozzle was of the variable-area type, a feature which was to become common on German axial turbojets. The object of such a nozzle was to provide as far as possible optimum operating conditions. For example, an increased nozzle exit area (and lower turbine back pressure) would be selected for engine starting and to avoid surging of the axial compressor at full rotational speed above a certain altitude. (One of the drawbacks of the axial compressor, as opposed to the centrifugal compressor, is that its optimum operating speed is close to its surging speed, and the two can be caused to coincide by the effects of altitude. The phenomenon of surging produces large variation in rotational speed, considerable vibration and loss of power.) In the case of the HeS 30, the exhaust nozzle exit area was varied by moving the sloping inner wall of the tail cone in or out. This inner wall was attached in sections to three streamlined struts attached to a central bullet. The bullet was held out by a spring and was pulled in by a system of pulleys and cable.

Three experimental engines were ordered under the official designation of 109–006, but facilities at Rostock were still not conducive to fast progress in building experimental engines, and the first HeS 30 progressed slowly. A major problem lay in trying to match the turbine to the compressor, which gave an air mass flow greater than designed for. The first engine was running under its own power by about April 1942, but in May that year Mueller fell into a dispute with Heinkel and resigned. However, by October the HeS 30 was running with full success and developing a static thrust of about 860 kp (1,896 lb) for a weight of only 390 kg (860 lb). Its thrust to weight ratio was far better than that of, say, the Junkers 109–004, and in terms of specific fuel consumption and frontal area per unit thrust, it is claimed that it was not equalled anywhere until 1947.

Unfortunately, this remarkable engine was not favoured by Schelp, who had some objection to the 50 per cent reaction-type blading of the compressor, while, from the production viewpoint, the engine's bearing arrangement was somewhat extravagant. He also considered the HeS 30 too small for future requirements and thought its place would soon be adequately filled by the more conservative Junkers 109–004 and BMW 109–003.

The official RLM line was that the 109–006 was excellent but too late, since the Junkers and BMW engines were further along the development path and they were not prepared to waste money just to help Heinkel catch up. According to von Ohain, Ernst Heinkel almost cried when he heard this. The RLM said it was his own fault for not listening in the first place when they were trying to run a national programme.

Schelp, in fact, wanted a Class II engine started, and by the end of 1942 orders had been given for Heinkel to concentrate on such an engine. (This, the HeS 011, is described later.) Although the HeS 30 lost RLM sponsorship after only a few test runs, it seems that tests were still made from time to time, probably in the hope that the engine would again be ordered into development. During a test in 1945, a static thrust was obtained as high as 910 kp (2,006 lb), but the basic data for the HeS 30 are as follows:

Maximum planned thrust	1,125 kp (2,481 lb) at 800 km/h (497 mph)
Thrust obtained	860 kp (1,896 lb) to 910 kp (2,006 lb) static
Speed	10,500 rpm
Weight	390 kg (860 lb)
Pressure ratio	3 to 1 approx.
Specific weight	0.44 approx.
Diameter	0.62 m (2 ft 0 in)
Length	2.72 m (8 ft 11 in)
Frontal area	0.30 sq m (3.23 sq ft)

The constant-volume HeS 40 (TL)

At some time during 1940–41 the interesting design of the HeS 40 turbojet engine was worked upon by Mueller's team. The special interest of this engine derives from the fact that it was designed to operate on the constant-volume principle as opposed to the constant-pressure principle universally employed for turbojet and gas turbine engines to-day. Indeed, during the period covered by this history, active protagonists of the constant-volume cycle were in the minority, and some explanation of the cycle may be worthwhile here.

In the conventional constant-pressure turbojet, the compressor delivers air at a constant pressure to the combustion chamber, where it is accelerated by the application of heat, the reaction to this acceleration providing the thrust. In such a turbojet, the combustion chamber is open at both ends. If, however, air at little more than atmospheric pressure is mixed with the fuel in a combustion chamber with the ends closed, combustion takes place at constant volume and density. The pressure then rises in proportion to the temperature, and release of the combustion gases through a nozzle gives a propelling impulse. Complication then arises when attempts are made to give a continuous thrust process by using a number of combustion chambers firing at different times to even out the impulses. Such a system, albeit not for a turbojet engine, was followed by Holzwarth (mentioned elsewhere). A simplified constant-volume jet engine is the pulsejet, whereby only one end of the combustion chamber (the intake) is closed during combustion. This type of engine is dealt with in its own section in this book.

Operating with a compression ratio of less than 2 to 1 and with a considerably increased combustion speed, the constant-volume, intermittent-combustion turbojet was regarded as a competitor of the constant-pressure turbojet having a high compression ratio and slower combustion. There were, however, a number of problems to solve, not least of which was the method of control. It was therefore decided to build an experimental engine, the HeS 40, but using parts such as the axial compressor (modified for a lower compression ratio), variable-area exhaust nozzle and other mechanical parts from the HeS 30 turbojet engine, which was at that time making slow progress. By having the HeS 40 and 30 as near as possible, similar, useful comparisons of the two systems were made easier.

The layout of the HeS 40 is shown in Figure 2.18. Compressed air was delivered to the combustion chambers by a five-stage axial compressor driven by a single-stage turbine. Six individual combustion chambers were provided, each being open at the exit to the turbine

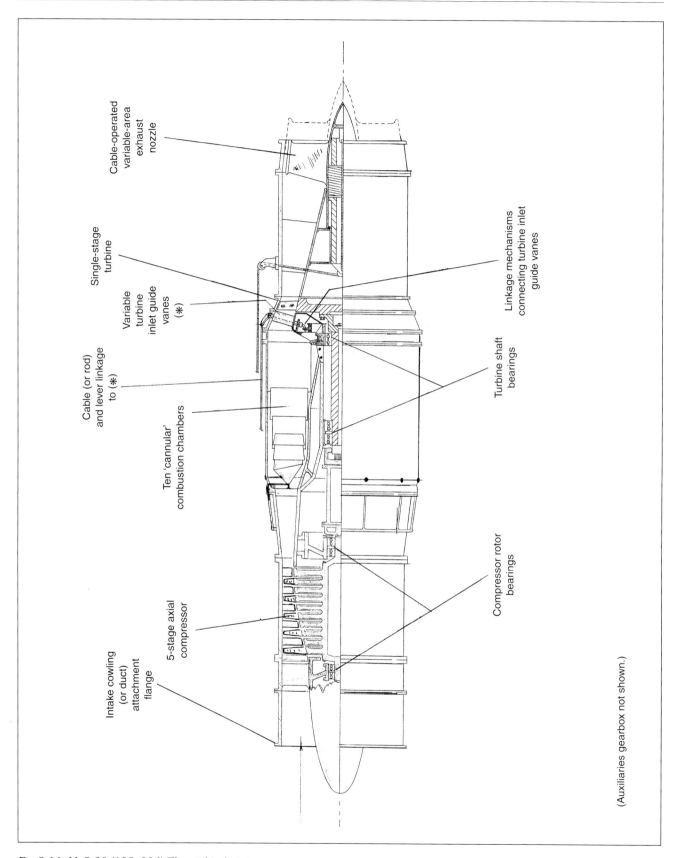

Fig 2.16 HeS 30 (109–006) TL axial turbojet. *(Author)*

GAS TURBINES FOR AVIATION

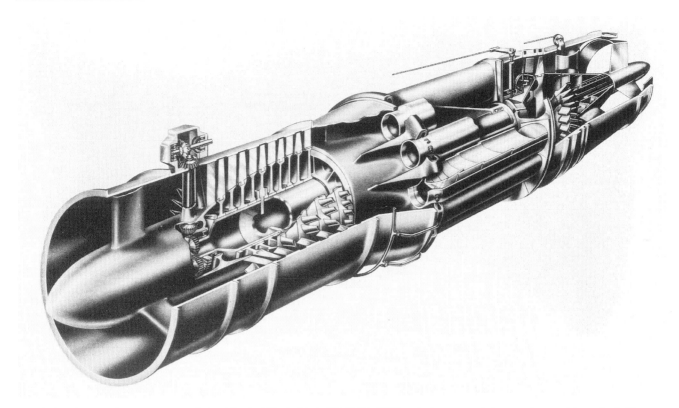

Fig 2.17 An artist's impression of the Heinkel-Hirth HeS 30 (109–006) turbojet, one of the most promising of all German turbojet engines. The simplicity and slimness of the engine are apparent, but note the novel feature of variable-angle turbine inlet guides. Official abandonment of this engine came about because Heinkel had not fitted into the national programme earlier on, and despite the engine's promise, it was seen as being always behind developments at Junkers and BMW. *(Richard T. Eger)*

and closed at the opposite, inlet, end during combustion by a poppet-type valve. Each inlet-valve stem extended forward into streamlined guides mounted in the diffuser section leading from the compressor. It is not known if the experimental HeS 40 was even built, but in any case it was abandoned by 1942 when the HeS 30 was making excellent progress. The HeS 40 was projected to have a thrust of 940 kp (2,070 lb) at 13,400 rpm at sea level. It would have had some weight increase because of the valve gear and larger combustion chambers, but this could have been offset by lightening the less-loaded compressor.

The Mueller ducted-fan engines (ML and MTL)

A particular interest pursued by Mueller's group at Rostock was the development of ML and MTL ducted-fan units powered by piston engines. The background for these had been studied during Mueller's former days at Junkers, where the stage of a wooden mock-up was reached. It was thought that these units, with their airflow rate somewhere between that of the conventional airscrew and the turbojet, would prove more advantageous than the piston-driven airscrew unit at speeds around 720 km/h (447 mph) on long-duration flights of up to ten hours. It was also thought that the ML unit would have about half the fuel consumption rate of the turbojet unit (TL). And such compound engines avoided the problems of developing a compressor.

The first ML unit (known at Heinkel as Marie Louise), HeS 50d, was designed to use a diesel engine since, despite its high specific weight, advantage was hoped for with the low specific fuel consumption during long-duration flights. However, the prime requirement of an ML or, later, MTL (combination of ML and TL) unit was seen to be an engine with a high power-to-weight ratio and very high rpm. The first requirement could be fulfilled by a two-stroke engine, but there were unsolved problems involved in increasing the rpm of such an engine. Thus, it became necessary to develop a piston engine specially suited to the ducted-fan units, but such specialised work was outside the scope of the Heinkel company and the Hirth company was not at that time controlled. Research was therefore put in the hands of Prof. Wunibald I.M. Kamm of the Forschungsinstitut für Kraftfahrwesen und Fahrzeugmotoren of the Techniches Hochschule Stuttgart (FKFS). This research reached the stage of having a single-cylinder unit built

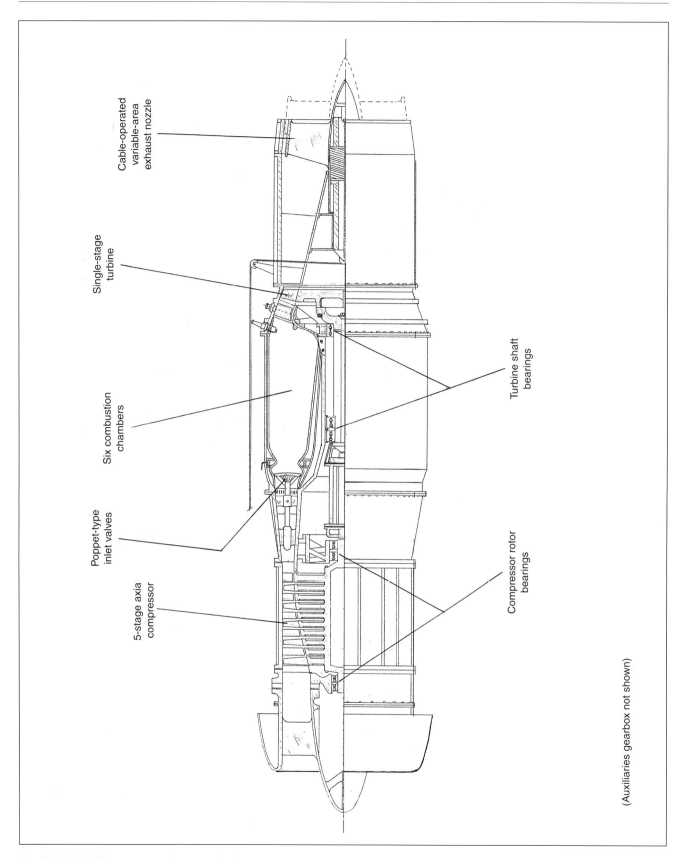

Fig 2.18 HeS 40 constant-volume axial turbojet. *(Author)*

and tested which formed the basis for designing multi-cylinder, two-stroke petrol engines. The piston engines subsequently designed for ducted-fan units, e.g. HeS 50z and 60, had a stroke shorter than the cylinder bore and were designed around a similar basic cylinder unit. Concerning the ducted-fan unit as a whole, it was calculated that the pressure ratio of the ducted fan should be between 1.5 and 1.9 to 1, whilst the piston engine should have a compression ratio of between 3 and 6 to 1.

Piston-driven ducted-fan units, however, did not progress beyond the research and mock-up stage at Rostock. Between 1940 and 1942, three designs were worked on, these being the HeS 50d, 50z and 60 described below. The two former designs, at least, had official backing. In a LAL report for general circulation, dated 31 January 1941, Mueller stated,

> 'It is to be expected that trial runs of the first ML full power unit will take place shortly. Already the results attained with the single-cylinder engine make the cherished hopes appear fully realisable. These results not only promise to be decisive for the development of ML power units but, beyond that, also for the whole range of use of high-speed piston engines of extremely large output per litre in the case of units which are not too small.'

In fact, the cherished hopes were no longer striven for by Heinkel after May 1942 since Mueller resigned then and ducted-fan development was immediately abandoned.

HeS 50 (ML and MLS)

Three prototypes of the HeS 50 series of ducted-fan engines were contracted for though not built. The first firm layout, designated HeS 50d, planned the use of a 24-cylinder, liquid-cooled diesel engine having a rated power of 1,000 bhp and a take-off power of 1,200 bhp. This engine had a bore of 95 mm (3.74 in) and a stroke of 90 mm (3.54 in), the cylinders being arranged in four rows of six in a horizontal H. A gear drive was taken from between and halfway along the dual crankshafts to a shaft driving a two-stage fan at the duct intake. A duct led from the intake, over the engine and curved around the rear of the engine section to a nozzle at the rear, the whole unit being enclosed in a streamlined nacelle. The air intake for the diesel engine was directly behind the ducted fan at the top of the engine, and was provided with a centrifugal impeller or supercharger. Another intake at the bottom of the engine (still inside the duct) was provided for the radiator. Some acceleration was given to the air flowing through the duct by adding heat from the diesel exhaust gases, but for boosting speed over a short period the possibility existed of adding an afterburner section. Such an afterburner reclassified the unit as an MLS unit, and such means of boosting entailed an enormous increase in fuel consumption. From the following data for the HeS 50d will be noted the large fall in thrust expected as the air speed increased:

Thrust	950 kp (2,095 lb) at 0 km/h
	325 kp (717 lb) at 800 km/h (497 mph)
Fan speed	4,600 rpm
Weight	678 kg (1,495 lb)
Diameter	0.85 m (2 ft 9 in)
Length	2.375 m (7 ft 9 in)
Frontal area	0.57 sq m (6.13 sq ft)
Thrust with afterburning	1,800 kp (3,969 ft) at 0 km/h
	2,000 kp (4,410 lb) at 800 km/h (497 mph)

Following the piston engine research of the FKFS, the HeS 50d design was superseded by a new design, the HeS 50z, the layout of which is shown in Figure 2.19. Compared with the previous design, the HeS 50z was simpler, lighter, smaller, and was expected to have a far more even thrust curve with more thrust at the high end of the speed range. The new two-stroke petrol engine beneficially had a higher speed and higher-temperature exhaust gases to heat the ducted air. This 16-cylinder engine had a rated power of 800 bhp and a take-off power of 1,000 bhp. The cylinders were arranged in four rows of four in an X formation, the bore being 100 mm (3.937 in) and the stroke 70 mm (2.756 in). Simplifications arose by employing air cooling and a single crankshaft for the engine.

A three-stage, lattice-type fan with a large-diameter hub was provided at the duct intake which led back between the engine cylinder banks. A top section of the duct diverted part of the airflow into the engine section at the rear, where it was drawn forward into the carburation system by a gear-driven centrifugal fan. (The HeS 50d differed in that air for the engine was drawn in at the front immediately behind the ducted fan.) The engine exhaust pipes were built up with a series of annular intakes to admit some of the ducted air for heating. Data for the HeS 50z are:

Thrust	550 kp (1,213 lb) at 0 km/h
	400 kp (882 lb) at 800 km/h (497 mph)
(Supercharging was estimated to increase thrust by about 300 kp (661 lb) at all air speeds.)	
Fan speed	6,000 rpm
Weight	370 kg (816 lb)
Diameter	0.62 m (2 ft 0 in)
Length	1.47 m (4 ft 9^m in)
Frontal area	0.30 sq m (3.23 sq ft)

HeS 60 (MTL)

The HeS 60 was projected to supersede the HeS 50 series from which it differed in having a turbine to extract a certain amount of power from the exhaust gases and air before discharge. In addition, a gear-driven supercharger fan was considered essential for the piston engine.

The idea was to provide a power unit with larger reserves of power than the ML unit, but with a small fuel consumption rate during normal cruising flight. It

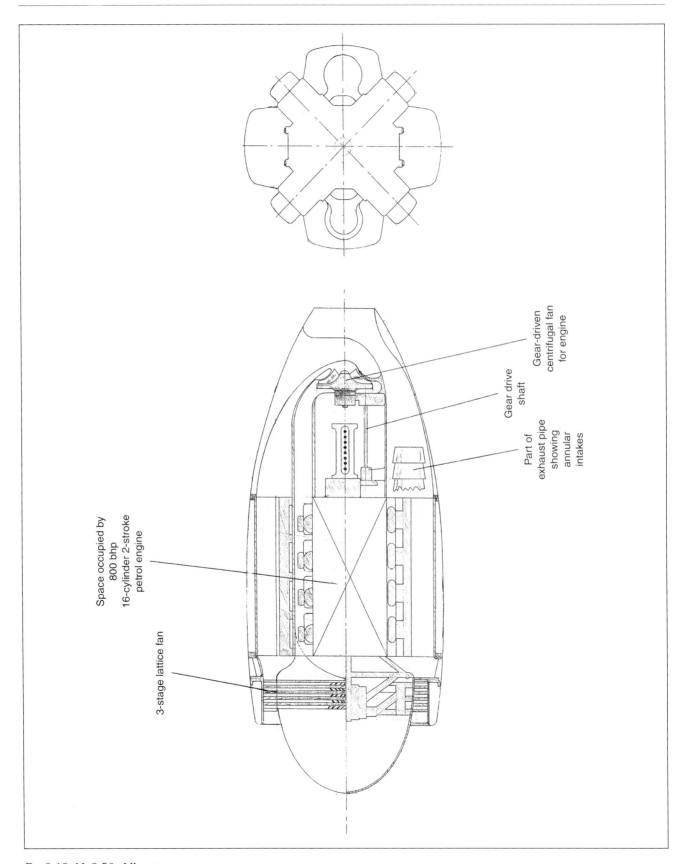

Fig 2.19 HeS 50z ML unit. *(Author)*

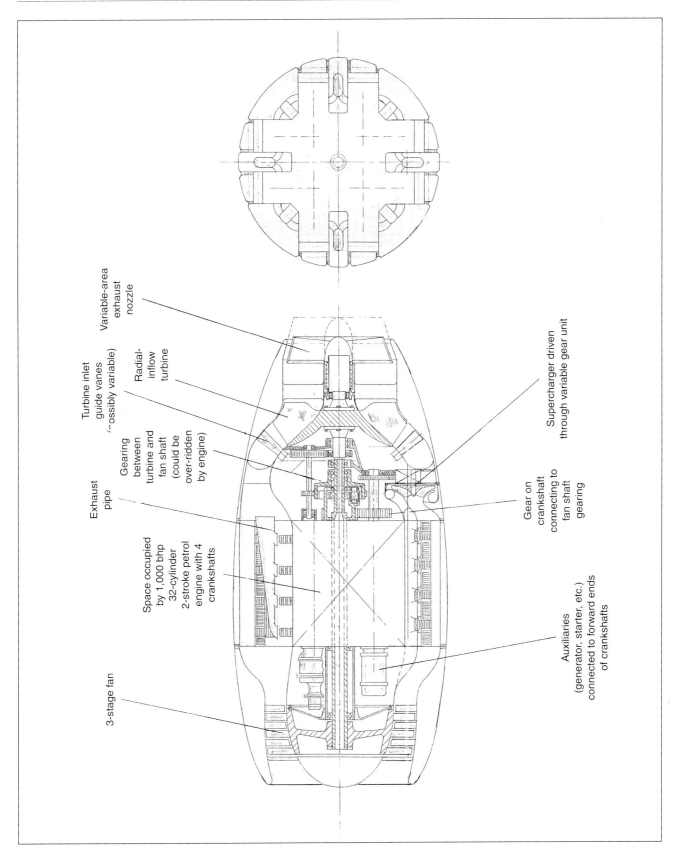

Fig 2.20 HeS 60 MTL Ducted-fan unit. *(Author)*

was thought that the resulting type of MTL unit would have a normal fuel consumption rate near to that of a similar-sized, conventional piston engine and airscrew unit, whilst, with increased output, the fuel consumption rate would be below that of a turbojet engine.

An increase in output of more than twice the normal was thought possible using the piston engine supercharger, the air output of the supercharger being governed by driving it through variable gearing.

Figure 2.20 shows the layout of the HeS 60. Its petrol engine had the same basic cylinder unit as planned for the HeS 50s, i.e. a bore of 100 mm and a stroke of 70 mm, but double the number of cylinders were used and in a different layout. Thus, for the HeS 60, the 32 cylinders were arranged in eight rows of four in a double X formation. This engine had a rated power of 1,000 bhp and a take-off or boosted power of 2,000 bhp. Its complex cylinder arrangement necessitated the use of four crankshafts which were geared to a central shaft driving the three-stage fan at the duct intake. The radial-inflow turbine at the other end of the duct was connected to the central shaft by gearing which allowed the piston engine to over-run the turbine if the rpm of the latter fell below that required to add power to the central shaft and fan. The four crankshafts provided ample means for compact connection of auxiliaries (such as a starter motor, pump and generator) to the front ends of the crankshafts, while the supercharger drive was taken from the rear end of one of the lower crankshafts. The supercharger drew in exhaust-free air from the lower part of the duct, and provision may have been made to cool the air output from the supercharger. Finally, a variable-area exhaust nozzle, using moveable, sloping, inner walls at the end of the duct, was provided. Data for the HeS 60 are:

Normal thrust	525 kp (1,158 lb) at 0 km/h
	425 kp (937 lb) at 800 km/h (497 mph)
(Supercharging was estimated to increase thrust by about 675 kp (1,488 lb) at all speeds.)	
Fan speed	6,000 rpm
Weight	800 kg (1,764 lb)
Diameter	0.90 m (2 ft 11 in)
Length	2.075 m (6 ft 9¾ in)
Frontal area	0.64 sq m (6.9 sq ft)

Focus on the HeS 011 or 109–011 (TL) at Zuffenhausen

On 18 July 1941 the GL/C Technisches Amt issued a specification for a bomber to be powered by a new turboprop power unit. The turboprop unit layout was planned by Helmut Schelp's department and consisted of two gas turbines (with their own compressors and combustion chambers) which acted as gas producers to feed a third power turbine driving a variable-pitch airscrew. Heinkels investigation of this project suggested that it would be necessary to first build a turbojet engine which could later be employed as one of the gas producers of the turboprop unit. The first turbojet layout considered for development by von Ohain's team at Rostock is believed to have been the HeS 9, already described as an HeS 8 development. However, before the end of 1941, Heinkel had been persuaded by Schelp to begin work on another turbojet which incorporated the latter's ideas on compressors. The compressor of the new engine consisted of a diagonal or oblique flow stage (a supposed compromise between axial and centrifugal flow), followed by three axial-flow stages. The efficiency of the diagonal stage could not be high, and its use meant an increase in engine diameter, but it was claimed that, having robust blades, it would be less susceptible to damage from dirt and stones and would be less likely to ice up. Von Ohain's team was not enthusiastic about the proposed compressor system, and in fact a similar system had already been studied and rejected some years before when Mueller was at Junkers. Nevertheless, Ernst Heinkel accepted the development since it was tied up with the eventual use of the Hirth Motoren works for Heinkel turbojet work. (In particular, he hoped to employ these works for the full development of the HeS 30 engine, but this aim was never accomplished.)

Although the dual turboprop unit originally proposed never materialised, work went ahead on the single turbojet engine as the HeS 011, or 109–011, which Schelp became keen to have developed into a Class II unit in its own right. This was considered all the more important by the end of 1942 when the Class I 109–004 turbojet being developed by Junkers was making rapid progress and mass production of that engine seemed quite near. Also by the end of 1942, experimental work for the HeS 011 had begun, preliminary studies had been completed in September that year and official orders were given for Heinkel to concentrate on that engine only. Thus, von Ohain's team at Rostock was eventually moved and consolidated with Mueller's former team to work together on the HeS 011 at the Hirth Motoren works at Stuttgart/Zuffenhausen, where about 150 engineers, designers, and research workers, etc., were concentrated on the development. Although von Ohain remained as chief engineer in charge of design and construction, another man, Harold Wolff, was brought from BMW and sent to the Zuffenhausen works as technical controller of Heinkel-Hirth turbojet development; and soon (March 1943), by order of Erhard Milch, he was in complete charge of the whole works. Some of the key numbers of the Heinkel-Hirth turbojet staff at this time are listed below with their departments:

Name	Job	Dept. Ref. Code
Director K. Schif	Technical direction	TK
Dr Stieglitz	Assistant technical direction	ATD
Dr von Ohain	Design and construction	MBA

Name	Job	Dept. Ref. Code
Dr Vanicek	Basic design and research	ENA
Dr-Ing. Max Bentele	Preliminary development, almost at project stage	VEW
Schmitz	Drawing office for construction of experimental engines	MKB
Hartenstein	Experimental division	VSA
F. Rees	Technical adminstration	TVW
F. Schäfer	Flight testing	FLV
I. Rees	Production shop	SKB
Dr Slattenschek	Material research	WKF

Particularly noteworthy in the above list is Dr Bentele, who was nationally renowned as an aeromechanical engineer specialising in turbine blade vibrations.

Existing staff and workers at Zuffenhausen were particularly suited to adapting to turbojet work since part of the work of Hirth Motoren, in conjunction with the DVL, concerned the production and development of turbo-superchargers for piston engines and gas turbines as auxiliary power plants, e.g. 19–518 and 109–051. Much of this work was performed under licence, but the company was working on a turbo-supercharger to enable the Daimler-Benz DB 605G piston engine to operate at a 15,000 m (49,000 ft) altitude. An order from the RLM in 1944 cancelled all this work in order to concentrate on turbojets. Turbojet facilities at Zuffenhausen included test beds of the overhead gantry and trolley type (whereby a suspended engine could be moved and positioned to and from the test cell) and a 1,600 kW compressor test rig.

Development of the HeS 011 or 109–011 to the production stage was planned as follows:

(a) 109–011 V1 to V5: Experimental static test engines, differing in major details, to establish the basic form.
(b) 109–011 V6 to V25: Experimental engines suitable for flight tests.
(c) 109–011 V26 to V85: A second series of experimental engines spread over a long period and mostly modified from production engines. (Reports were made for V86 and subsequent engines but no specific plans were made for these.)
(d) 109–011 A-0: First production engines.

Stages (a), (b), and (d) are examined below. Although a start was made on stage (c), this did not progress far because of the concentration made on putting the engine into production. The requirements laid down in September 1942 for the 109–011 engine were:

Static thrust	1,300 kp (2,866.5 lb)
Cycle temperature	750 °C
Pressure ratio	4.4 to 1
Airflow	30 kg/sec (66.15 lb/sec)

109–011 V1 to V5 series

Although the diagonal-flow compressor (or Kombinationsgebläse) for the 109–011 was virtually insisted upon by Schelp of the RLM, it is interesting to note the flow path development sequence for HeS turbojet engines shown in Figure 2.21. Before building the first experimental engine it was necessary to establish the dimensions and form of the diagonal-flow compressor more accurately than could at that time be calculated. Tests were therefore carried out with three different axial inducers in combination with three different diagonal-flow compressors, and the final combination selected was then tested with each of the three axial compressor stages. This work was performed at speeds below the design point on the inadequate 1,600 kW test rig at Zuffenhausen.

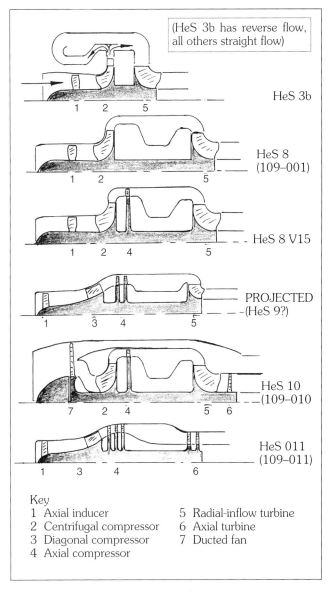

Fig 2.21 Flow path development of Heinkel turbojet engines.

When the first experimental, static engine, the 109–011 V1, was built, the compressor unit failed after less than one hour's running time. The diagonal stage of the compressor (machined from the solid) broke down through centrifugal stresses and poor material, but failure of the axial stage was attributed to fatigue caused by vibration. The problem was the responsibility of Dr Bentele, whose activities included the aerodynamic testing of compressors and special stressing work. His first report on the 109–011 V1 was VEW 1–139 issued in October 1942, and his solution of the problem was as follows. Various methods of stress calculation, checked when possible with evidence from the engine itself, were supplemented with frequency tests. Unfortunately, this first vibration failure caught Heinkel-Hirth unprepared, and these frequency checks were carried out using a piano and detecting by ear! (Later, a set of specially calibrated tuning forks was acquired, and later still, electronic and pneumatic equipment was borrowed from the FKFS.) From these calculations and tests it was found that the four-armed bearing support spider following the axial compressor stage (in the same plane as the inlet guide vanes to the combustion chamber) was the cause of the failure.

Dr Bentele was not in a position to redesign the compressor, but he redesigned the support spider with spacings of 60–120–60–120 degrees between the four arms, instead of the previous equal spacings. This altered the harmonic value of the engine speed and increased the running time to four or five hours without failure. This great improvement was but a start, however, and various experiments were made on compressor blades for later engines. To reduce root-bending stresses on the diagonal compressor blades, each blade was fitted with a bulb root allowing it to rock freely and take up a position where the air load was balanced by the centrifugal force. Friction damping plates were fitted between the blades. Also tested was the rolling of short grooves into the tips of axial compressor hollow blades to stiffen the trailing edge. These experimental blades are shown diagrammatically in Figure 2.22.

The layout of the 109–011 V1 is shown in Figure 2.23. Its essential features were as follows. Built up from steel castings and sheet metal, no attempt was made to save weight or meet aircraft design standards. The compressor set, drum-type rotor and turbine set were connected by a long tie rod to take out axial imbalance, and the whole assembly was carried by three sets of anti-friction bearings. The first bearing set comprised three ball races positioned in front of the diagonal compressor. Aft of the axial stages was the second bearing, comprising one roller race, in the plane of the exit guide vanes. Finally, the third bearing comprised another roller race aft of the turbine.

The diagonal stage of the compressor was milled out of a solid cheese of aluminium alloy, an operation which proved extremely difficult but was finally achieved by J.M. Voith GmbH of Heidenheim. Even so, an alarming

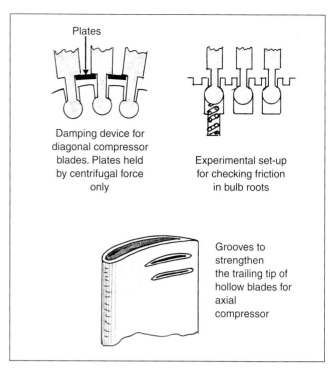

Fig 2.22 Compressor blade experiments for HeS 011 (109–011).

3,000 or so man-hours were needed on the milling machine to produce one diagonal compressor. This compressor was bolted to the front of the bolted assembly of axial rotor blade discs, these discs in turn being bolted to the front of the drum rotor. A similar construction was used for the turbine at the other end of the rotor. The two-stage, axial-flow turbine used solid blades which were held in their discs by means of pinned V roots. For the 109–011 V1, a fuel and combustion system was provided similar in design to that of the HeS 8 V15, and some secondary air was introduced into the annular combustion chamber through radial fingers or sandwich mixers. A two-position tail cone, operated by a hydraulic cylinder, was used to vary the exhaust nozzle area. Since the engine was not for flight, no accessories were fitted but an electric starter motor was mounted above the front support ring. A drive shaft led from the starter gearbox through one arm of the support spider. The same arms were set at an angle to act as guide vanes for the air flowing from the axial inducer to the diagonal compressor. Data for the 109–011 V1 engine are:

Static thrust	1,115 kp (2,459 lb)
Speed	9,920 rpm
Cycle temperature	660 °C
Airflow	21.65 kg/sec (47.7 lb/sec)
Nozzle area	0.14 sq m (1.506 sq ft)

According to the specification laid down in September 1942, a considerable advance in performance was to be aimed at by increasing the cycle temperature and air

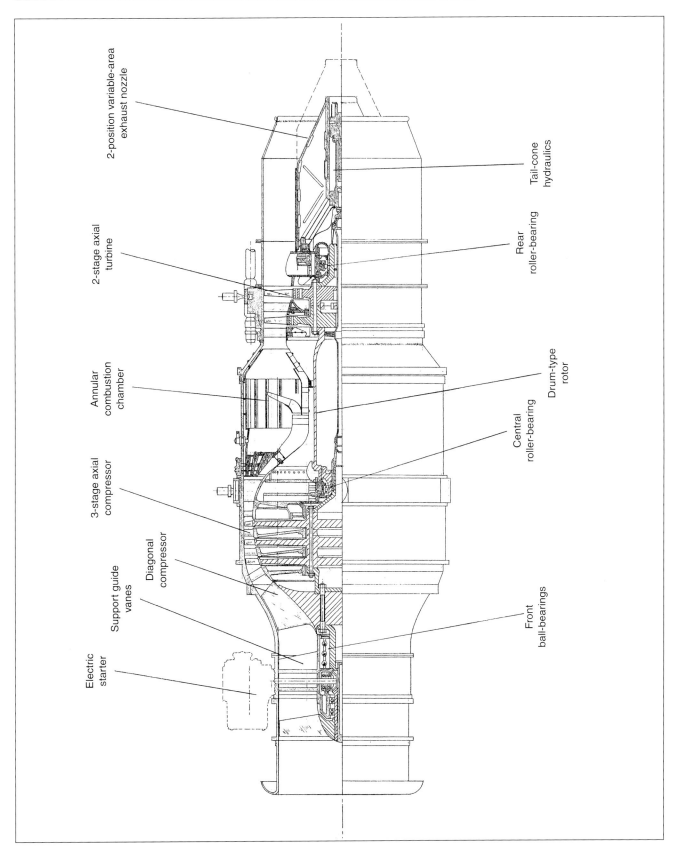

Fig 2.23 HeS 011 (109–011 V1) turbojet. *(Author)*

mass flow. The compressor set was not tested up to full speed (11,000 rpm) until December 1943, or a little later, when a DVL test bed at the Augsburg works of MAN was used. Specific details for the other four engines in the V1/V5 series are lacking, but some may have had adjustable compressor stator blades as described for the V6 and later engines. The V5 engine was tested with a DVL-designed, single-stage turbine having air-cooled hollow blades, this being the first attempt to supplant the expensive blades of solid heat-resisting alloy.

However, Dr Max Bentele's team was not satisfied with the DVL's work (which was carried out under Dr Fritz A.F. Schmidt) because they found that too much emphasis had been placed on the cooling system and not enough on mechanical and manufacturing considerations. Bentele therefore studied a dozen concepts of cooled turbine blades before conceiving a turbine blade made entirely from sheet metal – the Faltschaufel, or folded blade. Then, from two interested manufacturers, the famous Würtembergische Metallwaren-Fabrik (WMF) at Geislingen/Steige was chosen to produce blades. After some modification to suit WMF manufacturing methods, the final blade was known as the Topfschaufel (tubular, or bootstrap blade), and it became very successful in turbochargers and, finally, the 109–011 engine. (The patent application for this blade design was filed on 6 November 1943 under the names of the inventors, Max Bentele and Hans Braig (WMF)). Experimental methods of construction tried for the compressor set included forging and casting the axial inducer, fabricating the diagonal compressor and forging and sheet-metal-forming the axial stator blades. The five engines in this series were completed, as near as ascertainable, by the end of 1943 or early in 1944, but the running hours accumulated are not known.

109–011 V6 to V25 series

Work on this series of engines, following the previous broadly experimental engines, was aimed at refining the design and developing it to suit aircraft. This work phase started at Zuffenhausen early in 1944, but the threat from Allied bombing necessitated the dispersal of factories in the Stuttgart area and the continuation of work at safer locations. One such location was the salt mine known as Staatliche Saline Friedrichshall at Kochendorf. The preparation of this mine was begun in the spring of 1944 by the Todt organisation using SS prisoners, and the installation was code-named Eisbak. This installation comprised 144,250 sq m (1,552,130 sq ft), of which 36,100 sq m (388,436 sq ft) was allocated to Heinkel-Hirth and known as the Ernst Werk. Heinkel-Hirth began the move there in June 1944 and were ready to resume work on the 109–011 engine by about August/September 1944. (They were also to begin production again of exhaust-driven turbo-superchargers for piston engines.) By this time, almost two years had

Fig 2.24 This is the best photograph found of the Heinkel-Hirth 109–011 V1. Study it in connection with the drawing.

been spent on the protracted difficulties with the first five engines, and the compressor, in particular, was still giving trouble. Increasing demands were made to speed up development, and as early as the beginning of 1943, the specification of the 109–011 engine had been issued to aircraft companies to begin designing new, badly needed jet aircraft.

The major layout changes introduced with the V6/V25 engine series included a shortening of the engine by redesigning the combustion system and by supporting the rotor on two instead of three bearing sets. Figure 2.25 shows the layout of the 109–011 V6, and the following description applies to this engine. The compressor set and turbine were bolted to the ends of a much simplified drum rotor (unbroken in the middle for a centre bearing) which was mounted in an anti-friction bearing set at each end. The bearing sets comprised a ball and roller race in front of the diagonal compressor and a single roller race behind the turbine. Mounting of the front bearing housing was by three streamlined struts inside an annular casting, the struts of this spider being angled to act as guide vanes for the incoming air. Lubrication and cooling air were fed to the rear bearing through the arms of its support spider.

Composite construction was now used for the diagonal-flow compressor. Its 12 blades were machined from aluminium alloy forgings and each was held in the steel hub by a 33 mm (1.3 in) diameter bulb root of the type already described. The damping blocks between the blades at the bases also defined the inner wall of the air passage. Curvature of the diagonal-flow compressor blades was in the direction of rotation at the inlet and against the direction of flow at the outlet. At the rear of the hub was a flange to which were bolted succeeding stages, and at the front was a stub shaft for the front bearing. This shaft was internally splined for connection of a quill shaft leading to the auxiliary drive gears and axial inducer at the front. Mounting of the rotor blades of the axial-flow compressor was by U-shaped feet, the first and second stages being in aluminium alloy and the third stage in steel. Work was performed in both rotor and stator blades of the axial compressor (presumably this was the case with previous engines), but the amount of reaction is unknown. These blades followed a modified aerofoil section with a parabolic camber line.

Although the axial compressor was still a three-stage type, the stator blades were reduced from four to three rows, the last stage of the rotor blades being followed immediately by a single row of inlet guide vanes leading to the combustion chamber. These guide vanes were of steel, while the preceding stator blades were of aluminium alloy. Fixing of the stator blades was in the outer aluminium alloy casing, with mechanical means of adjusting the blade angles. Each stator blade had a shaft located axially in steel split rings keyed into the casing, the shaft having a boss with a slot for adjustment. A circular base to each blade fitted into a ring and labyrinth seal.

A complete change was made in the combustion system. In place of the former evaporative fuel system with 128 nozzles, there were 16 downstream injection nozzles. These fuel nozzles were set inside an annular shrouding ring which formed the front end of a primary air chamber. Secondary air was introduced into the combustion chamber through sandwich mixers set just aft of the primary chamber. In addition to the attempt to establish a primary combustion zone, an attempt was also made to replace combustion chamber parts (formerly of heat-resisting steel) with parts of aluminised, or aluminium-sprayed, mild steel. This was an essential move in view of the shortage of heat-resisting steels. At the exit from the combustion chamber were fitted nozzle guide vanes which were adjustable for angle, a similar feature having been used on some of the previous engines. The nozzle guide vanes were hollow for air cooling, the cooling air being carried to them from annular ducts surrounding the combustion chamber to holes at the top and bottom of each blade, the air leaving through slotted trailing edges.

Air-cooled turbine blades were apparently not tried on this engine. For obscure reasons, the former variable-area exit nozzle was dropped and a simple fixed exhaust nozzle introduced; inside this nozzle was a large, domed fairing at the centre of the rear bearing support spider.

Since the V6 engine (and those that followed) was to be suitable for flight tests, its external shape had to be made aerodynamically acceptable. Inlet ducting was attached to the front flange of the front annular casting. A sheet-metal housing led back from the intake duct to enclose the accessories at the top of the annular casting and an oil tank at the bottom. A vertical shaft passed through one of the arms of the front support spider to connect the quill engine shaft with the accessories gearbox, to which was fitted the starter motor and up to six accessories, including an impulse oil pump.

Little is known about engines subsequent to the 109–011 V6, but only this and three others in the V6/V25 series accumulated any running time. At least one engine had a cast main casing with diagonal, external bracing ribs in a large diamond pattern. The four engines tested in this series were mostly run on the ground, but one or two flights were made before the end of the war, using a Junkers Ju 88 as a test bed. No engine flew under its own power, however. Altogether, the four engines (V6 and three others) had accumulated 184 hours of running time by January 1945. Of these, 154 hours were at thrusts below 800 kp (1,764 lb), and only three hours of the remainder were at thrusts in excess of 1,100 kp (2,425 lb). Nevertheless, performance of this series of the 109–011 was quoted as:

Static thrust	1,300 kp (2,866.5 lb) at sea level
Speed	11,000 rpm
Specific fuel consumption	1.32

GERMAN JET ENGINES

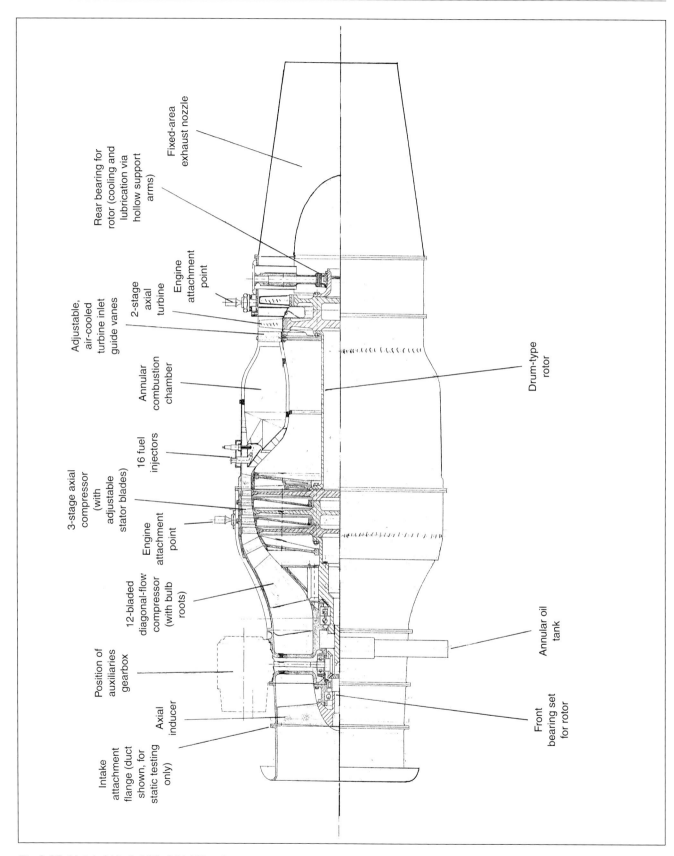

Fig 2.25 Heinkel-Hirth 109–011 V6 turbojet. *(Author)*

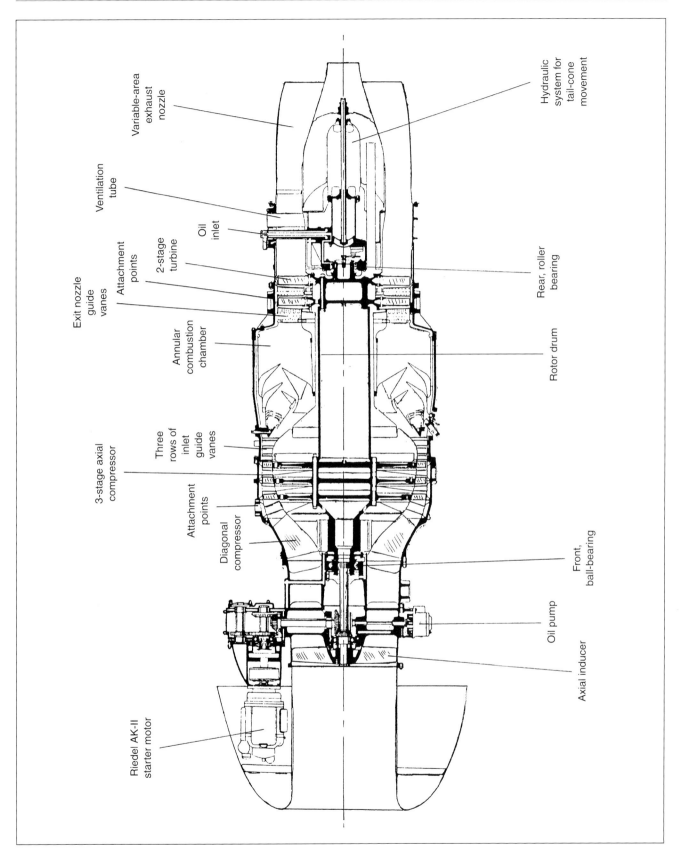

Fig 2.26 Heinkel-Hirth 109–011 A-0 turbojet.

Fig 2.27 A mock-up of the Heinkel-Hirth 109–011 A-0 turbojet built as a preliminary to production. The rod frame has been fitted to check on cowling clearances. Although not fully developed when production was initiated, this engine was the most powerful in Germany by the end of the war. *(German photograph taken on 14 March 1945)*

109–011 A-0

By the end of 1944, Heinkel-Hirth were forced to turn their attention towards the first production engine, the 109–011 A-0, despite the fact that their speeded-up efforts had still not produced a satisfactory experimental engine. As we have seen, the experimental engines were permutations of various layouts and features, making the final choice very difficult at an early stage in development. The compressor assembly was still imperfect and probably presented the major problem. Dr Encke of the AVA, Göttingen, conducted a test on the compressor assembly and pronounced its performance poor. He expected, in general, an 85 per cent stage efficiency from an axial compressor, but the flow efficiency of the 109–011 compressor was only 80 per cent because of the lowering effect of the diagonal stage. Trouble was also being given with the combustion system, there being burning of the metal mixing devices, but there were no carbon troubles. Runs were made with both B4 (paraffinic gasoline) and the cruder J2 fuel, the switch to the latter (essential because of the critical fuel shortage) having been accompanied by some starting problems.

A major redesign was undertaken for the production engine, partly as a result of the previous experiments and partly to adapt the design for mass production. Production was scheduled to begin in March 1945 at the factory already being used as a dispersal for BMW's Munich works, the dispersed factory being in a former cotton spinning mill at Kolbermoor (about halfway between Bad Aibling and Rosenheim, Bavaria). Smaller components were to be made at Ernst Werk (Kochendorf), with final assembly at Werk Waltersdorf (Berlin), while the specialised manufacturers made parts such as fuel injectors (L'Orange GmbH, Zuffenhausen) and starters (Norbert Riedel KG, Muggendorf). No complete engines were turned out during the war, however, since the dispersal to Kolbermoor was not completed until the first week in April 1945 and the war was over in the following month. The programme was affected very little by bombing after the dispersal from Zuffenhausen, and the lift shaft of the Ernst Werk was closed by artillery fire only in the final stages of the war.

The layout of the 109–011 A-0 is shown in Figure 2.26, while the photographs in Figure 2.27 show a mock-up of the engine, built as a preliminary to production. The compressor/turbine set was similar in layout to the V6/V25 series but simplified by the omission of the forward bearing roller race. Eleven blades were provided for the axial inducer on the quill shaft, each blade angle varying from 9° 30′ at the root to 48° 30′ at the tip. Little change was made in the diagonal-flow compressor which had 12 inserted blades. All blades of the three-stage axial compressor were now of hollow, sheet-steel construction, the stator blades being riveted into sheet-metal shroudings and the rotor blades being riveted into their discs. Where previous engines had used a single row

Fig 2.28 Inlet and impeller of the Heinkel-Hirth 109–011 A-0. *(Crawford Museum, via Richard T. Eger)*

of inlet guide vanes, following the last row of the rotor blades, there were now three rows of inlet guide vanes leading to the combustion chamber.

Air-cooled, hollow blades of Bentele/WMF (Topfschaufel) design were used for the two-stage turbine. Each blade (see Figure 2.29) was held by its hollow root by a single pin in the turbine disc, and was thus able to rock to avoid root stresses in a similar manner to the diagonal compressor blades. Cooling air from the annular space surrounded by the combustion chamber was led through holes in the turbine discs, up through the rotor blades and out through their tips. Part of the same cooling air also entered the exit nozzle guide vanes preceding the turbine as before. Cooling air for the single row of turbine stator blades was led into their roots from the outer annular duct surrounding the combustion chamber. The working temperature in the region of the turbine was about 750 °C, while the maximum temperature in the blades was kept below 600 °C by cooling.

A considerable re-design of the combustion system was made for the 109–011 A-0. Ideas on a primary combustion zone had crystallised further. Sixteen downstream fuel nozzles were provided, the fairing for each nozzle acting as supporting spokes for a V-sectioned ring shrouding the nozzles (similar to the V6). Primary air led around this shrouding ring and mixed with the fuel with the assistance of six rows of small mixing fingers set aft of the primary chamber. For the introduction of secondary air into the combustion chamber, a return was made to the radial finger type of sandwich mixer used in the first experimental engines,

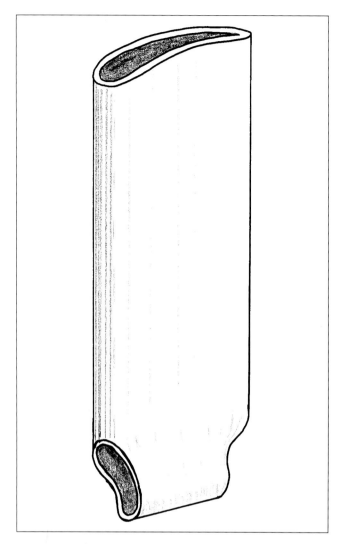

Fig 2.29 Topfschaufel hollow turbine blade for 109–011 A-0.

of it is ejected in unburnt droplets, and combustion efficiency is lowered. The duplex fuel injector for the 109–011 A-0 is shown in Figure 2.30. Two fuel manifold inlets were used, one to supply the primary stages of the atomiser and the other the secondary stages. Two fuel pumps were provided, one a service pump to draw fuel from the main tanks and to feed the inlet of the second pump, which delivered fuel under pressure to the two manifolds of each injector. A loaded relief valve permitted fuel to pass to the primary stage manifold at pressures up to about 1.55 kg/sq cm (22 psi), above which pressure both manifolds were fed. The service pump was made by Ehrich & Graetz of Hamburg and the pressure pump was made by Barmag of Remscheid. Maximum output of the 16 nozzles was about 1,770 kg (3,900 lb) of J2 fuel per hour, and fuel pressures ranged from 1.05 kg/sq cm (15 psi) up to 60 kg/sq cm (853 psi).

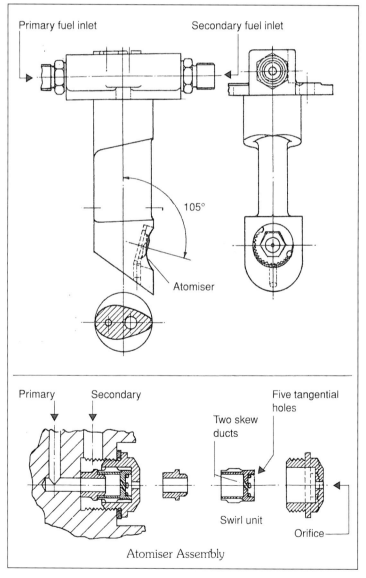

Fig 2.30 Duplex fuel injector for 109–011 A-0.

but now four rows of fingers were used. These fingers scooped air from the annular spaces surrounding the combustion chamber, which used aluminised steel sheet construction. Quite good mixing was achieved with the fingers for a pressure loss of about half an atmosphere, but oxidation of the metal was rapid.

Although the drawing shows simplex-type or single-nozzle fuel injectors, production engines were, in fact, to be turned out fitted with 16 fuel nozzles of the duplex type. This was an important feature and of special interest since no German turbojet was so equipped, those turbojet engines which went into service use (BMW 109–003 and Junkers 109–004) being fitted with single-nozzle fuel injectors. Since a fuel injector not only sprays but meters the fuel, a single nozzle cannot supply the correct amount of fuel under all conditions. Thus, when the engine is running slowly or is operating at high altitudes, i.e. when the combustion chamber pressure is low, the fuel leaves the injectors in a coarser spray, much

A variable-area exhaust nozzle was again introduced, movement of the open-ended tail cone being by means of a hydraulic cylinder. An increase in the overall length of the engine occurred as a result of a rearrangement of the forward end. Where, previously, the auxiliary drive shaft had passed through the support spider and guide vanes just aft of the axial inducer, auxiliary drive shafts for the 109–011 A-0 now passed out through their own two hollow fairings, the guide vanes being positioned immediately behind these two fairings. The system of bevel gear drives from the engine quill shaft to the radial auxiliary shafts was essentially as before, as was the accessory gearbox. The additional, second auxiliary shaft drove an impulse oil pump at the bottom of the engine intake, engine lubrication being on a closed-circuit system using a 50/50 brand of an engine oil and a light oil. The starter motor, attached to the front of the accessory gearbox, was a 10 hp Riedel AK11 (9–7034) two-stroke petrol engine which could be started by hand or by a small electric motor. A larger starter motor, Riedel AK1 (9–033), could be fitted but necessitated a new engine intake cowling which lengthened the engine by 0.15 m (6 in). (A more complete description of the Riedel type of starter is given in the Junkers section.)

Data for the 109–011 A-0 are:

Thrust	1,300 kp (2,866.5 lb) static
	1,040 kp (2,294 lb) at 900 km/h (559 mph) at sea level
	500 kp (1,002.5 lb) at 900 km/h (559 mph) at 10,000 m (32,800 ft)
Max. speed	11,000 rpm
Idling speed	6,000 rpm
Specific fuel consumption	1.31
Compressor efficiency	80 per cent
Weight (±2%)	950 kg (2,095 lb). Engine was being lightened and was expected to weigh 865 kg (1,907 lb) finally
Specific weight	0.73 approx.
Height	1.08 m (3 ft 6 in)
Width	0.864 m (2 ft 10 in)
Length	3.455 m (11 ft 4 in)

Planned application of the 109–011 A

One of the first aircraft designed to use the Class II 109–011 A turbojet was the Heinkel He 343D, which was planned as a four-engined reconnaissance version of

Fig 2.31A The Heinkel-Hirth 109–011 A-0 turbojet mock-up demonstrating the under-wing fitting.

Fig 2.31B Heinkel-Hirth 109–011 A-0 turbojet at the Cranfield Institute of Technology, Bedford, March 1970.

(Richard T. Eger)

Fig 2.31C and D A Heinkel-Hirth 109–011 A-0 turbojet in American hands. *(Official USAF photo)*

the fairly conventional He 343 jet bomber project. During 1944, Focke-Wulf made project studies to ascertain the best bomber layout to utilise two 109–011 A engines, the common basis for each bomber study being a 1,000 kg bomb load carried at 1,000 km/h on an action radius of 1,000 km. Since the 109–011 A was planned to replace Class I turbojets, there were naturally projected developments of jet aircraft, already built or in service, which used Class I turbojets. The Arado Ar 234, the world's first jet bomber, was projected in various versions (P-3, P-5 and D-2) to use two 109–011 As, while the Ar 234 C-7 night-fighter was to use four such engines. Similarly, there were the projects for the Messerschmitt Me 262 fighter (in service) and Junkers Ju 287 B bomber (experimental), each to be fitted with two 109–011 As, and the Heinkel He 162 fighter (just entering service) and Gotha Go 229 fighter-bomber (experimental), each to be fitted with one such engine.

Most of the new aircraft planned to use the 109–011 A were fighters. When details of this engine were released in 1943, aircraft firms (especially Messerschmitt and Focke-Wulf) began planning single-engined jet fighters which could more efficiently equal or surpass the performance of twin-engined jet aircraft such as the Me 262. However, official faith was laid with existing jet aircraft (powered by Class I turbojets) until near the end of the war, when the problem of unreliable operation of Class I turbojets above about 11,000 m (36,080 ft) was recognised as a serious shortcoming. This was all the more serious when it was realised that even the German speed advantage could be challenged by the imminent appearance of Allied jet aircraft. It was hoped that the 109–011 A turbojet would operate at ceilings in excess of Allied aircraft, especially if the duplex fuel injectors (already described) came up to expectations. At the end of 1944, therefore, the OKL was demanding a new jet fighter to defend a reeling Germany, and an emergency fighter competition was instigated by Col. Siegfried Knemeyer (Chef/TLR). The specification issued to all the main aircraft companies called for the new fighter to be powered by a single 109–011 A turbojet, have a level speed of 1,000 km/h (621 mph) at 7,000 m (22,960 ft), operate at altitudes up to 14,000 m (45,920 ft), carry about 1,500 litres (330 gallons) of fuel, and be armed with four 30 mm MK 108 cannon. Eight projects from five companies had been received by February 1945, and were considered at an official meeting on 27 and 28 February. Briefly, these eight projected aircraft were:

Type	Layout	Air intake/s
Blohm und Voss P.212	Tailless	Nose
Heinkel P.1078C	Tailless	Nose
Junkers EF 128	Tailless	Fuselage sides
Messerschmitt P.1101	Tailed	Nose
Messerschmitt P.1110	Tailed	Annular, above fuselage
Messerschmitt P.1111	Tailless	Wing LE roots
Focke-Wulf Ta 183/I	Tailed	Nose
Focke-Wulf Ta 183/II	Tailed	Nose

From these designs, Focke-Wulf's Ta 183/I was chosen for immediate development and production. The first prototypes were to be powered by Junkers 109–004 turbojets pending delivery of the first 109–011 A turbojet, but despite prodigious effort, even detail design was not completed when all Focke-Wulf factories suitable for production were over-run by Allied troops in April 1945. To enhance its interception performance, the Ta 183 was to have a 1,000 kg (2,205 lb) thrust rocket motor fitted beneath the turbojet, rocket fuel for a 200-second burn being carried in underwing drop-tanks.

The aircraft intended to result from the emergency fighter competition was the most important one planned for the 109–011 A engine. Even before the competition was held, early in 1944, Messerschmitt designed the P.1101 and, in July 1944, began construction of a prototype on their own initiative. This prototype could have the sweep-back angle of its wings varied on the ground, and its construction was about 80 per cent complete when it was captured by the Americans (see Figure 2.32). Only a mock-up 109–011 A engine was fitted, however, and the P.1101 prototype was scheduled to begin tests with a 109–004 engine.

Other fighters projected to use a single 109–011 A turbojet included the Messerschmitt P.1116 and P.1106, Heinkel P.1079 A-1, Blohm und Voss P.215 and Dornier Do 435, the latter being planned to use a DB603 piston-engine in addition to the turbojet. Space precludes more than this brief mention of some of the aircraft designed around the 109–011 A, which, almost to the end of the war, officialdom was expecting to see in service.

The Tuttlingen engine

Work on a novel gas-turbine engine, attributed to von Ohain and H. Wolff, began after the early stages of the 109–011 turbojet development were over. The name of Tuttlingen was given to the new engine after the location of the dispersed site where the work was performed, Tuttlingen being on the river Donau approximately 100 km (62 miles) SSW of Stuttgart.

The aim of the designers was to provide an improvement of the Holzwarth constant-volume combustion turbine by avoiding the intermittent combustion of the cycle with its associated mechanical difficulties. In the Holzwarth cycle, cold air at low pressure is heated at constant volume to give intermittent high pressure and temperature gas for the turbine. Von Ohain proposed that a portion of the temperature drop available from the hot compressed gas should be used to compress the cold air to the higher pressure. For this purpose, the Tuttlingen engine had the essential feature of a single rotating element which performed the function of a compressor and a turbine. Also, there was fixed valving arranged around the rotor element to ensure an intermittent hot and cold gas stream. The scheme had many variations proposed, but the most favoured is described in the following.

GAS TURBINES FOR AVIATION

Fig 2.32 Three views of the Messerschmitt Me P.1101.

Top and centre views: The aircraft with a Heinkel-Hirth 109–011 A turbojet engine and mock-up cannon. The engine has an axial alignment pole fitted in the outdoors shot.

Bottom view: Later, in the USA, the damaged P.1101 had a trial fitting of an Allison J35 axial turbojet and was extensively studied by Bell Aircraft Corporation prior to design of the Bell X-5 research aircraft. Two X-5s were built, the first flying on 20 June 1951 from Edwards AFB.

For the mode of operation, reference should be made to Figure 2.33, which gives the arrangement of the Tuttlingen engine and four cross-sections. Cross-sections BF and GH show stationary sector valves at the ends of the rotor element, the outlet area being greater than the inlet area. Cross-section AB (at the forward end of the rotor element) shows the guide vanes and combustion chamber outlet directing the exhaust gases down one side of the rotor element, while cross-section CD (at the rear end of the rotor element) shows the guide vanes and inlet leading air up from the other side of the rotor element into the mixing and combustion chamber. It will be noted that the rotor element drives a precompression fan at the engine air intake. After passing through this fan, the air passes into the rotor element blades at the bottom and is drawn up into the mixing and combustion chamber. At the same time, a fuel/air mixture is being burned in the combustion chamber, and the resulting hot gases exhaust down the other side of the rotor element and give up some of their heat to the incoming air as they pass out of the rotor end into the power turbine section. The exhaust gases are partly directed onto this turbine by fixed nozzles behind the rear stationary sector valve.

In the arrangement chosen, the power turbine was free from the rotor element, thus necessitating four bearing sets for the complete engine. The planned application of this engine is not known and probably could not be decided upon until extensive tests showed the full characteristics of the engine. The arrangement shown, however, was inconvenient for aircraft use unless the power turbine drove a pusher airscrew. On the other hand, the Tuttlingen engine, without the power turbine, could be treated as a gas producer, and two such units in parallel could drive a separate power turbine connected to an airscrew. This arrangement would have fitted in with Helmut Schelp's idea for a dual turboprop unit in 1941. Advantages claimed for the Tuttlingen engine were a small overall diameter, simple construction and the use of a high working temperature, since the rotor material was subjected to the high temperature for only a part of the working period. However, the engine appeared unattractive to Allied investigators after the war, since, even by using a combined compression-expansion efficiency of 80 per cent, a cycle efficiency of only 27 per cent appeared possible.

According to von Ohain, a small test model (60 hp) was built and run at Tuttlingen, but the engine either blew up or was dismantled and the parts dispersed. In addition, detailed drawings were lost and little evidence remained by the end of the war.

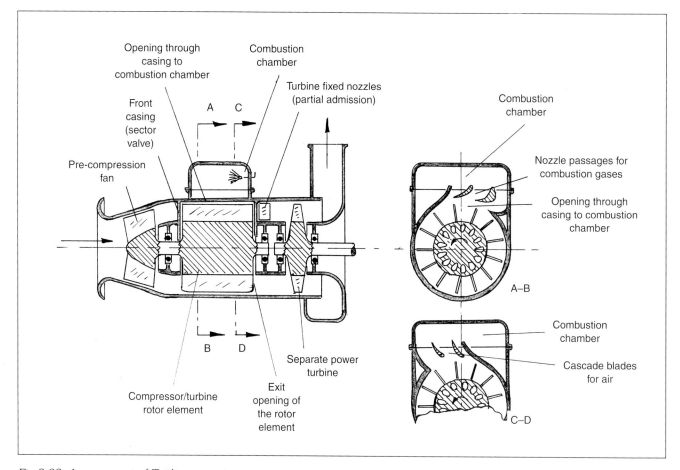

Fig 2.33 Arrangement of Tuttlingen engine.

109-021 (PTL)

In January 1945, a turboprop (PTL) version of the 109-011 turbojet was projected, designated 109-021. This was based as far as possible on the 109-011 A-0, but with the shaft extended forward to drive a single propeller via planetary gearing, though no extra turbine stages were added. There was a short ducted spinner forming the air intake.

Little else is known about this project apart from the details shown in the Heinkel-Hirth drawing 109-021.6008 dated 29/1/45 (see Figure 2.34). This was the simplest way of adapting the 109-011 turbojet to give propeller power, but strangely the project was also given to Daimler-Benz, which company projected a more complex version of the 109-021 (see under Daimler-Benz, page 150).

Conclusion

Despite the early start made in 1936 by the Ernst Heinkel AG, the Heinkel, and later Heinkel-Hirth, turbojet development programme was a failure, since after nine years of work no engine was in production. The reasons for this failure, all the more poignant in view of the early pioneering, are not hard to find. Principally, far too few engine specialists could be recruited, and their efforts were dispersed over a great many developments and projects. This situation changed in 1942 when the order was given for all effort to be concentrated on developing a single turbojet, the 109-011. However, another weakness was then shown up in that the Heinkel-Hirth organisation was geared more for research than for developing an engine up to production status, and the layout chosen for the 109-011 proved to be a difficult one. Although this engine was the most powerful turbojet in Germany by the end of the war and was about to go into production, it was by no means fully developed. Following an official request, BMW looked into the questions of giving technical assistance to Heinkel-Hirth and the experimental manufacture of their 109-011 engine, and held meetings to discuss these questions on 15 and 19 February 1945. BMW were then invited to give their opinion of the 109-011 design at a conference held in March at which von Ohain and Schmidt (or Schmitz?) of Heinkel-Hirth met Oestrich, Biefang and Hagen of BMW. A film strip was shown of the engine after an eight-hour test, and the combustion chamber was in a very bad condition. The opinion was formed that performance indicated a thermodynamically poor match between the compressor and turbine, and the BMW personnel were not impressed with the 109-011. Thus, Heinkel-Hirth found themselves in a difficult position, but the story might have been different had officialdom allowed Heinkel to complete the development of the promising HeS 30 or 109-006 turbojet.

With the end of the war, a wealth of documentation remained for the study of Allied investigators. Held in custody for interrogation at Kolbermoor were von Ohain, Hartenstein, Tetzloff (a production engineer) and numerous other Heinkel-Hirth turbojet personnel. Most of the engineers soon found employment abroad, Dr von Ohain and Dr Bentele, for example, working for a time with the Curtiss-Wright Corporation in the USA. It was the Americans who expressed chief interest in Heinkel-Hirth turbojet work. Under the auspices of the US Navy, six examples of the 109-011 A-0 turbojet were ordered. Parts for ten engines were already on hand at the Kolbermoor plant, and final assembly of the engines took place at Stuttgart/Zuffenhausen at the end of 1945. At least one of these engines went to the National Gas Turbine Establishment, Farnborough, England. Of the others, it appears some were tested to destruction at Kolbermoor, while some went to the USA for flight tests. It is said that these tests did not warrant continuing development, especially as turbojet development was by then becoming very rapid in the USA and elsewhere. However, a political card also came into play when the Soviet press made a big fuss about the apparent encouragement the US Navy was giving to German turbojet development. Therefore, the whole programme was very quickly closed down and co-operation was ended in 1946. One source said that Ernst Heinkel, as a last throw of the dice, tried to sell the 109-011 to the Egyptian government. Other engines which interested the US Navy were the 109-006 turbojet and the Tuttlingen engine, but these interests were not pursued far. Two preserved examples of the 109-011 turbojet are about all that remains of Heinkel-Hirth turbojet work today.

From 1951, contracts signed between the Spanish government and certain German companies resulted in the establishment of a group of German and Spanish engineers at the Instituto Nacional de Industria (INI) to develop the 109-011 turbojet up to 2,500 kp (5,512 lb) static thrust as a first step and then put it into production. Three examples of the engine were built in 1956 but, unfortunately, interminable technical difficulties and lack of machine tools resulted in the closure of the programme around that time.

The general consensus of opinion was that Heinkel-Hirth had been overtaken technically by BMW and Junkers in the development of turbojets. The earlier Heinkel engines relied to a great extent on the use of heat-resisting steels, which, in view of the German shortage of such metals, became quite unacceptable as the war progressed. They therefore made a late start on

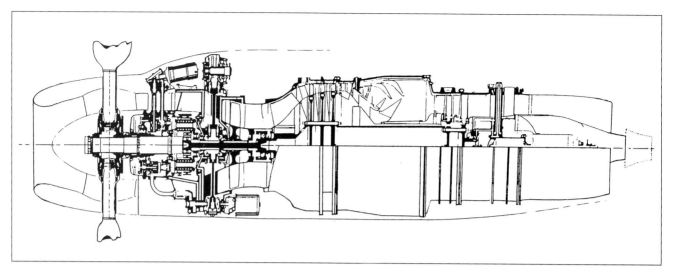

Fig 2.34 Heinkel-Hirth projected 109-021 (PTL) turboprop.

the development of more-economical, air-cooled components which the other firms eventually had down to a fine art. This is but one example of where Heinkel-Hirth lagged behind, but to be fair it seems probable that more progress would have been made once sufficient testing of the 109–011 A-0 turbojet had revealed all the areas where further improvements were indicated. Other considerations aside, however, credit for the early Heinkel pioneering will stand the test of time; in particular will be remembered the world's first all-turbojet flight performed by the He 178 powered by an HeS 3b turbojet engine, both designed and built by the same company. Furthermore, the pioneering work of von Ohain, nurtured by Ernst Heinkel, was instrumental in kick-starting Germany's turbojet programme.

An interesting postscript comes from the talented Helmut Schelp, who instigated a national programme out of the RLM, when he was brought to England for interviewing in 1945. He said: 'Immediately after the war, the British flew me to London for interrogation. They put me up in an apartment house and let me run around London on my own. I was only to be available on twenty-four hours' notice for talks with British officials. I was treated royally. As I was walking around London one day I noticed a display of the Whittle turbojet plane. I guess it was the Gloster E.28, which was the first one they flew. It had a sign on it saying: *This aeroplane made the first turbojet flight in the world*. The next time I saw Lord Banks* I kidded him about it. I said, "You know that's not true". I told him that certainly he knew that Ohain and Heinkel flew their turbojet on 27 August 1939 and, well, hell, I knew that was true because I was there. He said, "I believe we overlooked that fact." '

That was to do with early confusion over the first flight, but the same confusion can still occur today. It's the same story with the engine itself. For many years, in the interests of national pride, the words 'Who was first with the turbojet engine, the British or the Germans?' could be heard and still can. Finally, when Dr von Ohain received the AIAA Goddard Award in January 1966, he met the British turbojet pioneer Sir Frank Whittle for the first time. When the subject of the first demonstration turbojet engine, the HeS 1, was discussed, Whittle laughingly said 'Then you did beat me in the date of running the machine prior to my first demonstration'. Von Ohain, however, generously discounted this and stated that his simple, sheet-metal model operating on hydrogen gas could not be compared with Whittle's first engine, which had a high degree of engineering perfection and had already solved the problem of combustion with liquid fuel.

Von Ohain was, from the early 1960s, working in the USA as the Director of the US Airforce Aeronautical Research Laboratory when, later, Whittle also settled in the USA and the two men became great friends. Amongst the great number of honours that both men received was the Charles Draper Prize for Technology, which they received jointly in 1992. Whittle died in August 1996, followed by von Ohain who died in April 1998.

For more details of post-war Heinkel work on turbojets see pp.247/251 of *Turbojet – History and Development 1930–1960*, Vol.1 (A. L. Kay).

*Schelp was referring to Air Commodore F.R. Banks.

Junkers Flugzeug- und Motorenwerke AG

Junkers turbojet development facilities and personnel at Dessau — Evolution of 109–004 A (TL) — The 109–004 B-0 (TL) — The 109–004 B-1 (TL) — Description of the 109–004 B-1 — The 109–004 B-2 and B-3 (TL) — Hollow turbine blades by Prym and Wellner — The 109–004 B-4 (TL) — Production of the 109–004 B-1 and B-4 — Testing the 109–004 — Utilisation of the 109–004 B — The 109–004 C (TL) — The 109–004 D-4 (TL) — The 109–004 E (TLS) — The 109–004 F — The 109–004 G (TL) — The 109–004 H (TL) — The 109–012 (TL) — The 109–022 (PTL) — Initial Allied intelligence on Junkers turbojets — Conclusion

The Junkers Flugzeug- und Motorenwerke AG, Germany's largest aeronautical manufacturer during the Second World War, was formed from the merging of the Junkers Flugzeugwerk AG and the Junkers Motorenbau GmbH on 15 July 1936 to give separate airframe and engine divisions within the same company. Hugo Junkers (who became a professor in thermodynamics at the Technisches Hochschule, Aachen, in 1897) founded the Junkers Motorenbau in Magdeburg (near Dessau) in 1913 and the Junkers Flugzeugwerk in 1918, although he had much to do with aviation prior to this time. Prof. Junkers's main concern in the engine field was the development of his stationary, double-piston engine in which two pistons worked against each other. This engine was developed for trucks and aircraft and, finally, as a free-piston diesel compressor which used no crankshaft or connecting rods and proved exceptionally reliable. (Raul P. Pescara also worked on such engines and took out some of the earliest patents in France and Switzerland. One of his patents, 509,758 granted in October 1930, covered the design of a closed-cycle gas turbine with a free-piston compressor.) The Junkers free-piston engine was granted a patent (629,222) as a gas generator in April 1936 under the name of Theresa Junkers, while control equipment for this generator was filed (709,066) in May 1938 under the name of Franz Neugebauer, who worked at the Forschungsanstalt with Prof. Junkers.

Even before 1935, however, the Junkers Motorenbau had made studies of the free-piston gas generator to power a turbine driving an airscrew, this representing the first Junkers proposal to utilise a gas turbine to power an aircraft. Other studies, with a far more modern approach, were also made at this time, and concerned turboprop engines with constant-pressure combustion. These were rejected because of anticipated high fuel consumption.

While Hugo Junkers was a most capable engineer and had many technical and business achievements to his credit, he was anti-Nazi and refused to collaborate, and by 1933 had been forced by the Nazis virtually to relinquish control of his businesses and leave Dessau. He retired to Munich to continue work on his free-piston engine, and there died in February 1935, aged 76. For a short time, Klaus Junkers filled his father's former position, but he knew little about aircraft or engines and was soon replaced by Generaldirektor Dr Koppenberg.

Despite the 1936 Junkers merger, the engine division (sometimes known as Jumo) and the airframe division (sometimes known as JFA) continued for some time to work on projects without interchanging ideas. Thus, Dr Anselm Franz of the engine division was considering schemes for piston engines to give more power in the form of jet thrust (and less to the airscrew shaft), while Herbert Wagner of the airframe division was interested in aircraft power plants other than piston engines. In both cases, the efficient attainment of higher speeds than were possible with a conventional piston engine was the aim.

Herbert Wagner, chief of airframe development, suggested in 1936 that a turboprop engine (PTL) might prove worthwhile, and before 1938, special development facilities, independent of the main Jumo engine factory at Dessau, were set up at Magdeburg. Here, Max Adolf Mueller set about studying the turboprop with a team of about thirty designers and draughtsmen. As the work expanded, studies also covered the turbojet (TL), the piston-driven ducted-fan (MTL) and sundry other schemes. One of the early schemes for an MLS unit, shown in Figure 2.35, projected the use of a multi-bank X-formation, two-stroke engine to drive an axial compressor. Air from the compressor flows into passages at the top and bottom of the engine and passes between the cylinders into mixing chambers at the sides of the engine. The air in the mixing chambers, already partially heated by the engine cylinders, is then further heated by the ejected cylinder exhaust gases, and the mixture in each chamber moves rearwards to the two exhaust nozzles. Before each exhaust nozzle is a combustion chamber in which extra fuel can be burned to give extra thrust.

During 1938, a prototype of an axial-flow turbojet engine was built, but was only persuaded to run up to partial speed by using an external compressed-air supply. This engine was persevered with, still unsuccessfully, into the early months of 1939, by which time construction of a turboprop engine and wooden mock-ups of a piston-driven ducted-fan engine was under way. A description of the later development of some of Mueller's engines has already been given in the Heinkel section (see under HeS 30, 40, 50 and 60) since, by the summer of 1939, Mueller and some of his staff had resigned from Junkers and gone to work for the Ernst Heinkel AG. It is interesting to note, however, that, apparently after

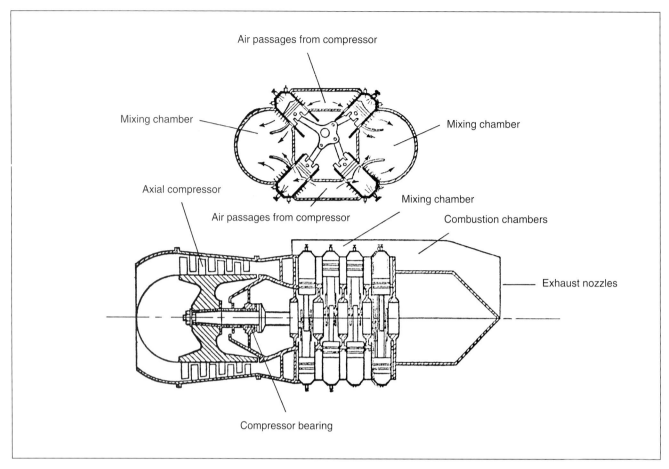

Fig 2.35 Early Junkers scheme for an MLS unit. *(G.G. Smith)*

Mueller had left Junkers, a patent was filed in his name and that of the Junkers Flugzeugwerk on 29 July 1939 for *turbine rotor cooling*.

The enterprising developments undertaken by the Junkers airframe division received no official sponsorship, although the RLM was approached by Wagner in the summer of 1938. As we have seen, the view of Hans Mauch of GL/C3 was that engine development was no job for airframe companies, and he wanted the Junkers engine division to take over all Mueller's engine work. Already, at the end of 1938, Mauch and Schelp had visited the head of the Junkers engine division, Prof. Otto Mader, and persuaded him at least to accept a study contract concerning gas turbine and jet propulsion. Mader was not, however, enthusiastic over developing new propulsion systems, but he was an outstanding engineer and wanted first to fully develop the piston aero-engine. Also, he was another man, like Hugo Junkers, who was not at all keen to collaborate with the Reich authorities. However, unlike Junkers, he managed to maintain his position until his death late in 1943, when he was succeeded by Dr Richard Lichter, who was also a capable man on piston engines. August 1938 saw a start made on the propulsion study contract, this work being led by Dr Anselm Franz, an Austrian engineer, who directed work on piston-engine aerodynamic and thermodynamic problems and the development of gear-driven superchargers and piston-engine jet exhaust systems.

Following a general survey of jet propulsion possibilities, Franz turned to more specific studies early in 1939. With the wealth of the Junkers experience with free-piston engines, it was natural that a study was first made of free-piston jet engines, but Franz soon concluded that such engines would have a power-to-weight ratio far too low for aircraft use. Following the official proposal that the Junkers engine division should take over Mueller's engines, Franz then inspected Mueller's work at Magdeburg in about May 1939, but formed the opinion that none of the engines was worth pursuing. In any case, after an earlier visit of his own to Magdeburg, Mader was willing to take on the development of only one type of engine, and this led to Mueller's negotiation and eventual employment with Heinkel.

The outcome for Junkers was that Franz favoured making a fresh start to develop a turbojet engine, and in the summer of 1939 the company received an RLM contract to develop such an engine, which was later designated 109–004.

Junkers turbojet development facilities and personnel at Dessau

Work began on the turbojet development at the Jumo engine plant, Dessau, with only a small addition to Franz's piston-engine design staff, as it was in 1939, while constructional work was initially performed in the experimental piston-engine workshops. Although Otto Mader was still anxious to concentrate available resources on piston-engine development in order to catch up with foreign developments, separate facilities and increased staff for turbojet development began to grow in 1940.

Fig 2.36 The control board of one of a line of Junkers 109–004 turbojet test cells, probably at Dessau. *(Richard T. Eger)*

A full-scale altitude test plant was built giving twin cells for engine tests, and in 1942 a second, larger, altitude test plant was built which could be run in parallel with the first one. With both plants in operation, about 7 kg/sec (15.4 lb/sec) of air could be supplied at an equivalent altitude of 10,000 m (32,800 ft), which was a sufficient air mass flow for the 109–004 turbojet engine. The power available was 6,500 hp, and ICAN temperature and pressure could be maintained up to 13,000 m (42,640 ft) altitude. The air supply for these test cells was cooled by expansion turbine and dried by means of calcium chloride, while the exhaust gases from the engines under test were cooled by water sprays and heat exchangers in the cooling towers. The Junkers full-scale altitude test cells in no way compared with the very fine BMW high-altitude chamber at Oberweisenfeld, but nevertheless provided a valuable aid for research and test work and supplemented flight tests with long hours of ground running.

All the combustion chamber test rigs were each for a single, can-type combustion chamber only. Two of these test rigs were served by a 1,800 hp compressor which supplied 3.5 kg/sec (7.7 lb/sec) of air at 4 atmos. pressure to each chamber. One altitude test rig was later available for combustion chamber tests; in this the hot gases were ducted away from the combustion chamber, which was mounted flexibly on rubber joints, the thrust being measured by reaction.

By early 1943, the Dessau turbojet staff and workers had grown from the original handful to something like 500 engineers, designers, scientists and mechanics. When Mader died late in 1943, Dr R. Lichter became technical director of the plant, while Franz headed all engine development as chief engineer. Principal members of his turbojet staff included:

Name	Job
Dr Decher	Chief technician
Dr Stein	Chief designer
Dr Beck	Design engineer – turbines
Dr H. Adenstadt	Metallurgical research
Dr Bates }	Aerodynamic research, e.g.
Dr Hagen }	compressors
Dr T.T. Schmitt	Experimental testing. Burner and turbine research
Ing. H. Timm	Test engineer

Only a few of the previous Magdeburg turbine group stayed on at Junkers to take jobs in the new group under Franz, one such notable being Dr Heinrich Adenstadt.

Evolution of the 109–004 A (TL)

In the design of the new turbojet, Junkers were given an almost free hand, the only official requirement being that the engine should develop 600 kp (1,323 lb) thrust at 900 km/h (559 mph) at sea level, which is equivalent to about 680 kp (1,500 lb) static thrust at sea level. (The RLM rated the thrust of a jet engine at 900 km/h at sea level purely as a common reference point for all engines, and not to designate a maximum speed. It will be noted that much of the thrust data already given for Heinkel engines are in non-official terms, being calculated according to that firm's earlier practice.) Having fixed the thrust, the next important requirement was that the 109–004 should be rapidly developed up to production status, even, if necessary, at the expense of the engine's performance, which could be improved once production was under way.

In choosing the essential features of the engine it was decided to use an axial-flow compressor (instead of the centrifugal type familiar from the Junkers supercharger work) to gain the benefits of a less devious flow-path through the engine and an engine with a smaller frontal area. The compressor blading was designed at the AVA, Göttingen, while the single-stage turbine to drive the compressor was designed largely by Prof. Kraft of the

Allgemeine Elektrizitäts Gesellschaft (AEG), Berlin. For the combustion system, individual can-type combustion chambers were chosen, instead of the preferred annular combustion chamber, largely on the grounds of the more rapid development that these promised, i.e. one can-type chamber could be more easily tested and modified on its own.

To obtain data for the first engine design, it was decided to construct and test a scaled-down model, the scale being such that the compressor was to absorb only about 400 hp at a speed something over 30,000 rpm and to allow the use of testing facilities then extant. The model was completed late in 1939, but the small scale gave inadequate combustion and the engine suffered from vibration. Independent tests were then made on the compressor alone, but this burst during one test at high rpm. With the failure of this model and the fresh impetus brought on by the start of the Second World War, it was decided that studies would be as well if performed on a full-size engine, and work on this, the 109–004 A, was begun in December 1939.

The 109–004 A had an eight-stage axial compressor, six separate combustion chambers and a single-stage turbine. This model was intended to be rapidly developed into a pre-production engine only, and used solid turbine blades in heat-resisting steel, but since great economies in such steel were already foreseen as necessary, the development of hollow air-cooled turbine blades was simultaneously begun for future, production engines. The prototype of the 109–004 A was first test run on 11 October 1940, without an exhaust nozzle, and was giving 430 kp (948 lb) thrust at 9,000 rpm by the end of January 1941. All was not well, however, and this engine was almost destroyed following a vibration failure of some of the compressor stator blades. To solve the problem, the renowned blade vibration specialist Dr Max Bentele was brought in. He changed the design of the stator blades, which had previously been cantilevered from their outer ends. Also, the material of the stator blades was changed from aluminium alloy to steel. A good six months of work and the testing of several engines produced a partial remedy to the vibration problem. Although these changes gave only a partial increase in life, the design thrust of 600 kp (1,323 lb) was reached during a test on 6 August 1941. At one time, consideration was given to using an axial compressor with contra-rotating stator blades, but the resulting overall diameter was considered too large for aircraft use, and there were so many other problems needing solution in the standard design that the contra-rotating compressor scheme was not pursued very far.

On 15 March 1942, a 109–004 A was first flight-tested, using a piston-engined Messerschmitt Bf 110, and soon after, two of these turbojets were sent for fitting to a prototype of Messerschmitt's Me 262 fighter. Earlier, in 1941, attempts had been made to fly the Me 262 V1 (PC+UA) with two Heinkel HeS 8 (109–001) turbojets and then with two BMW 109–003 turbojets, but without success. The first all-jet flight of this fighter type was therefore made on 18 July 1942 by the Me 262 V3 (PC+UC) fitted with two 109–004 A-0s rated at 840 kp (1,848 lb) static thrust each (see Figure 2.38). By this time, the engine was in pilot production, and about thirty engines were finally built; a pair of these were used to power the Me 262 V2 (PC+UB) on 2 October 1942, but the rest were used by Junkers for intensive experimental work. On the test stand, the V5 prototype of the 109–004 A developed as much as 1,000 kp (2,205 lb) thrust by operating at an unacceptably high temperature for a short time, but average data for the type are:

Thrust	840 kp (1,848 lb) static
Speed	9,000 rpm
Weight	850 kg (1,874 lb)
Specific fuel consumption	1.4 per hour
Specific weight	1.01
Diameter	0.960 m (3 ft 1¹ in)
Length	3.80 m (12 ft 5¹ in)

The 109–004 B-0 (TL)

While experience was being gained with the 109–004 A, the design of a production model, the 109–004 B, went ahead. Also, at an early stage, some of the larger parts were being made, since these were expected to differ little from the pre-production engines of the A-series. The layout of the 109–004 B was chosen in December 1941, and its detail design was completed by October 1942. December 1942 saw the first runs on the test stand, and the first pre-production 109–004 B-0s were ready for delivery in January 1943. The chief alterations made for the B-0, mainly in the interests of mass-production, were:

(i) Modified compressor construction using a rotor with separate discs.
(ii) Replacement of castings with sheet metal, where possible.
(iii) Substitution of more than half the weight of strategic material used in the 004 A model (although solid turbine blades were still employed).
(iv) Improved entry to air intake.

The 109–004 B-0 model was not entirely satisfactory, but the small number built were released for use with prototype aircraft.* Thus, the Me 262 V1 (PC+UA), which had first flown on 18 April 1941 with a piston-engine only, was refitted with two B-0 engines only and flew on 2 March 1943. The similarly powered Me 262 V4 (PC+UD) flew in the next month, and on 22 April this aircraft was flown by Gen. Lt. Adolf Galland as part of the effort to convince the non-technical Luftwaffe and official personnel of the desirability of jet aircraft. These first prototypes of the Me 262 fighter, unlike subsequent examples, had tailwheel undercarriages which directed

---

*Initially the Me 262 V1 to V5 prototypes used 109–004 A engines.

GAS TURBINES FOR AVIATION

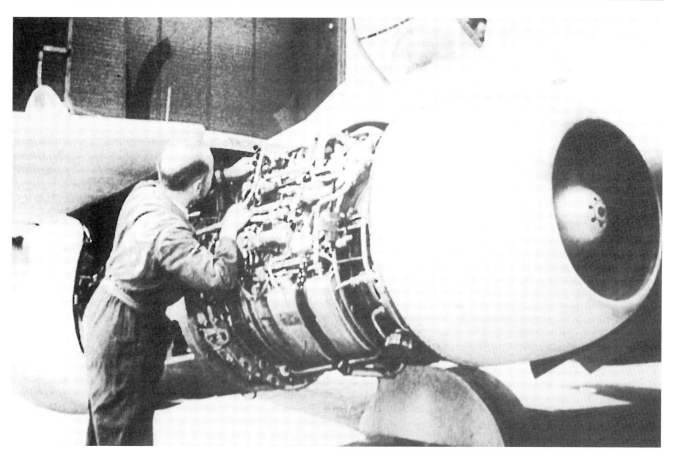

Fig 2.37 A mechanic working on a Junkers 109–004 A-0 turbojet fitted to the Messerschmitt Me 262 V3 (PC+UC). Note that there is no ring-pull in the inlet fairing for a starter engine. This version of the engine was first flight-tested under a Messerschmitt Bf 110 on 15 March 1942, and first powered the Me 262 (V3) some four months later.

Fig 2.38 The first all-jet flight of the Messerschmitt Me 262 was with the V3 prototype (PC+UC) on 18 July 1942 at Leipheim when two Junkers 109–004 A-0 turbojets were used. The early Me 262 prototypes had tailwheel undercarriages, and the pilot, Fritz Wendel, had to tap the brakes at the right speed during take-off in order to lift the tail off the ground.

the line of the jet exhausts towards the runway, with the result that, during take-off, chunks of tarmac and even concrete were torn up. April of 1943 saw the He 280 V2 (GJ + CA) and V7 (NU+EB) flying on B-0 engines, while on 15 June 1943 the Arado Ar 234 V1 (TG+KB), prototype of the world's first jet bomber, made its maiden flight with two 109-004 A engines*. By the end of 1943, two other similarly powered prototypes of this bomber had flown. Data for the 109–004 B-0 are:

Thrust	840 kp (1,848 lb) static
Speed	8,700 rpm approx.
Weight	730 to 750 kg (1,610 to 1,654 lb)
Specific fuel consumption	1.4 per hour
Specific weight	0.88 approx.
Diameter	0.960 m (3 ft 1^i in)
Length	3.80 m (12 ft 5^l in)

These data show, compared with the A-series, that the chief gain made was a considerable reduction in weight (and strategic materials) while still maintaining the engine size and performance. The large tolerance in weight shown was because of sand-cast parts, which were only machined where necessary.

*The Ar 234 V1 to V4 prototypes had 109-004 A engines, the V5 being the first with 109-004 B-0 engines.

The 109–004 B-1 (TL)

Since this engine was the major production model, it will be illustrated and described in the most detail as representative of the 109–004 turbojet. Modifications to the compressor and turbine entry nozzles decreased vibration and increased the thrust to 900 kp (1,984 lb). Although steel compressor blades had been tried previously, failures still occurred owing to resonance in the third row of stator blades. Thus, while Prof. Encke preferred to use compressor blades of thin aerofoil section, e.g. Göttingen 684, throughout, it was found better in the existing design to use thin-sectioned, wide-chord blades for the first two stages, and comparatively thicker-sectioned, narrower-chord blades for the last six stages. This did not cure the resonance problem, although the increased rigidity of the blades ensured a longer life, and other changes were made in later engines.

The first 109–004 B-1 production engines were delivered in May or June 1943, and the first aircraft to fly on the power of these engines was the Me 262 V6 (VI+AA, Werk Nr. 130 001). This aircraft was the first in the series to have a fully retractable nosewheel undercarriage, and made its maiden flight in October 1943. The same aircraft was demonstrated on 26 November to Hitler, who set back the jet fighter

Fig 2.39 The Me 262 V3 (PC+UC) after it was rebuilt following a take-off accident on 11 August 1942. It returned to flight testing on 20 March 1943. Note the Heinkel He 177 Greif bombers on the airfield perimeter.

programme by about six months by insisting that the aircraft be developed to carry a bomb load. The 109–004 B-1, although in slow production by then, was also in trouble with vibration of the turbine blades. To gain data, Anselm Franz decided to discover the ringing natural frequency of the turbine blades by asking a musician to stroke the blades with his violin bow and use his trained musical ear to determine the frequencies. Dr Max Bentele was again consulted and he discovered that the turbine rotor blade vibration problems originated with the six combustion chambers and also the three struts of the jet nozzle housing of the turbine. By early 1944 the problem was cured by increasing turbine blade taper, shortening the blades by 1 mm, modification of the turbine entry nozzles and reducing the maximum speed from 9,000 rpm to 8,750 rpm. By April that year, 13 pre-production Me 262 A-0 fighters had been manufactured in addition to some 12 developmental prototypes (see Figure 2.40), and the first of these was delivered in the summer to Erprobungstelle Rechlin for testing. At the same time, the first pre-production Arado Ar 234 B-0 bombers were leaving the assembly line, some also going to Rechlin. Although jet fighters and bombers, using the 109–004 B-1 turbojet, were soon in production and service, it was not until the war situation deteriorated further for Germany that the full priority was given to jet aircraft over conventional airscrew aircraft. A short résumé of jet operations is given later under the heading 'Utilisation of the 109–004 B'.

Description of the 109–004 B-1

A section through the complete engine is shown in Figure 2.41, while three views of an engine taken from a captured Arado Ar 234 are shown in Figure 2.42. The following describes the 109–004 B-1, the principal operational German turbojet, from nose to tail.

Nose cowling and intake casting:
The streamlined aluminium alloy nose cowling was designed to support and fair into cowling for the whole engine when used in an exposed mounting, e.g. underwing. Inside this cowling was an annular fuel tank divided into two sections; the upper section carried about 3.5 litres (0.77 gallons) of a petrol-oil mixture for the Riedel starter motor, and the lower section carried about 17.0 litres (3.75 gallons) of petrol for initiating the combustion only of the turbojet. Once started, the engine was switched to the main J-2 fuel supply. Behind the annular fuel tank was an annular oil tank of 18.0 litres (3.97 gallons) capacity. Warm oil was fed into the top of the tank beneath a baffle close to the inner surface, flowed around this inner, cooling annulus and then into the main section of the tank at the bottom. Heat from the oil was transferred to the inner surface of the oil tank, which was kept cool by air flowing into the engine.

Fig 2.40 A pre-production Messerschmitt Me 262 A-0 as a test pilot lights up the port Junkers 109–004 B-1 turbojet. Bad velocity distribution of the combustion gases was in the process of being cured with this version of the engine.

The central cowling inside the intake housed the Riedel two-stroke starter motor which was attached to the inner shell of the compressor intake casting. The form of the outer and central nose cowlings was such as to maintain an approximately constant-area intake passage of about 1,420 sq cm (220 sq in) up to the intake guides. The compressor intake casting is seen in Figure 2.43, which shows the engine with nose cowling, petrol and oil tanks and the upper half of the compressor casing removed. Four faired struts can be seen attaching the inner shell to the outer shell of the intake casting, which was of magnesium alloy. Housed in the inner shell was the bevel gear assembly for the auxiliary drives, followed by the front compressor bearing assembly. This bearing assembly consisted of three DKF single-row ball races mounted in steel liners and carried in a hemispherical light alloy housing. Several springs kept this housing in contact with the female hemisphere which was part of the intake casting. The spherical mounting was considered necessary to accommodate any misalignment of the rotor shaft, but experience showed

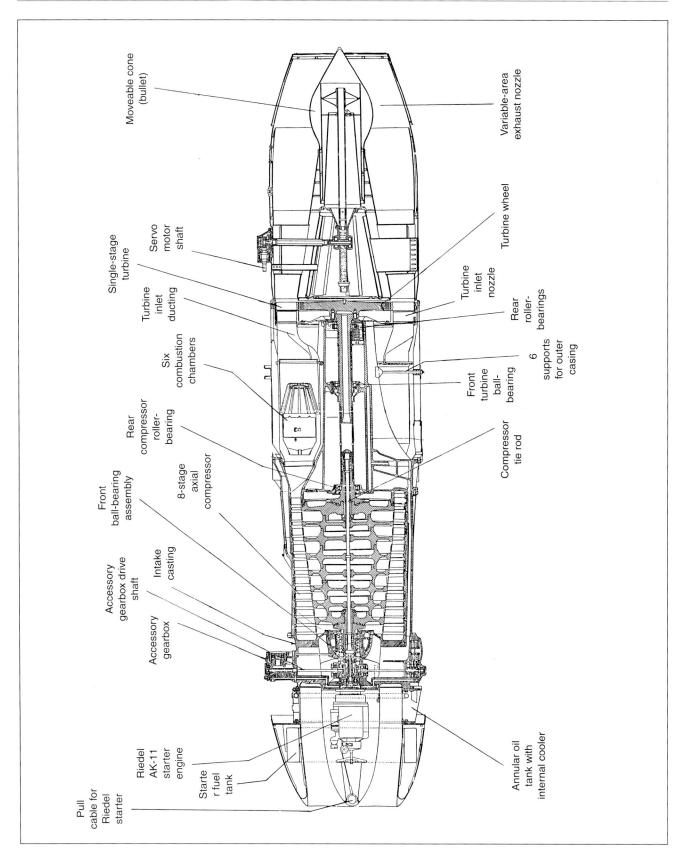

Fig 2.41 Junkers 109–004 B-1 turbojet.

GAS TURBINES FOR AVIATION

Fig 2.45 Components of the Junkers 109–004 B-1.
TOP: Turbine inlet ducting and combustion chambers.
ABOVE: Turbine inlet nozzles and inlet ducting.

Fig 2.46 Components of the Junkers 109–004 B-1 turbojet.
TOP: View of the exhaust cone or bullet from the turbine. End cover removed to show the rack and pinion drive for the cone.
ABOVE: The exhaust system.

Turbine and exhaust system:
The single-stage turbine (and the compressor it drives) is described more fully later. The turbine and compressor rotors were connected by means of a splined torque tube which allowed for axial movement. A forged molybdenum-steel disc carrying the turbine rotor blades was attached to the stub shaft by six bolts, one ball race and one roller race carrying the stub shaft in the central casting. The bearings were cooled by oil only.

Figure 2.46 gives two views of the exhaust system components. The inner and outer cones, supporting struts and outer casing were all built up from mild-steel sheet and then aluminised. The bullet fitted to the rear of the inner cone was movable in an axial direction in order to vary the area of the exhaust nozzle to suit varying operating conditions. Movement of this bullet was by means of the rack and pinion mechanism shown in Figure 2.46. The pinion shaft was driven through bevel gears by a long shaft (outside the engine) leading from a gear-type oil motor. An estimated force of about 225 kg (496 lb) was needed to move the bullet. Cooling air from the six ducts, already mentioned, led through the double skin on the outer case and through the supporting struts to the inner cone.

67

Compressor:
The eight-stage axial compressor used impulse-type stator blades whereby practically the whole of the stage pressure rise occurred across the rotor blades. Such a compressor allowed simple construction for the stator blades and did not require fine axial clearances between the stator shroud rings and the disc rims. Thus, although production was eased, a relatively heavy construction resulted. The compression ratio was 3:1 or more, and the efficiency was between 75 and 80 per cent, although a special compressor had obtained an efficiency of 83 per cent at full speed. (For such a performance, a centrifugal compressor could have been built having less weight but about twice the frontal area. As we have seen, the smaller frontal area was one of the reasons for the German choice of the axial compressor.) Air entered the compressor at about 130 m/sec (426 ft/sec) and left it at about 80 m/sec (262 ft/sec) and a temperature of about 150 °C. Since the exit velocity was low, no diffusing passage was needed between the compressor and combustion chambers. The critical speed of the compressor rotor was between 10,000 and 11,000 rpm, with a critical Mach number of 0.76 at the tips of the first rotor compressor blades at full speed on the test stand. Compressor stall occurred on the low-pressure stages along the full length of the blades.

Figure 2.47 shows the compressor mounted in the engine, while the removed upper half of the stator casing showing the stator blades is seen in Figure 2.48. The compressor rotor consisted of eight forged duralumin discs fastened together by overlapping flanges and pressure from a central tie rod, each disc, when complete with blades, being dynamically balanced before assembly with the other discs. A steel stub shaft for the bearings was attached to the disc at each end. Projections on the rear stub shaft and bolts through each disc flange ensured torque transmission from the turbine. The tie rod force pulling the compressor assembly together was of the order of 8 tonnes.

The periphery of each rotor disc had staggered grooves into each of which the dovetail root of a forged duralumin rotor blade was fitted and held in place (axially) by a grub screw. The tip stagger of the rotor blades was fairly constant on the first six stages but increased in the last two stages, while the rotor blade chords decreased in steps through the compressor. The widest-chord rotor blades of the first two stages had similar aerofoil sections of small thickness/chord ratio, while the reduced blade chords of the subsequent six stages gave comparatively thicker sections. Each of the first two stages had 27 rotor blades, while each of the other stages had 38.

Rings to carry the stator blades were made in two halves to secure into the longitudinally halved compressor casting by means of studs and nuts at the ends of each ring half. Figure 2.48 illustrates the assembly of the stator blades. The inlet guide vanes (top of the picture) and the first two rows of stator blades were of light alloy, with the ends of the blades fitted into slots in the shroud rings and secured by brazing. From the fourth row onwards, the stator blades were of steel sheet and were attached by three tabs at each end which fitted through slots in the shroud rings and were then folded over and spot welded. The third row of stator blades (shown in black) varied in both blade form and method of attachment; in some cases these blades were of light alloy sheet and brazed to the shroud ring, while in other cases the blades were of steel sheet and spot welded as for succeeding blades. The inlet guide vanes and the last row of stator blades acted as airflow straighteners, but the other (impulse) stator blades were set at almost zero stagger merely to redirect the air into the next row of rotor blades. A slot pointing upstream was set into the compressor casting over the fifth row of rotor blades to bleed off the air for cooling purposes.

Combustion chambers:
Air from the compressor was delivered at a maximum velocity of about 75 m/sec (246 ft/sec) to the six circular combustion chambers. The components of one combustion chamber are shown in the four illustrations of Figure 2.49, while a diagram of the flow path through one combustion chamber is given in Figure 2.50.

Each combustion chamber consisted of three main pieces, these being a mild-steel outer air casing, 523 mm long × 219 mm maximum diameter (20.6 in × 8.6 in), inside which was an aluminised mild-steel liner and a flame tube. The liner was spaced from the air casing by a corrugated sheet to allow cooling air to pass between the liner and casing. The flame tube consisted of a mild-steel cylinder (black enamelled) and a mild-steel, grid-like stub pipe assembly which carried a dished, circular baffle plate some 100 mm (4 in) in diameter. The stub pipe assembly consisted of 10 chutes welded to a ring at one end and to the baffle plate at the other, the ring being riveted to the black enamelled cylinder. To help force air down the hollow stub pipes, small semicircular baffles were riveted to the support ring. One spark plug was fitted to every other combustion chamber, combustion being initiated in the other three chambers by virtue of the interconnecting tubes between the chambers. The weight of one complete combustion chamber was 8.6 kg (19 lb).

Primary air for fuel combustion entered each combustion chamber through a six-vane swirler, the fuel being injected upstream into this swirling primary air from a fixed nozzle on a long stem. As the diagram shows, mixing of the hot gases of combustion with the cooler secondary air occurred at the stub pipe section. Some of the secondary air passed down the stub pipes and was forced back into the hot gases by the baffle plate at the end, but most of the mixing was done by deflecting the hot gases into the secondary air, because if the reverse was done the air deflectors for the cool air had their tips burnt away. (Heinkel-Hirth problems in this respect will be recalled.) An experimental

Fig 2.47 View looking down on the compressor rotor of the Junkers 109–004 B-1. The air inlet end is on the right.

Fig 2.48 Upper half of compressor stator casing for the Junkers 109–004 B-1.

Fig 2.49 Components of a combustion chamber for the Junkers 109–004 B-1 turbojet.

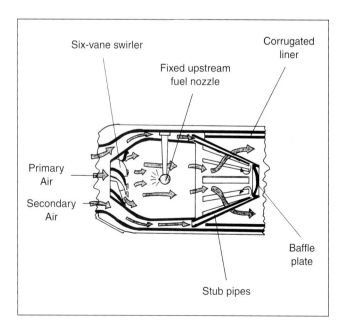

Fig 2.50 Flow path for one combustion chamber of 109–004 B-1.

combustion chamber was tried whereby secondary air was deflected into the hot gases, but not much development work was performed on it. The best combustion efficiencies with the production design (at sea level) were 97 to 98 per cent at full speed, dropping to 92 to 95 per cent at 80 per cent full speed, the pressure loss being between 0.1 and 0.15 atm.

The standard fuel used was the J2 oil-type grade, but with the fixed-components injection nozzle, atomisation was poor during starting. Thus, to give easy starting, B4-grade petrol was first fed by a small booster pump to the nozzles to initiate ignition. However, the chief weakness of the system was instability of combustion at altitude, which, because of the difficulty of relighting, imposed an inadequate service ceiling on operational aircraft. Instability became marked around 10,000 m (32,800 ft) altitude, and the maximum altitude reached by an Me 262 fighter using two 109–004 B engines was 12,700 m (41,656 ft). To permit safe operation at higher altitude, a duplex fuel injector was being developed.

The life of the combustion chambers was one of the chief limiting factors on the life of the engine. Combustion chamber parts usually had to be replaced at the 25-hour overhaul. Unfortunately, changing a chamber meant a complete engine strip, owing to the inaccessability of the chambers. It will be recalled that these individual can-type chambers were initially chosen to speed development, but in view of the short life the poor accessability of the engine design is curious. The fuel injectors or burners were the components with the shortest life (field spares were 400 per cent of operational requirements), but these at least could be changed quite easily working through the six large holes in the steel shell surrounding the combustion chambers. It was merely necessary to undo a fuel-pipe union, remove two bolts and then draw out a complete injector.

Turbine:

A view of the engine with exhaust system removed to show the turbine wheel is given in Figure 2.51. Since AEG collaborated in the design of the turbine, contemporary steam practice was followed. A certain amount of reaction was chosen for the blades to obtain the requisite amount of work from them with a relatively low turbine speed and without too much swirl in the exhaust gases. Also, the intention of later using afterburning (see later) was kept in mind. The amount of reaction chosen was 20 per cent, although Junkers wanted more and AEG wanted none at all. This low reaction gave an end thrust on the turbine rotor sufficiently low for one ball thrust race to cope with, and also enabled cooling air to be drawn up over the blade roots. Although solid turbine blades were first used (004 B-1 and B-2) for quick results, air-cooled, hollow blades were put under development from the beginning so that higher temperatures and lower fuel consumption could be realised in later engines. All versions of the engine used air-cooled, hollow turbine inlet nozzles. The general design of the turbine gave an unspectacular efficiency, somewhere around 80 per cent, and metallurgical examination (by X-ray) was carried out at the 25-hour overhaul stage. If passed, the turbine was refitted for another 10 hours' service, but 35 hours was the absolute maximum life for the turbine.

Cooling air for the 35 turbine inlet nozzles was tapped off the compressor and passed through three of the holes beneath the ribs of the central casting. The air emerged into the space between two diaphragms (supporting the inlet nozzle ring and inlet ducting), and from there out to the hollow nozzles and out through their trailing-edge slots. Each nozzle was formed from manganese-steel sheet (18.0 Mn 9.0 Cr) bent round and welded to shaped spacers. Due to its low scaling resistance, the steel of the nozzles would not have given adequate service without air cooling.

Most of the 109–004 B-1 engines were fitted with 61 turbine blades, each being a solid forging with a forked blade root secured by two rivets in the turbine disc. One blade weighed 346 gm (12.2 oz), and the heat-resisting material was known as Tinidur (0.13 C, 0.8 Si, 0.7 Mn, 2.1 Ti, 29.2 Ni, 14.9 Cr, blce Fe), a high-alloy, non-magnetic steel made by Krupp of Essen. Tinidur was similar to the British Nimonic 80, but its creep strength fell off sharply at about 580 °C, and an increase in its nickel content to counteract this was impractical because of the German shortage. The blade roots and tips, however, ran fairly coolly at about 450 °C, due largely to a cooling airflow over the blade roots and the nature of the temperature distribution from the combustion chambers. On the other hand, the centres of the blades ran at a higher temperature, nearer the limit, and centrifugal and gas bending stresses on the blades were accordingly kept low.

The solid turbine blades gave varying results at first, some lasting as long as 100 hours while others were finished after only five hours' running, without any obvious differences. However, Junkers found that it was necessary to exactly control all stages of blade manufacture to obtain the necessary uniformity. In particular, the forging operation needed special care, since at that stage the basic heat-resisting properties of Tinidur were formed. Most of the solid blade problems were solved by the end of 1943, and more concentration was then given to the full development of hollow turbine blades, described for later engines.

The turbine inlet nozzles also gave much trouble and failed in many different ways at first, their life varying between 90 and two hours. Various combinations of inlet nozzles to turbine blades were tried, some with very bad results, but the combination 35/61 appeared to most effectively stop resonance. The problem then resolved itself into one of detail design in order to obtain sufficient rigidity in the shroud ring and good inlet nozzle cooling. Even the 109–004 B-1 engines in production often showed detail changes in the troublesome nozzle assembly, and the scrap pile at Dessau was witness to the difficulty.

Auxiliaries:

The main gearbox was fitted to the top of the intake casting (as shown in Figure 2.41) and drove a tachometer, air-oil separator, hydraulics pumps and main engine fuel pump on its forward face and a generator and governor on its rear face.

The oil pumps at the bottom of the intake casting had a separate drive shaft. On the starboard side of the compressor stator casing was an electric petrol pump (for engine starting), while on the port side was a servo motor for the variable-area exhaust bullet.

The starter engine, located in the intake central fairing, was of a Riedel type AK 11 (9–7034) air-cooled, two-stroke, two-cylinder, petrol engine (see Figure 2.52). Also known as the RBA/S10, this starter unit weighed 16.5 kg (36.4 lb) complete with an integral 4.8 to 1 reduction unit, and was started up by means of a pull-cable or its own electric starting motor. The Riedel starter had a maximum crankshaft speed of 10,000 rpm and gave 10 bhp at a dog speed of 1,200 to 1,500 rpm. From the weight and size viewpoint, the starter was excellent but it could not restart a turbojet at altitude,

Fig 2.51 View with exhaust system removed, showing location of the turbine wheel of the Junkers 109–004 B-1.

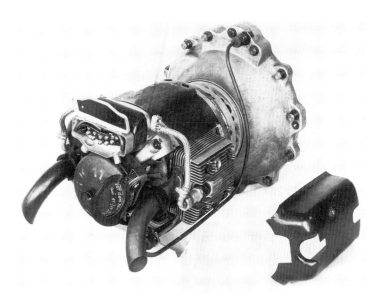

Fig 2.52 The compact, reliable Riedel type AK 11 (9-7034) air-cooled, two-stroke, two-cylinder petrol engine used to start German Class I jet engines. Note the ring-pull for starting the engine.

and its performance soon fell off and became unreliable in normal service.

Fuel system:

The main fuel pump was a single-stage gear pump with a full rpm capacity of about 2,275 litres (501 gallons) per hour at 70 kg/sq cm (996 psi) pressure, the fuel being delivered to the injectors for combustion via an all-speed governor. This governor was very complicated (see Figures 2.53 and 2.54), and consisted of the following five parts, which were interconnected:

(i) An oil pump which gave the oil circulation and pressure for operating the servo mechanism of the governor and the motor moving the variable-area exhaust bullet.

(ii) A centrifugal governor which in essence consisted of flyweights, a pilot piston coupled to the flyweights and a follow-up mechanism. When the flyweights moved away from the neutral position they moved the pilot piston which controlled the flow of servo fluid to the pistons operating a spill valve. The throttle varied the spring load on the governor flyweights and hence the governed engine speed.

(iii) A spill valve which controlled the flow and pressure of fuel delivered to the throttle. It was servo-operated by oil fed to the ends of a piston, and the flow of servo fluid was chiefly controlled by the centrifugal governor.

(iv) A throttle which gave a variable restriction in fuel line between the spill valve and the fuel injectors (or burners), and could be fully closed to act as a high-pressure shut-off cock. Opening of the throttle decreased the resistance to fuel flow to the burner line and so more fuel was instantly fed to the burners.

(v) A pressure control valve (or safety valve) which limited the pressure drop across the throttle to 10 kg/sq cm (142 psi) by opening the spill or fuel return needle. This prevented the fuel pump building up too much pressure.

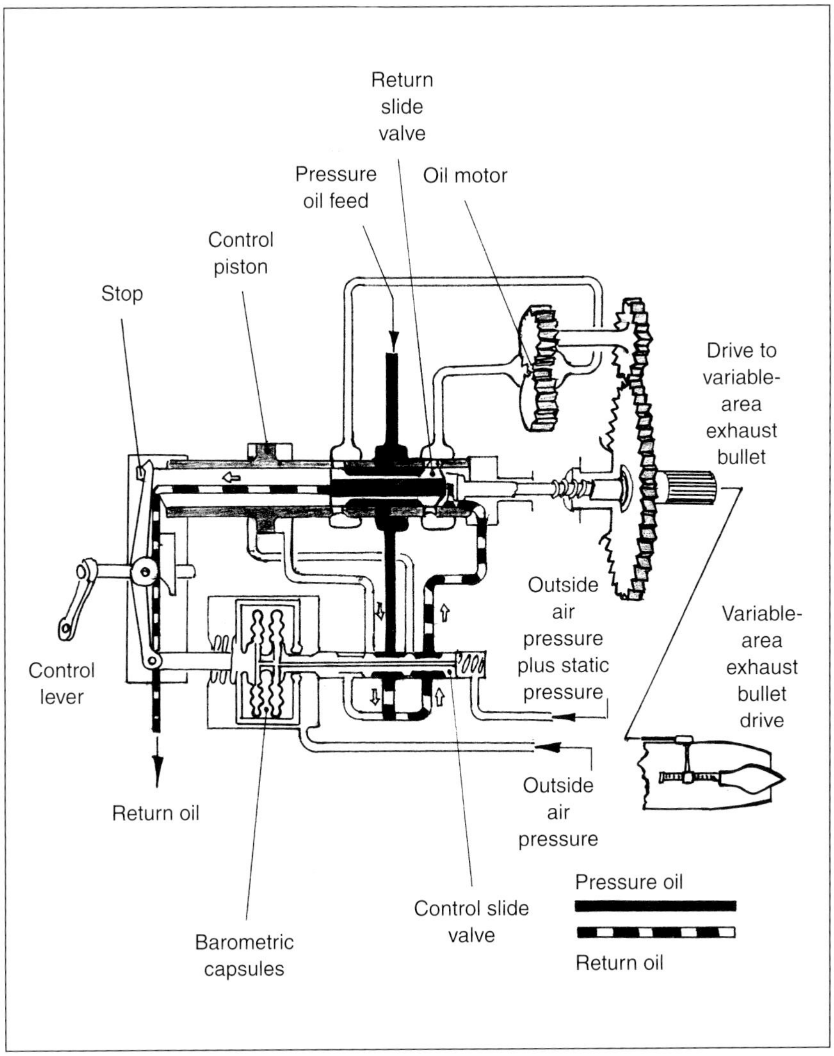

As a modification to the above fuel system, Junkers were performing work on fuel regulation by means of a temperature-responsive element such as a thermocouple operating through an electric relay, but no work was carried out with electronics.

For starting, the engine was run by the Riedel starter and petrol was fed to the burners by the small, electric booster pump. Once accelerated to 2,000 rpm on petrol, the throttle was moved to the idling position to permit the main fuel system to cut in. The combustion chambers could be burned up by over-rapid throttle movement which could boost the temperature by over 200 °C. This danger was greatest between 3,000 and 6,700 rpm because of the low air speed through the engine, and work was still proceeding on time interval control over the throttle.

To ease the start and give reasonable acceleration without compressor surge or overheating, the exhaust nozzle was set fully open at low rpm. When the throttle was moved to fully open, the exhaust nozzle bullet moved rearward to decrease the nozzle exit area to an appropriate value, say for take-off conditions. The bullet was moved by a gear-type servo motor fed by oil from the gear pump in the governor. The control for this servo motor was operated through a cam, mechanically linked to the throttle, and by the pressure difference across a barometric capsule. Whether the barometric control was effective or not is uncertain, but some engines had the pressure points going to the capsule blanked off and thus needed more attention from the pilot.

Typical data for the Junkers 109–004 B-1 follow:

Thrust	900 kp (1,984.5 lb) static
Thrusts at 900 km/h (559 mph)	730 kp (1,609.7 lb) at sea level
	320 kp (705.6 lb) at 10,000 m (32,800 ft)
Speed	8,700 rpm ± 50
*Weight	720 kg (1,588 lb) + 3% (See weight breakdown)
Pressure ratio	3.0 to 1 up to 3.5 to 1
Specific fuel consumption	1.38 per hour (this could increase to 1.45 within a few hours by build-up of dirt on compressor)
Airflow	21 kg/sec (46.3 lb/sec)
Specific weight	0.85 approx.

*The weight of 720 kg given above was the net dry weight of the engine (without cowling and aircraft auxiliaries) given in the official manual D.(Luft)T.g.3004 B-1, whereas the following weight breakdown for the 109-004 B-1 is based on the weighing of components of an actual captured engine:

Fig 2.53 Scheme of thrust regulator (oil motor) for the Junkers 109–004 B turbojet engine.

Gas temperature   700 °C max.
Diameter   0.8 m (2 ft 7 in)
Length   3.864 m (12 ft 8 in) or, with exhaust bullet extended, 4.144 m (13 ft 7 in)
Frontal area   0.586 sq m (6.3 sq ft) cowled

	kg	(lb)
Intake section:		
Casting w/oil pumps and filters	26.0	
Bevel gear assembly and drive shafts	8.2	
Gear box and drives	15.9	
Front compressor bearing assy	11.4	61.5 (135.6)
Compressor section:		
Stator casing and stator blades	90.8	
Rotor w/stub shafts and tie bolt	100.0	190.8 (420.7)
Centre section:		
Centre casting and fittings	74.0	
Outer casting and fittings	45.5	
Rear compressor bearing assy	3.0	
Front turbine bearing assy	3.4	
Bear turbine bearing assy with scavenge pumps	4.1	130.0 (286.7)
Combustion section:		
Six combustion chambers w/burners, igniters and interconnectors	52.8	52.8 (116.4)
Turbine section:		
Inlet ducting and joint rings	19.0	
Nozzle assembly	19.5	
Diaphragm plates	4.5	
Disc with SOLID blades	68.5	
Shaft, sleeve and fittings	13.5	125.0 (275.6)
Torque tube:		3.5 (7.7)
Exhaust nozzle assembly:		86.5 (190.7)
Bare weight:		650.1 (1,433.4)
Auxiliaries:		
Oil tank	12.3	
Fuel pump	4.0	
Governor	7.7	
Tachometer	0.7	
Centrifugal separator	1.8	
Nozzle control servo motor	8.0	
Drive shaft for nozzle bullet	1.8	
Fuel filter	0.9	
Fuel non-return valve	0.45	
Throttle and dash linkage	3.5	
Pipes, bolts and fittings	11.4	52.55 (115.9)

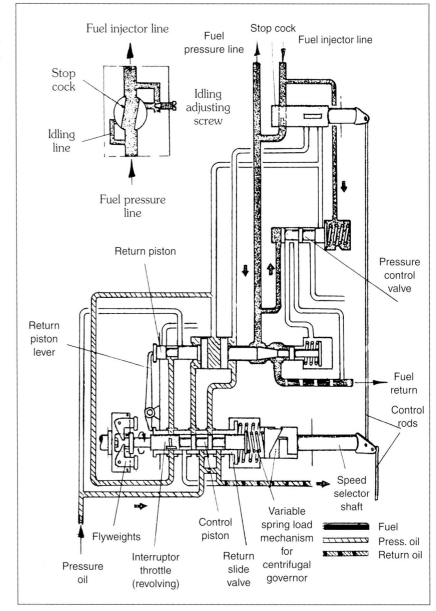

Fig 2.54 Scheme of all-speed governor for the Junkers 109–004 B turbojet engine.

	kg	(lb)
Engine mounting brackets	7.0	(15.4)
Net dry weight :	709.65	(1,564.7)
Starter:		
Riedel 2-stroke starter	16.5	
Petrol tanks and support flange	9.1	
Petrol pump	2.7	
Igniter coils	1.4	29.7 (65.5)
Net dry weight with starter:	739.35	(1,630.2)

	kg	(lb)
Aircraft auxiliaries:		
Generator and fittings	16.5	
Hydraulic pump	3.7	20.2 (44.5)
Cowling:		
Riedel starter cowling	1.8	
Front cowling (around tanks)	7.7	
Remainder of engine cowling	39.0	48.5 (107.0)
Total dry weight of fully cowled engine		808.05 (1,781.7)

The total dry weight of the fully cowled engine given above can be compared with that of 805 kg (1,775 lb) for the 109–004 B-1 (numbered 1106) weighed at Wright Field, Ohio, in the USA.

The 109–004 B-2 and B-3 (TL)

The design of the 109–004 B-2 was completed by about mid-1943 with a new compressor design resulting from co-operation between Junkers and the AVA. By this time, Junkers were performing their own aerodynamic studies, and their Dr Bates worked closely with Prof. Encke of the AVA. The new compressor design aimed at finally eradicating vibration failures while at the same time giving an improved altitude performance. Firstly, the number of rotor blades in each stage was reduced to put the critical rotational speed (when trouble occurred) outside the engine's speed range. For better altitude performance, the rotor blades were redesigned with increased blade chords to give six stages of small thickness-chord ratio plus two stages with comparatively thicker sections (instead of the previous two stages of thin blades plus six stages of thicker blades). The profiles of the stator blades were unaltered, but the blades were reset at a slightly different angle, while the axial clearances between rotor and stator blades remained unaltered. Despite the effort, the new compressor suffered its own form of blade vibration, so that the 109–004 B-2 did not go into production. Scant evidence indicates that it was about 20 kg (44 lb) heavier than the B-1 model. Research continued in order to find improvements for the compressor with the space limitations of the engines in production, and various modifications were introduced from time to time. One late scheme aimed at altering the frequencies of the rotor blades by introducing a slot cut down the centres of the blades of the first, third, fifth and seventh stages, but it is not known if this was introduced in time for production.

Only one questionable source was found which mentioned the 109–004 B-3 engine and thought it was the first model with air-cooled, hollow turbine blades. The B-3 could well have been a developmental model prior to putting into production the B-4 with hollow turbine blades. In any event, we must now look at the development of these hollow turbine blades for Junkers turbojets.

Hollow turbine blades by Prym and Wellner

The need to develop hollow air-cooled turbine blades arose, of course, because of the German shortage of metals used in heat-resisting alloys, and the best results had to be obtained with the inferior alloys available. Compared with solid turbine blades, hollow blades needed less material, while the air cooling of them permitted either a higher working temperature with increased efficiency and thrust or the temperature as before but with an increased engine life. Junkers experiments with high-temperature ceramic blades failed to produce sufficiently robust blades (other work in this field is described in Section 6 and elsewhere), so hollow metal blades had to be the answer. Looking more to the future, it was realised that, for a given limiting stress in a turbine wheel, a higher rotational speed and thrust was possible with the lighter, hollow blades.

Using Tinidur sheet metal, Junkers first tried making hollow blades by folding and welding, but Tinidur proved unsuitable for welding. Junkers therefore sought the assistance of outside specialists, and in February 1943 asked the William Prym firm of Stolberg/Rhld (near Aachen) to develop a process for the manufacture of hollow turbine blades for the 109–004. Already, between November 1942 and February 1943, Prym had carried out work on the hollow blades for the Junkers Jumo 207 exhaust turbo-supercharger, the firm having been brought to the attention of Junkers by AEG. Prior to the war, Prym had made pins, needles, buttons, etc., but in order to expand, had established a separate deep-drawing department (known as Abteilung Spanlose Formung) under the direction of Hermann Köhl. Success was such that Prym eventually became one of the major German manufacturers of deep-drawn items such as cartridge cases.

After a couple of months' work, Prym delivered the first 70 hollow blades for the 109–004 on 24 April 1943, but these had an error in root dimensions. Nevertheless, it was obvious that a satisfactory job could be done, and a meeting to discuss blade problems was held at the Stolberg works on 11 May 1943. This meeting was of some importance and illustrates the official keenness to iron out problems and get the desired turbojets into full production. Not only were RLM officials present at the meeting, but also personnel from Junkers, Dessau; AEG, Berlin; Krupp, Essen; Brown Boveri, Mannheim; DVL; BMW, Berlin; Heinkel-Hirth and Daimler-Benz, Stuttgart.

A month after the meeting, Prym received an RLM contract to develop turbine blades for the 109–004, and by July 1943 had delivered 150 blades. These blades, however, were based on wooden models, since Junkers had still not finalised the blade form, and it was not until August 1943 that a master blade became available. According to the production schedule, 800 blades should have been made by then. For 1943, 140,000 blades were

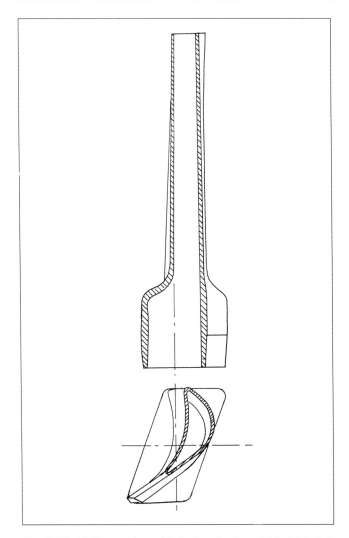

Fig 2.55 Hollow turbine blade for Junkers 109–004 B-1 turbojet.

the manufacturing process being as follows. Using 3 mm thick Tinidur sheet, a 97 mm diameter blank with a 9.5 mm hole in the centre was punched out. Cold drawing operations then began, the blank first being shaped as a cup and eventually as a closed-end tube. The closed end was then cut off and the tube progressively widened, flattened and formed to the correct blade section with a 'rectangular' root. Initially, forming from tubular stock was tried, but when the circular blank was settled on, considerable experimenting was needed to establish the exact size of the blank. At one stage, Junkers changed the length of the blade root and Prym had to begin the work of establishing the blank size all over again. The final blade weighed 173 gm (6.1 oz), or half the weight of a solid blade.

Whilst Prym were developing their deep-drawing process for blades in Tinidur, the Saechsicher Metallwarenfabrik Wellner Soehner AG of Aue were working on a process for producing hollow turbine blades in another heat-resisting alloy known as Cromadur (18 Mn, 12 Cr, 0.65 V, 0.5 Si, 0.2 Ni, less than 0.12 C, blce Fe). Cromadur was produced by Krupp as a substitute for Tinidur and used manganese in lieu of the scarce nickel. Since Cromadur proved easy to weld (unlike Tinidur), Wellner produced the hollow blades by folding and welding down the trailing edge. Although the creep strength of Cromadur was inferior to Tinidur, the Wellner welded blades actually proved superior to the Prym drawn blades in service. However, neither company could achieve the requisite production rate, so that both Cromadur and Tinidur hollow blades were kept in production for the 109–004 B-4.

scheduled, and by mid-1944, 225,000 blades, by when production was to be at the rate of 500 per day. These targets were never reached, production for August 1944, for example, being only 5,000 blades.

Clearly, expansion of facilities was needed, and a new Prym factory specifically for blade production was begun in 1943 at Zweifall (approximately 15 km SE of Stolberg). The Zweifall factory was to be ready in October 1944 with forty or fifty presses (compared with fifteen to twenty presses at Stolberg) and a floor space of 2,250 sq m (24,210 sq ft). Its initial production schedule was 100,000 blades per month, rising to 300,000 per month by the end of 1943, but the factory was not ready by the end of the war.

For the development of the hollow blade process, the RLM assigned to Prym's deep-drawing department some forty Luftwaffe technicians, and the whole project was classified secret. Figure 2.55 gives a drawing of the type of hollow turbine blade produced for the 109–004 B-4,

Fig 2.56 A partially machined turbine wheel for the Junkers 109–004 turbojet to test the fitting of the air-cooled turbine blades. *(Richard T. Eger)*

The 109-004 B-4 (TL)

By the end of 1944, the 109-004 B-1 was replaced in production by the B-4 version, which differed chiefly in having air-cooled hollow turbine blades. Other minor differences included changing the cooling air bleed after the last compressor stage from an incline to a right angle with the engine axis.

Although, by a minor modification of the combustion system, this engine was capable of a higher working temperature and thrust, the previous thrust of 900 kp (1,984.5 lb) static was maintained in order to gain greater service life and reliability at the same temperature as the B-1. Apart from their economy, a possibly unforeseen benefit of the hollow turbine blades was their faster and easier production compared with solid blades.

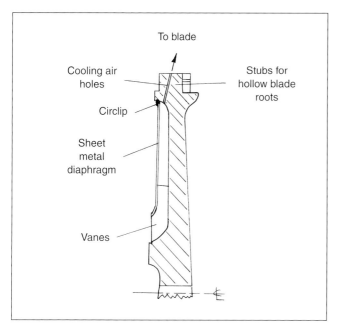

Fig 2.57 Sketch through turbine wheel of 109–004 B-4 showing cooling-air system.

The hollow blades were pressed over stubs on the turbine wheel and brazed on. To allow cooling air to pass up into the blades, holes were drilled through the rim of the turbine wheel, the cooling air being forced out into the blades by centrifugal vanes attached to the upstream face of the wheel (see Figure 2.57). This turbine wheel was lighter than that for the solid blades, and owing to the lower stress, the pull of one blade on the wheel was about 3.4 tons. About 3 per cent of the total airflow through the engine was used for cooling purposes at the expense of engine efficiency, but although cooling the turbine blades brought about a further slight decrease in efficiency, this effect was offset by advantages already mentioned. Unlike BMW, Junkers did not find an inner sleeve necessary for their hollow blades.

The brazing medium for the blades was either Degussa Flussmetal (85% silver, 15% manganese) melting at 1,000 °C or Silma solder (86% silver, 14% manganese) melting at 960 °C. Lithium fluoride was used as flux. The following notes on attaching the blades with Silma solder are based on German instructions:

(a) Preparation of wheel and blades
Grind down radii on wheel stubs, taking care not to chamfer. Blades must fit easily with a mallet.
Insert solder wire ring into soldering groove (solder consumption per rotor 250 gm).
Sand-blast blade feet, degrease wheel stubs and apply flux to same areas.
Fit blades with a wooden mallet before flux solidifies.
Bore holes for fixing pins using soda-water cooling (no oil).
Insert pins. (These were merely for holding during soldering.)

(b) Soldering
Place whole turbine wheel in oven heated to 600/800 °C but turned-off.
Warm for 20 minutes.
Light up oven.
Heat up to 1,050 °C ± 15 °C in about 40 minutes, turning the wheel. Work with excess gas to avoid oxidation.
After heating whole wheel, remove at soldering temperature and allow to cool in still air.

(c) Hardening
Hardening performed in a gas oven.
For blades of Tinidur – five hours at 650–80 °C.
Hardness required (measured at blade feet) for Tinidur – HP 20 220 kg/sq mm. If this hardness not obtained, repeat process but for a shorter time.

The soldered and hardened turbine wheel was then mechanically finished, inspected and balanced. The efficacy of solder in holding the turbine blades depended largely upon correct assembly and solder flow (see Figure 2.58), but blades came off in service owing to

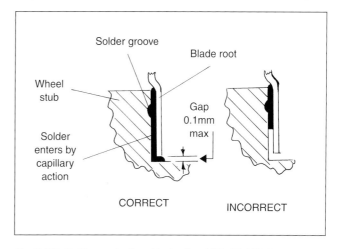

Fig 2.58 Soldering hollow blades for 109–004 B-4.

both incorrect soldering and engine overheating which weakened the solder. To ensure that blades were well fixed with solder down the whole root, each blade was bowed with a violin bow, experience recognising the right note. This process was known as Prufenweisung. Another trouble sometimes found with the hollow blades was splitting of the trailing edge tips because of vibration, and it became necessary to reinforce the trailing edge tip with two rivets.

Specific data for the 109–004 B-4 are lacking, but performance and external dimensions were generally as for the B-1, while the weight was slightly less. The complete engine could scarcely have been more economical with the scarce metals such as nickel, tungsten, molybdenum and chromium. A 109–004 B-4 with Tinidur turbine blades used 4.6 kg (10.14 lb) of chromium, 6.5 kg (14.33 lb) of nickel and 0.2 kg (0.44 lb) of molybdenum, while the same engine with Cromadur turbine blades used 4.7 kg (10.36 lb) of chromium, 3.5 kg (7.71 lb) of nickel and 0.2 kg (0.44 lb) of molybdenum. No tungsten was used at all. These figures can be compared with those for the 109–004 B-1 (with solid Tinidur turbine blades), which had 6.35 kg (14.0 lb) of chromium, 9.85 kg (21.72 lb) of nickel and 0.205 kg (0.45 lb) of molybdenum.

Production of the 109–004 B-1 and B-4

Towards the end of 1943 the RLM finally decided to put jet aircraft into production, and 109–004 B-1 turbojets were being produced in quantity. The first production engines were made at Dessau, where all the experimental engines were also built, but from August 1944 production was taken over by Junkers factories at Köthen (25 km SW of Dessau) and Muldenstein (across the river from Dessau), the latter factory eventually achieving the highest total production. The vast underground factory known as Mittelwerke GmbH in the Harz mountains near Nordhausen joined in the programme in October 1944, followed by a factory at Zittau (about 75 km ESE of Dresden) the following month. The final production centre in Germany was known as Mitteldeutsche, located at Leipzig, but this had barely begun 109–004 B production when the war ended. One other plant had also just begun producing engines outside Germany, and this was near Prague, possibly at the Letov aircraft factory. These were the factories at which completed engines were produced, total production for the 109–004 B-1/B-4 and experimental engines, 6,010 in all, broken down as shown in the following table.

The time taken to produce one 109–004 B was about 700 man-hours, which compares with about 600 man-hours for the BMW 109–003 turbojet engine. In some cases, of course, engines took longer than this to produce. At the Mittelwerke, where the Junkers section occupied the rear portion of a tunnel about 3 km long

Plant	1944					1945		
	Aug.	Sep.	Oct.	Nov.	Dec.	Jan.	Feb.	Mar.
Dessau	96*	–	–	–	–	–	–	3
Köthen	25*	60	170	200	210	100	20	40
Muldenstein	189*	220	430	448	467	334	335	307
Mittelwerke (Nordhausen)	–	–	10	20	50	430	630	815
Zittau	–	–	–	12	53	86	115	120
Mitteldeutsche (Leipzig)	–	–	–	–	–	–	–	5
Prague	–	–	–	–	–	–	–	10
TOTALS	310	280	610	680	780	950	1,100	1,300

(* Totals *up to* August 1944)

NW of the town, the programme was rather disorganised because of the heavy load of work in the machine shops on parts for V-weapons and other aero-engines. The large number of measuring gauges indicated a lack of jigs for production, at least at the Mittelwerke. Nevertheless, it seems that the later total production rate for the 109–004B was somewhere near the scheduled rate. A point on which complete information is lacking is that some of the plants in the above production schedule were dispersed.

Other production centres were being set up apart from those already mentioned, but were either too late or were abandoned to advancing Allied forces. The largest such centre was probably situated at Strasbourg in France, where, from June 1940, Junkers had taken over the Matford plant which had formerly built Ford cars. Renamed as the Junkers-Meinauwerke, this plant was re-equipped to produce 250 new piston engines, e.g. Jumo 211, and repair 500 piston engines per month. Production of 109–004 B engines from Strasbourg was scheduled at 20 engines in September 1944, increasing to a maximum of 1,000 per month by July 1945, but bombing of the area on 27 May 1944 set the programme back. Dispersal was then begun around the general Strasbourg area, not for protection but to make use of the existing facilities. The chief sites and the 109–004 B parts they were to deal with were as follows:

Hasserode: Centre casting.
Mutzig: Machining of centre casting.
Rotham: Machining of turbine shaft.
Illkirch (Graffstaden Mill): Stores and machining of turbine wheel. Brazing of blades elsewhere.
Lutzelhausen: Hollow shaft and bevel gear assembly for exhaust bullet.
Pfaffenhausen (Omega works): All sheet-metal work.

Other parts, such as compressors and turbine blades, were to come from Germany, and final assembly of engines at Strasbourg was to be in the Junkers-Meinauwerke. Completed engines were then to be sent to Muldenstein and the Mittelwerke for testing and release. Using largely Alsatian and Russian labour, work began in the Strasbourg area in July 1944, but no complete engines were turned out before the liberation. Presumably in an attempt to confuse the inquisitive, the 109–004 was referred to in France as the Jumo 203, this number being in line with the Junkers Jumo piston engine designations 9–204, 205, etc.

Testing the 109–004

In addition to the functional testing of production engines before release, regular performance tests on both production and experimental engines were carried out by E-Stelle Rechlin and Junkers. Official static testing and flight testing by E3 Rechlin and E2 Rechlin respectively have already been mentioned in the Introduction.

E2 Rechlin performed its turbojet flight tests at constant weight, which necessitated a separate flight for each point on the performance curve in order to eliminate the effect of the high fuel consumption of the whole flying test-bed. Jet thrust was measured in flight using a Junkers Ju 52/3m test-bed aircraft, and the results so obtained were then applied in subsequent tests with the same turbojet engine on specific aircraft. To measure thrust, the engine was mounted in a parallel-motion frame under the fuselage, and an electro-induction method (load cells) was used to measure the compression of a suitable spring.

Junkers used a different method of measuring jet thrust in flight by reading the pressures before and after the turbine and at the exhaust nozzle. Fuel consumption was measured by using flowmeters. A typical flight test installation (Figure 2.59) shows a 109–004 turbojet fitted beneath the port wing of a Junkers Ju 88 (GH+--).

Flight tests at Rechlin using an Me 262 A powered by two 109–004 B-1s showed that, for a 15 °C difference in air temperature, a difference of 30 per cent in thrust and 25 per cent in fuel consumption resulted. This meant that the average maximum level speed of the Me 262 fighter at 7,000 m (22,960 ft) varied from 868 km/h (536 mph) in winter to 820 km/h (508 mph) in summer. Examples of Me 262s used by Rechlin for 109–004 flight tests are:

Me 262 A	Werk Nr. 130 018	Code: E3+01
Me 262 A-1a	Werk Nr. 130 163	Code: E2+01
Me 262 A	Werk Nr. 130 168	Code: E3+02
Me 262	Werk Nr. 170 095	Code: KD+EA

Flying carried out under icing conditions gave no trouble with the compressor, but to counteract compressor icing, should it occur at a later date, plans were made to divert hot gas tapped from downstream of the turbine or to employ a fluid de-icer. No trouble was encountered with sand or grit erosion of the compressor blades, but then, German airfields were not too bad in this respect.

Fig 2.59 A Junkers Ju 88 A-5 (GH+FQ) with underslung Junkers 109–004 turbojet to perform flight tests. The engine is mounted at a hard point normally used to carry bombs. Other Ju 88 test beds include DE+DK.

Fig 2.60A The first pair of Junkers 109–004 B-0 turbojets were delivered to Arado's Warnemünde factory in February 1943 and fitted to the V2 prototype of the Ar 234 as shown. This view illustrates the useful size of the inspection panels.

Fig 2.60B The Arado Ar 234 V3 taking off with rocket assistance. This prototype, fitted with Junkers 109–004 B-0 turbojet engines, made its first flight on 25 August 1943 and had a pressurised cabin and ejector seat. It was destroyed on an early flight. *(Phillip Jarrett)*

Fig 2.61A Arado Ar 234 V5 (GK+IV) powered by two Junkers 109–004 B-0s and showing the type of take-off trolley used prior to the development of a retractable undercarriage.

Fig 2.61B Messerschmitt Me 262 A-1a (probably KD+EA, Werk Nr. 170 095, at E-Stelle Rechlin) with a partially uncowled Junkers 109–004 B-1 turbojet.

Utilisation of the 109–004 B

Although many aircraft were planned, the two jet types which went into operational service using the 109–004 B turbojet were the Me 262 and Ar 234. According to one source, the 109–004 B was known unofficially by the name Orkan (hurricane), which at least seems apt. From July 1944, the 1st Staffel of the Versuchsverband Ob.d.L., followed by the 3rd Staffel, began using Me 262s and Ar 234s on experimental reconnaissance missions. Servicing of the new aircraft was initially performed with the assistance of personnel from Messerschmitt, Arado, and Junkers. The aircraft were found to be excellent in these trials, being easier to fly and much faster than piston-engined fighters, although they were less manoeuvrable and required long take-off and landing runs. The 109–004 B-1 engines, however, were prone to premature failure and would stand no mishandling from the pilot. In particular, failures occurred in the compressor, turbine inlet nozzles and the turbine wheel, and new engines were eventually given a general overhaul before being put into service. Jet aircraft types being prepared for mass-production and service were the Me 262 A-1a Schwalbe (swallow) fighter, the Me 262 A-2a Sturmvogel (stormbird) bomber which Hitler had ordered, and the Ar 234 B reconnaissance and bomber aircraft. In addition, various sub-variants were to follow for specialised roles such as night-fighting.

By September 1944 the first semi-operational jet fighter unit, Erprobungskommando 262, had finished evaluating the Me 262, and a special Ar 234 test unit, Sonderkommando Götz, was formed. The first fully operational jet unit, Kommando Nowotny, began operations on 3 October with about twenty Me 262 A-1a fighters. These first operations, against USAAF bomber formations, proved disappointing, and a number of Me 262s were shot down when they cut their speed because of the pilots' inability to attack with accuracy at high speed. Other aircraft were lost in accidents, of which 34 per cent were due to undercarriage failures, 33 per cent to turbojet failures, and 10 per cent to tailplane failures. In the latter case, resonance from the turbojet exhausts set up oscillation leading to structural failure in the Me 262 tailplane.

From November 1944, two further small Ar 234 reconnaissance units were set up, but were replaced in the new year by 1.(F)/100, 1.(F)/123 and 1.(F)/33, which were able to carry out reconnaissance over the British Isles and other areas with impunity. Another unit, Sonderkommando Sommer, performed similar work over northern Italy. Also in November 1944, a new fighter wing, JG 7, was established and attempts were made to solve the problems of high-speed attack, chiefly by fitting rocket armament to the Me 262 fighters. In order to familiarise pilots more safely with the peculiarities of jet aircraft, training units with two-seat

Fig 2.62 Messerschmitt Me 262 A-1a/U3, the photo reconnaissance version of the jet fighter, showing the nose bulge made to accommodate two vertical cameras. It was captured by the Americans and given the Foreign Evaluation number FE4012. A Junkers Ju 88 is behind. *(US Air Force photograph taken on 30 September 1945)*

Fig 2.63 A Messerschmitt Me 262 B-1a/U1 interim, two-seat night-fighter (Werk Nr. 121 442) with nose antennas for its SN-2 Lichtenstein radar equipment. It was captured by the Americans and given the Foreign Evaluation number FE610. *(US Air Force photograph taken 1945)*

versions of the Me 262 were set up for both fighter and bomber units. (Me 262 A-2a bomber units had begun forming in September 1944 and were known as Kommando Schenk and Kommando Edelweis.) Training for the first Ar 234 bomber unit began in November 1944 at Alt Lönnewitz, where the main Ar 234 production centre was. During the Ardennes Offensive of December 1944/January 1945, the Ar 234s of 6./KG 76 made the first bombing attacks with this machine on Allied positions, and soon other units of KG 76 became operational with the Ar 234 bomber. Operations were limited owing to a severe fuel shortage, but from early March, sorties by KG 76 increased, often supported by the Me 262s of I. and II./KG 51 bomber units. Some of the most desperate, and often suicidal, of these operations were against bridges such as the Remagen bridge over the Rhine. By the end of March, however, Ar 234 bomber operations had dwindled to a low level and had almost petered out by the end of the war.

In the meantime, the Me 262 was performing in its most successful role as a short-range reconnaissance aircraft, it being virtually immune to interception while flown fast and without deviation. For this work, Sonderkommando Brauegg was formed in December 1944 with Me 262 A-1/U3s. On New Year's Day 1945, the Luftwaffe made its last big effort by attacking Allied airfields in Europe with all available aircraft, but following this, a rapid decline followed, and soon the jet aircraft were amongst the few Luftwaffe aircraft still flying. April saw a few jets being used in the night-fighting role, Kommando Welter using the Me 262 and Kommando Bonow using the Ar 234. On the 10th of that month, the Luftwaffe carried out its last sortie over the British Isles when an Ar 234 of 1.(F)/33 flew from Stavanger in Norway on a reconnaissance mission over northern Scotland (see Figure 2.64).

Although the new year saw various former conventional bomber units preparing to begin operations

Fig 2.64 Sola airfield (Stavanger, Norway) 6 August 1945. From here, the Luftwaffe flew its last reconnaissance mission over the British Isles in April 1945. This Arado Ar 234 B-2/1p (9V+CH Werk Nr. 140 493), captured there, is being prepared for flight while British personnel watch on. Previously of Einsatzkommando 1.(F)/5, this aircraft was flown to RAE Farnborough.
(Imperial War Museum)

with the Me 262 fighter, of these only 1./KG(J)54 became operational. By the end of the war, only III./JG 7 and Jägdverband 44 were still operational as Me 262 fighter units. The famous Jägdverband 44 was formed by the ace pilot Gen.Lt. Adolf Galland as a last-ditch effort to concentrate the cream of the Luftwaffe's fighter pilots with the best fighter aircraft into a single unit. As one of Germany's final attempts to hit back at the Allied bomber streams which had already crushed her, it showed what the new jet fighters could do once the tactics for their use were fully grasped. JV 44, formed on 10 February 1945, began operations on 31 March from München-Riem with about 25 Me 262s and 50 pilots. A surfeit of almost 100 jet fighters was soon collected from other disintegrated units. However, only about six Me 262s were kept serviceable at a time, but by the time JV 44 surrendered on 3 May 1945, the unit had destroyed some 45 enemy aircraft. The only Allied jet fighter operational before the end of the war was the Gloster Meteor I, which began operations on the Continent on 16 April 1945, but no contacts with Luftwaffe jet aircraft were made.

This résumé of the utilisation of the 109–004 B turbojet in service must perforce be brief, but mention should be made of other, non-operational, aircraft designed to use the engine also. At about the time when the prototype of the world's first jet bomber, the Ar 234 V-1, began flying in July 1943, a study was made at Junkers for another jet bomber. Whereas the Ar 234 was rather small for a bomber and was limited in speed by its unswept wing, Junkers planned a much larger bomber with swept-forward wings to reach higher speeds in the region of mach 0.8. This was the Ju 287, which was to use more turbojet power than hitherto by flying with four 109–004 B-1 engines fitted one at each side of the forward fuselage and one beneath the trailing edge of each wing. Using this power, a low-speed test aircraft designated Ju 287 (RS+RA) made its first flight on 16 August 1944 from Brandis. As with the Ar 234 in certain instances, the Ju 287 V-1 required to use jettisonable Walter rocket units to assist take-off, but more turbojet power was planned for subsequent aircraft, e.g. six BMW 109–003 A-1s for the Ju 287A. Other projected bombers were to use turbojets as short-time speed boosters by supplementing the normal airscrew power.

Such a bomber was the large Focke-Wulf Ta 400 of conventional layout and with six radial piston engines plus two 109–004 B turbojets.

An interesting fighter-bomber project was the Blohm und Voss P.178, which was of asymmetrical layout, with the pilot's gondola situated towards the port side of the wing and a single 109–004 B engine towards the starboard side. Two such engines were planned for the Blohm und Voss P.202 projected fighter, the most interesting feature of this aircraft being its curious wing, which was to be at right angles to the fuselage for take-off and landing but was to pivot about its vertical axis for high-speed flight. Thus, the port wing was swept back and the starboard wing swept forward, which still gave, apart from stabilising problems, a fully swept lifting surface.

Less startling in concept but nonetheless advanced, was the Horten Ho IX (later Gotha Go 229) tailless, or flying-wing, fighter which was designed originally in 1942 to use two BMW 109–003 turbojets. However, following an unpowered prototype, the Ho IX V-2 was flown in 1944 at speeds up to 800 km/h (597 mph) using two 109–004 B-1 turbojets because of their greater availability. After only a few flying hours, this aircraft was destroyed during an enforced single-engine landing, but other prototypes were under construction when the war ended.

Naturally, other Axis powers were interested in German turbojet work, and Japanese delegations made regular visits to the Dessau works to inspect the turbojet work there, the last such visit being as late as 5 March 1945, when all the latest developments were seen. Drawings for the latest versions of the 109–004 B were to go to Japan by U-boat, but probably never arrived there. In any event, the prototypes of Japan's only Second World War jet fighter (the Nakajima J8N1 Kikka) were powered by scaled-down Japanese versions of the BMW 109–003 turbojet. In Italy it was purported that the projected Caproni Regianne Re 2007 fighter was to be powered by a single Junkers 109-004 turbojet but this was just a myth.

By the end of the war, hundreds of German aircraft had flown on the power of the Junkers 109–004 B turbojet. Most of the aircraft used twin underwing-mounted engines, the most prolific type being the Me 262, of which 1,433 were built, although fewer than 200 or so reached the operational squadrons. The care with which the 109–004 B had to be used in service is perhaps best illustrated by the following notes on the pilot's operating instructions, extracted from the Luftwaffe manual L.Dv.T.2262 (for the Me 262 A-1 and A-2) published in January 1945. Only the instructions relating to the engines follow:

(a) *Preparation for take-off*
  (i) When ATO units are fitted, only connect up the detachable plug immediately before take-off.
  (ii) Fuel cocks to be closed. Throttle lever in the 'stop' position.

(b) *Starting*
  (i) Place chocks in front of main landing wheels but not in front of nose wheel.
  (ii) Place protecting baskets in front of engine air intakes.
  (iii) Push in the red-ringed automatic switch. The switch labelled 'Fernselbstschalter' should be off when an external source of current is connected.
  Warning: Do not switch on the ignition button, as otherwise there will be a danger of fire.
  (iv) Press starting handle for three to five seconds. Starting fuel runs into the air intake of the Riedel starter.
  (v) Pull starting handle to switch on the ignition and starter of the Riedel starter.
  (vi) The Riedel must start within five seconds, otherwise recommence operations.
  (vii) As soon as the Riedel is running, press the change-over switch which selects lower scale on the rpm counter.
  (viii) Read revolutions on the inner scale of the rpm counter.
  (ix) At approximately 800 rpm, press the button on the throttle lever. This causes starter fuel to be injected and ignited by means of sparking plugs. Engine rpm increases.
  (x) Watch the temperature. If there is too much flame or the temperature rises above the highest permissible point, release the button and allow the Riedel starter to continue running, i.e. go on pulling the starter handle.
  (xi) At about 2,000 rpm, release the starter handle and rpm counter scale switch. Open fuel cock on the port instrument bank.
  (xii) Slowly move the throttle from the 'stop' position to the 'idling' position. The ratchet lever clicks into the stop. Continue pressing the ignition button until 3,000 rpm is reached. Do not move the throttle past the 'idling' position.
  (xiii) At 3,000 rpm, release the ignition button. The engine is now running on J-2 fuel.
  (xiv) Switch labelled 'Fernselbstschalter' to 'Ein' (on). Engine is now idling.

(c) *Running up*
  (i) Gradually move throttle lever forward from 3,000 to 6,000 rpm. Temperature may not rise above highest permissible point. The control unit only operates above 6,000 rpm. Therefore, if the throttle is moved forward too quickly, there is a danger of considerable flame and thus of fire.
  (ii) As from 6,000 rpm, gently move throttle lever to fully open. Max. revs: 8,700 + 200. Gas temperature must remain steady after one minute.
  (iii) Bring the throttle lever sharply back to the 'idling' position. This must not cause the engine to peter out.

(iv) With low oil temperatures, the rpm may exceed the highest permissible by 250. The rpm will drop again when the oil reaches operating temperature.

(d) *Proceeding to take-off point*
(i) The aircraft is towed to the take-off point. Use of towing gear essential.
(ii) According to ground conditions, the aircraft begins to move between 4,500 and 6,000 rpm. Keep engines running at equal speeds; in any event, first allow the aircraft to get under way and turn by braking on one wheel, i.e. engines not used for taxiing or manoeuvring.

(e) *Take-off*
Head aircraft into take-off direction. Keep brakes on until 7,000 rpm reached. At commencement of take-off run, select 'Full load' position for throttle. With ATO units, observe also the following points:
(i) On the sign 'Ready' the mechanic connects up the ignition cable.
(ii) At the commencement of the run, the tumbler switch to 'Ein' (on), the indicator light on the port instrument bank lights up.
(iii) Press the button. This will ignite the ATO units which burn for about six seconds. (A table of speeds at which to switch on ATO units for specific take-off weights is given.)
(iv) After take-off, jettison the rockets but not above 400 km/h (249 mph). To do this, pull the cable which is underneath the port instrument bank.

(f) *Cruising*
(i) Maximum rpm of 8,700 permissible for 15 minutes. This may only be exceeded during climb, combat, and at a great height.
(ii) Maximum continuous cruising 8,400 rpm with:
Fuel pressure: 40 to 60 atmospheres
Oil pressure: 1 to 3 atmospheres
Gas temperature: Not over the red mark

(g) *High-altitude flight*
(i) Over 8,000 m (26,247 ft) switch on both main fuel tank pumps.
(ii) Between 4,000 m (13,123 ft) and 8,000 m (26,247 ft) only open or close throttle very slowly. Above 8,000 m do not throttle engines below 8,000 rpm.
(iii) If the revolutions rise above 8,700 when the throttle is closed, select 'Full Load' again.
(iv) If a power unit stops above 4,000 m (13,123 ft), move throttle lever to 'Stop' position and do not attempt to re-start. Re-starting is only possible at 4,000 m or lower.

(h) *Diving*
Throttle lever to 'Idling' position and rpm about 5,000. Highest permissible diving speed is 950 km/h (590 mph).

(i) *Landing*
Do not close throttle beyond the 'Idling' position.

(j) *Switching off*
(i) Throttle engines down to 'Idling' position.
(ii) Close fuel cocks.
(iii) Press the ignition button and lift the throttle lever over 'Idling' stop and close fully.

(k) *Behaviour in special cases*
Single engine flight:
(i) The throttle of an unservicable engine must be immediately fully closed. Also close the fuel cock.
(ii) Trim the aircraft. A slight bank towards the running engine means less rudder and greater range.

Re-starting an engine in flight:
Only possible under 4,000 m (13,123 ft). Do not use the Riedel starter since switched-off engine remains running.
(i) Reduce speed to between 300 and 350 km/h (186 and 217 mph) or about 3,000 rpm on the engine.
(ii) Open fuel cock.
(iii) Switch on ignition and, when temperature has risen slowly, move throttle lever to 'Idling' position

Cavitation in the compressors:
At great height, above 4,000 m (13,123 ft), many power units are inclined to develop cavitation in the compressor which is noticeable by a falling in the thrust and a turning movement around the vertical axis. The temperature also rises. To overcome this, close the throttle until rpm drops to 8,400 and, as the temperature falls slowly, open the throttle again.

From the above instructions it can be seen that particular care was needed in handling the throttle. If the throttle lever was moved too quickly to increase power (pilots were used to a rapid throttle response with piston engines), the turbojet would either cut out or fail to reach full speed. Attempts were made to introduce a time delay in the governing system to avoid this difficulty, but the governor gave trouble at high altitude and the pilot had to be relied on to a great extent for successful operation.

Because of the slow throttle response with these early turbojet engines, a pilot was committed to land his aircraft once on final approach, and at that stage he was a 'sitting duck' should an enemy fighter 'bounce' him.

The protecting hemispherical baskets for the engine intakes, referred to in the above starting instructions, were, when used, removed once an engine was running, but tests were made with the baskets retained as guards against debris caused during aerial combat, and these allegedly did not affect performance.

An incident connected with the 109–004 B movable exhaust bullet is of interest and concerns several Me 262s which crashed following the failure of one engine in each case. At the time, an explanation was not forthcoming as to why a crash should follow immediately after one of the engines stopped, but the answer was found after the war when the German test pilot Ludwig Hoffman was ferrying an Me 262 for the Americans. The flight was planned from Lechfeld to Cherbourg, but an engine cut out en route and the aircraft immediately went into a dive which Hoffman was unable to correct. Fortunately the pilot was able to bale out on this occasion, and his evidence enabled the phenomenon to be explained. Apparently the moveable bullet in one of the engines had become detached, blown back and had blocked the air flow through the engine. The resulting severe effect on the airflow then caused the aircraft to sideslip and dive, following which the tail controls became ineffective because the airstream over them was blanketed off by the bulk of the aircraft. The first Me 262 on which this sequence of events occurred was the Me 262 V3 (PC+UC) which crashed and killed the pilot after it had been rebuilt following a previous crash.

Towards the end of the war, German turbojets (specifically the 109–004 B) in service were being run on unrefined fuel oil (centrifugally cleaned only) owing to refinery destruction. The oil was heated before filling the aircraft tanks, and apart from sooting up of the spark plugs no trouble was experienced. This success was very largely because Junkers had developed the 109–004 from the beginning to run on J-2 or diesel-type oil, though this choice was made on technical grounds (safety, economy, etc.), and not because of an envisaged shortage of regular aviation petrol.

The 109–004 C (TL)

The 109–004 C was a re-design of the B-4 model with detail refinements to give more thrust for less weight and size. Applications of this engine were to be as before plus those aircraft projects which were coming along in the meantime, such as the Blohm und Voss P.188 bomber (four engines), Messerschmitt P.1100 bomber (two engines), and P.1099 fighter (two engines). Production of the 109-004 C was just beginning when the war ended, and peak production was scheduled for July 1945*. An example of the engine is shown in Figure 2.65, and known data for it are as follows:

Thrust	1,000 kp (2,205 lb) static
Thrusts at 900 km/h	775 kp (1,709 lb) at sea level
(559 mph)	380 kp (838 lb) at 10,000 m (32,800 ft)
Speed	8,700 rpm approx.
Weight	720 kg (1,588 lb)
Specific weight	0.72
Diameter	0.755 m (2 ft 5^i in)
Length	3.830 m (12 ft 6^i in)

Either two 109–004 Cs or two 109–004 Ds were used in an experimental flight to give the Me 262 its highest level speed. The aircraft was probably the Me 262 V12 (VI+AG, Werk Nr. 130 007) and the speed attained was

*A Junkers document of January 1945 states that the 004 C was not being proceeded with and the information previously issued was to be destroyed.

Fig 2.65 A rare picture of the Junkers 109–004 C turbojet. This had detail refinements over the B-4 model to give more thrust. Production was just beginning in 1945.

BELOW:
Fig 2.66 The Horten Ho IX V2 (Werk Nr. 39) at Oranienburg prior to its first flight in late December 1944. Powered by two Junkers 109–004 B-1 turbojets, it was destroyed in an enforced single-engine landing in February 1945, killing the pilot, Ziller.

GERMAN JET ENGINES

Fig. 2.67 A Gotha Go 229 all-wing fighter with two Junkers 109–004 C turbojets under construction. This plant, at Friedrichsrode, was captured by troops of the VIII Corps of the US Third Army on 14 April 1945. *(US Army photograph)*

Fig 2.68 Another Gotha Go 229 jet fighter prototype, further advanced in construction, captured at the Friedrichsrode plant. Again, the abandoned state of the plant is apparent, but the time cards on the right seem to indicate that about thirty personnel worked there. *(US Army photograph)*

930 km/h (578 mph). Figures 2.67 and 2.68 show Gotha Go 229 prototypes under construction at the war's end, one fitted with 109–004 C engines.

The 109–004 D-4 (TL)

Modification of the combustion system of the 109–004 B-4 permitted hotter combustion and increased thrust at some expense in engine life. This appears to have been merely an interim thrust uprating, and the engine incorporating the modification, the 109–004 D-4, was not for large-scale production. The thrust of this engine was 1,050 kp (2,315 lb) static, and it was to fit the same nacelle as the engines in use. Aircraft projected to use two of these engines included the Arado Ar 234 P-4 night-fighter and the Ar 234 C-8 bomber.

The 109–004 E (TLS)

The 109–004 D-4 engine was to be introduced with after-burning as the 109–004 E in July 1945, and a 20 per cent increase in thrust was expected. To test this, a 109–004 was fitted with a long exhaust section, but 13 to 14 per cent thrust increase was the most obtained without excessive exhaust section temperature and deterioration of the structural work. Two fuel injection methods were tried, as follows:

(i) Injection at turbine inlet nozzle level, which gave perfect stability but badly affected the turbine blades.
(ii) Upstream injection, without baffling, immediately behind the turbine, which did not give perfect stability but was the preferred method since it did not increase the turbine temperature.

These tests were performed on the test bed only, and no flight tests were made. One source states that about 100 hours of running were performed on experimental engines both with and without the afterburner section. The 109–004 E was planned for a thrust of 1,200 kp (2,646 lb) static with afterburning, but since only about 1,140 kp (2,514 lb) static was practical, the type was not worth pursuing in the light of other developments.

The 109–004 F

No details are known concerning the F designation, and it is only mentioned as being in the 109–004 engine model sequence A to H under description. However, the most likely possibility is that the 109–004 F was projected to use water injection (F possibly standing for 'flüssig', or fluid). In any event, Junkers tested a 109–004 engine with water being injected radially, both before and after the turbine, and allegedly obtained a 15 per cent increase in thrust by virtue of the increased cycle temperature or increased energy level. This work was not carried far because, at the time, experiments were handicapped as the test beds were needed for production engine type tests.

The 109-004 G (TL)

Based on the 109-004 C, the 109-004 G was designed to have an 11-stage axial compressor, eight combustion chambers and the usual single-stage turbine. With the air mass flow thus increased, the engine was expected to give a Class II thrust of 1,700 kp (3,749 lb) static, but was not built.

The 109-004 H (TL)

As a complete re-design and virtually a new engine, the 109-004 H was planned to give double the thrust of the B model production engines, with only slight increases in the overall dimensions. The new engine was to have an 11-stage axial compressor and a two-stage turbine, and fell into Class II. By the end of the war, no parts had been made, and the engine was still in the final drawing stage since the RLM allotted a higher priority to the 109-012 turbojet (described later). Data for the proposed 109-004 H-4 production engine are:

Thrust	1,800 kp (3,970 lb) static
Speed	6,600 rpm
Weight	1,200 kg (2,646 lb) approx.
Pressure ratio	5.0 to 1
Specific fuel consumption	1.2 per hour
Airflow	29.5 kg/sec (65.05 lb/sec)
Specific weight	0.67 approx.
Diameter	0.860 m (2 ft 9^m in)
Length	4.00 m (13 ft 1 in)

Relatively few aircraft were designed to use the 109-004 H. The projected Horten Ho XVIII B long-range tailless bomber was to use six of these engines grouped beneath the wing and to cruise at 800 km/h (497 mph) over an 8,000 km (4,968 mile) range.

The 109-012 (TL)

Once the Class I 109-004 turbojet was in production and its main shortcomings were overcome, the GL/C Technisches Amt urged Junkers to develop a Class II turbojet. The 109-004 G and H designs at first put forward were not considered powerful enough, and so an entirely new, larger engine, the 109-012, was begun for Class III.

The design of the 109-012 was presided over by Dr Stein, but was hampered by a lack of facilities. In layout, the engine followed previous practice to a large extent. The compressor was an 11-stage axial type with solid blades enclosed in a neatly designed casing fabricated from mild-steel sheet with band strengtheners. Some initial difficulty was expected with distortion of this compressor casing, but a cast casing was ruled out on the grounds of weight and also because difficulties were being experienced in obtaining castings in quantity. The compressor blades, designed by Encke of the AVA with a critical mach number of 0.8, were thinner than those for the 109-004 and had much sharper leading edges, with maximum thickness occurring at approximately the 55 per cent chord point. Provision was made for an air bleed after the fifth or sixth compressor stage, to provide, if required, an auxiliary compressed-air supply. A fabricated compressor exit assembly led into eight combustion chambers housed inside a sheet-metal casing. The turbine was of the two-stage axial type with hollow, air-cooled blades, while a moveable cone was used to vary the exhaust nozzle area.

Ten 109-012 turbojets were ordered for initial development. Some parts for the first three were made at Dessau, where nine exhaust cones were also ready, the cones having been made elsewhere (probably by the firm of Opel). None of the parts was tested, however, since the RIM ordered a stop to the development in December 1944 (reason unknown), and the design work was dispersed from Dessau to Brandis, Aken and Mosigkau. A mock-up of the 109-012 is shown in Figure 2.69 and a drawing in Figure 2.70, and data for the engine are:

Thrust	2,780 kp (6,130 lb) static
Thrusts at 900 km/h (559 mph)	2,200 kp (4,851 lb) at sea level
	1,100 kp (2,425.5 lb) at 10,000 m (32,800 ft)
Speed	5,300 rpm
Weight	2,000 kg (4,410 lb) approx.
Pressure ratio	6.0 to 1
Specific fuel consumption	1.2 per hour
Airflow	50 kg/sec (110 lb/sec) approx.
Specific weight	0.72 approx.
Diameter	1.063 m (3 ft 5^m in)
Length	4.862 m (15 ft 11 in), excluding tail cone

Fig 2.69 A Junkers 109-012 turbojet engine. This is a partially completed mock-up. Designed to have three times the thrust of the 109-004 B, its size dictated a sheet-steel, fabricated compressor casing to minimise weight.

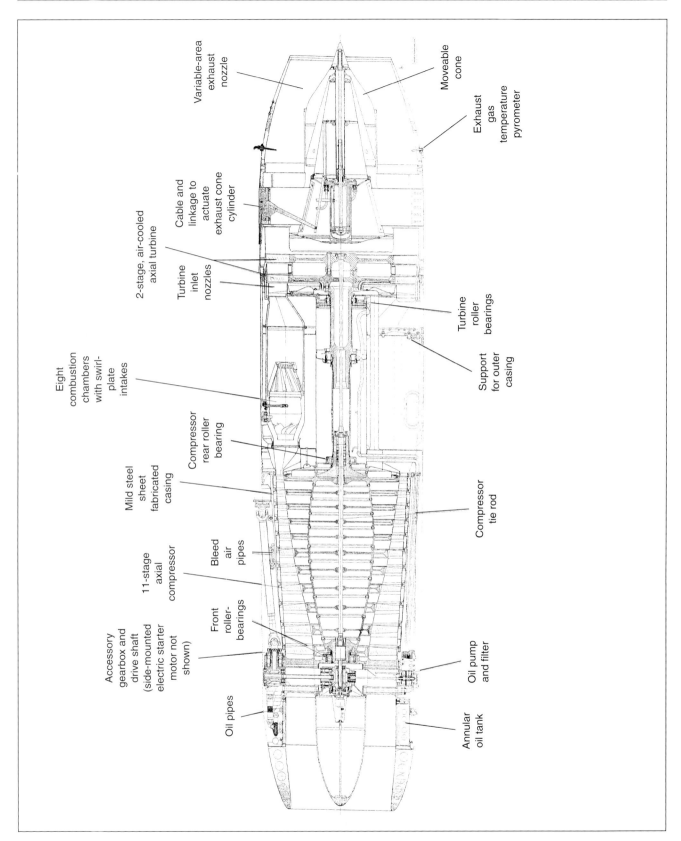

Fig 2.70 Junkers 109–012 A Class III turbojet engine. Based on Junkers drawings dated April 1944, this drawing has not been seen since the end of the war.

The Junkers Ju 287 B-2 bomber with swept-forward wings was originally designed to use two 109–012 (or two BMW 109–018) turbojets beneath the wings, and the projected Heinkel He 343 bomber was also to use two such engines.

The 109–022 (PTL)

In line with the official jet engine development programme, turboprop versions were to be developed from the new turbojets of Class II and larger so that the performance of large aircraft could be considerably improved. The projected turboprop version of the 109–012 turbojet engine was the 109–022, which Junkers were to develop as a unit intermediate in size between the smaller Daimler-Benz 109–021 and the larger BMW 109–028. Basically, the Junkers turboprop was to consist of the 109–012 turbojet, with an extra turbine stage added so that 3.50 m (11 ft 5^i in) diameter contra-rotating, variable-pitch, three-bladed airscrews could be driven through reduction gearing. Between 40 and 60 per cent of the unit's power was to be produced by the airscrews, depending on speed. By the end of the war, the design of the 109–022 was not completed, and the problems of designing the airscrew reduction gear sufficiently compact were still being wrestled with. In any event, work probably stopped in December 1944 when the turbojet version was shelved. Figure 2.71 shows the outline only of the 109–022, data for which follow:

Power rating at sea level, static	4,600 equivalent shaft hp (Separate figures for jet thrust and shaft hp not available.)
Power rating at sea level and 800 km/h (497 mph)	Jet thrust 1,200 kp (2,646 lb) approx. Shaft power 3,500 hp approx. (9,500 equivalent shaft hp)
Airscrew speed	1,200 rpm

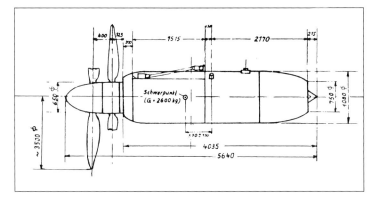

Fig 2.71 The Junkers 109–022 turboprop. Only this original outline has been found, giving projected overall size and centre of gravity.

Weight	2,600 kg (5,733 lb)
Diameter	1.063 m (3 ft 5^m in)
Length	5.640 m (18 ft 6 in) including 1.605 m (5 ft 3 in) for airscrew spinner
Diameter of airscrew spinner	0.650 m (2 ft 1^1 in)

Because the 109–022 was only at the project stage, few applications were planned.

Initial Allied intelligence on Junkers turbojets

Until the end of 1944, Allied technical intelligence on German turbojet work was fragmentary, sources in France, for instance, being of little help since the Germans had been careful to reveal no more than was necessary. Some time before the event, however, the appearance of jet aircraft in Luftwaffe service was expected. In April 1943, for instance, a senior officer of the Versuchsverband Ob.d.L. was captured and interrogated by the British. Amongst other interesting information, he revealed that his Commanding Officer, Colonel Rowehl, had recently been informed by Hitler that new weapons, including missiles and jet aircraft, were to be launched against Britain that summer. Actual Allied examination of these weapons depended on the Germans losing them during tests or operational sorties.

The first examples of Junkers turbojets were captured by the Allies in September 1944 when a Messerschmitt Me 262 crashed in Belgium. From the wreck, the US Army recovered two Junkers 109–004 B-4 turbojet engines and shipped them to England on 11 September for examination by Power Jets Limited at Pyestock. Unfortunately, these engines were badly damaged, especially at the forward ends, and test-bed running was not possible. Almost six months passed before engines capable of being put in running order were captured, and the circumstances surrounding this event are worth recording.

At 11.00 hours on 24 February 1945, Republic P-47 Thunderbolt fighters of the US Ninth Air Force shot down the first Arado Ar 234 to be captured by the Allies. This jet bomber (belonging to III./KG 76, which had become operational in February 1945 and was based at Rheine), suffered a flame-out in one engine and crash-landed near the village of Segelsdorf, which lay very near the front line at that time. The pilot was injured, and since the Ar 234 had suffered only minor damage, the Germans attempted to destroy it. However, in the face of advancing Allied forces, they had time only to recover the aircraft's documents, and after some skirmishing, the aircraft was in the hands of the US Ninth Army by 11.00 hours on 25 February. Further attempts were then made by the Germans to destroy the Ar 234 by artillery fire, and during one salvo, an officer of the US Air Technical Intelligence Section had a narrow escape when a shell lodged in the trailer of his jeep but failed to explode.

After dismantling, the Ar 234 was shipped to England for detailed examination. The two 109–004 B-1 engines, numbered 1163 and 2006, were received at Pyestock on 2 March 1945. Each was burned on the upper side, mainly around the gearbox and auxiliaries, and each was damaged by the penetration of 0.5-inch bullets. Engine 1163 was completely overhauled and first ran at Pyestock on 4 April, total running time in that month being nine hours 31 minutes. Apart from the failure of a few components, including the governor, the engine functioned very well. There was very little mechanical vibration, but roughness of the combustion was particularly noticeable. By British turbojet standards at the time, performance of the Junkers engine was considered unspectacular; engine 1163 gave a thrust of 839 kp (1,850 lb), a specific fuel consumption of 1.40 and a jet temperature of 660 °C, all at 8,800 rpm on the test bed.

Conclusion

Almost all the turbojet development effort of Junkers was concentrated on the 109–004 engine, but by the end of the war the production versions of this engine still had a high wastage rate through component failures; piles of both scrapped and bomb-damaged engines and components often greeted the Allies when they captured Junkers factories (see Figure 2.72). The performance and life of production 109–004 engines was also low and compared only with the first British Power Jets flight engine W.1 of 1941. Nevertheless, the aim of getting the then radical turbojet engine into service was achieved by Junkers, and their policy, in this case of conservative engineering being the road to moderate success, paid off. Examination of the Junkers achievement virtually means examination of the 109–004 engine, and in this respect the best evidence comes from the following extracts taken from the Power Jets Report No. R.1089 on the 109–004 B-1 engine:

> 'General engine design is such that performance appears secondary to production. This has resulted in a heavy engine with its major components, i.e. compressor, turbine, and combustion chambers, being of a simple design but having only a moderate performance. The design had to be made using a minimum of heat-resisting steels. By much ingenuity in design and by making full use of cooling air tapped off from the compressor, an engine has been built in which the only parts made in heat-resisting steel are the turbine nozzles and the turbine blades.
>
> Since the component efficiencies have been sacrificed for production requirements, a good fuel consumption could hardly be expected. The high consumption on this engine can be said to be partly due to poor component efficiencies and partly to the low compression ratio of the compressors . . .
>
> From a consideration of the turbine and compressor only, it seems that the development of this engine would proceed on similar lines to our own; that is, by increasing the throughput and the gas temperature. To increase the compressor and turbine throughput just means a change in the blade and nozzle settings and a 20% increase should easily be possible within the present scantlings. The combustion system, however, is less easily altered and the mean velocity through the combustion system is already some 10% higher than that corresponding to British designs. Also, the performance and life of the combustion chambers under the present design conditions is poor, as shown by the atmospheric rig tests. It therefore seems probable that combustion troubles would prevent any increase in thrust being obtained by simply increasing the throughput.
>
> The enemy has striven to keep down the temperature of the hot metal parts but, should a source of nickel and chrome become available so that better heat-resisting steels could be used, there is no reason why the engine should not run some 50 °C hotter. This would give a 10% increase in thrust but at the expense of a further rise in the already high fuel consumption.
>
> Without considering any change in the general scantlings or in the materials used, the engine could best be improved by reducing weight and improving combustion chamber efficiency. This would not, of course, give the engine a higher thrust rating.
>
> For general future design of gas turbine engines, there does not seem much to be learned from this engine. The enemy has always tended to sacrifice everything for production and has made strenuous efforts to overcome his shortages. In consequence, performance has suffered but it still shows that a useful jet engine can be built even when heat-resisting steels are in short supply.'

Fig 2.72 Junkers 109–004 B turbojet engines destroyed by bombing at the Magdeburg factory. Besides bomb damage, these early production engines had high failure and wastage rate.

The engine on which the above remarks were based did, of course, have uncooled, solid turbine blades. Apart from introducing air-cooled turbine blades to allow for possible higher temperatures, we have already seen the various methods Junkers resorted to in order to considerably improve their engine. Nevertheless, British and American turbojet engineers showed little interest in pursuing the Junkers work further, while in France the main emphasis was on actually using captured 109–004 engines to power various aircraft as an interim measure.

France's first jet aircraft, the Sud-Ouest S.0.6000 Triton, made its maiden flight on 11 November 1946 on the power of a single 109–004 B engine, and a similar engine was also used to power the Arsenal VG 70 which first flew on 23 June 1948. The Triton had a fuselage of large cross-section in order to flight-test various turbojet engines, but the VG 70 was a small research aircraft and achieved a level speed of about 900 km/h (559 mph).

However, until captured German jet engines became available in France, groups of French engineers studied jet engine problems in isolation and without knowledge of British and German work. One such group, which worked clandestinely during the German occupation, was the Société de Construction et d'Etudes de Matériels d'Aviation (SOCEMA), which, from 1941, began designing and building their TGA-1 (Turbo-Groupe d'Air) turboprop engine. From January 1944, the British Broadcasting Corporation mentioned aircraft without propellers, and the SOCEMA group suspected they used turbojet units derived from Whittle's known patents. Only rumours of German work in this field were heard. However, in August 1944, immediately after the liberation of France, partially destroyed Junkers 109–004 turbojet engines were found in a burnt-out German lorry near Orleans. SOCEMA were then asked by the Société Nationale de Constructions Aéronautiques de Sud-Est (SNCASE) to design a turbojet engine slightly larger than the Junkers engine but with double the thrust. The resulting engine was designated TGAR-1008, and benefited from a close study of the 109–004. Its eight-stage axial compressor was similar in construction (but used 100 per cent reaction), while an air-cooled, single-stage turbine and movable exhaust bullet were used. The engine regulator system was developed by Société Bosavia based on the Junkers regulator. However, the TGAR-1008 was an exercise in learning, not copying, and had many novel features, such as a duplex annular combustion chamber and a special method of air-cooling the turbine inlet nozzles. The TGAR-1008 passed a 150-hour type test at 2,100 kp (4,630 lb) static thrust, while a later development, the TR-1008, gave a static thrust of 2,500 kp (5,512 lb).

Meanwhile, in Czechoslovakia, where sub-contract work for the Germans was carried out during the war, manufacture of the Junkers 109–004 B-1 turbojet and Messerschmitt Me 262 was continued after the departure of German forces. Starting in 1946, parts were brought together and made to build 18 Me 262 As (as the S.92 fighter and CS.92 two-seat trainer) at the Avia plant. The 109–004 B-1 turbojets were constructed at the Letecke factory at Malesice under the designation M-04. A squadron of these jet aircraft stayed in service with the Czech Air Force until the introduction of the Soviet MiG-15 in the 1950s.

For the Soviets it was a different story. Towards the end of 1945, the first Soviet jet fighter, the Yak-15, appeared as an adaption of the Yak-9 piston fighter airframe with a single 109–004 B turbojet attached to the fuselage below the wing. During October 1946, trainloads of aeronautical equipment, turbojet engines, and 240 German gas turbine specialists began arriving in the USSR from Eastern Germany. This was a small part of the Soviet utilisation of German industry and know-how for which various isolated establishments were set up in the USSR. Most of the Junkers material came from the Dessau and Bernberg plants and was sent to the GAZ-19 establishment at Kuibyshev on the Volga, where German and Austrian engineers worked, mostly without choice, on turbojets under the Russian N.D. Kuznetsov. As part of a large training programme, the German work was followed closely by Soviet technicians. For a start, the Junkers 109–004 was put into production in the winter of 1946–7 as the RD-10 (RD-Reaktivny Dvigatel), and the Yak-15, followed by the improved Yak-17, went into squadron service with this engine (See Figure 2.73). The next stage was to put copies of British engines (Rolls-Royce Derwent and Nene) into production, but some attention was also given to the Junkers 109–012 turbojet and 109–022 turboprop engines. As we have seen, neither of the latter two engines had been run by the end of the war, and their estimated performance was not outstanding, but their comparatively large size was perhaps noteworthy. By September 1950, the team at Kuibyshev, headed by the Austrian engineer Ferdinand Brandner (formerly of Junkers), had developed a large turboprop engine of 6,000 shp based on the 109–022. Other developments followed (leading to the huge NK-012 which ran in 1951), and the surviving German and Austrian engineers were finally sent home in the summer of 1954.

Two examples showing Soviet continuance of German jet aircraft ideas shortly after the war will suffice. The Junkers Ju 287 jet bomber, with swept-forward wings, was continued as the EF 131, first at Dessau then at Stakanovo (now Zhukovskii) airfield in Russia. It was to have six RD-10 engines but was not flown (only the Ju 287 V1 was flown). This development was supplanted by the more advanced EF 140 (later 140 R) under Brunolf Baade's direction; it used two VK-1 turbojets (Soviet Rolls-Royce Nenes) and first flew on 12 October 1949. The second example was the design by Oleg Antonov of the so-called M fighter, which was based on the Heinkel He 162 and was powered by a single RD-10. Later changed to be powered by two RD-10s (and later still two RD-45s) with side intakes, the prototype was almost ready for flight when the project was cancelled in July 1948.

Fig 2.73 An example of post-war utilisation of Junkers turbojets is this Soviet Yak-17 fighter with an RD-10A turbojet (109–004). The similarly powered Yak-15 had entered service on 24 April 1946 and the Yak-17 entered service in 1948. Both were capable of about 500 mph at 3,000 m (9,840 ft). *(Author)*

Fig 2.74 An example of the Soviet-built version of the Junkers 109–004 turbojet, the RD-10. *(Jan Hoffmann)*

Most of the better-known Junkers technicians seem to have evaded Soviet efforts to deport them at the war's end. As an example, Dr Anselm Franz left Dessau a week or so prior to the arrival of US troops and went to live in Allenstadt (Harz), where he was interviewed by Allied personnel on 16 May 1945. Later, in 1954, Franz set up a gas turbine development team for Avco Lycoming in Connecticut, USA, and eventually rose to become vice-president and assistant general manager of that company. Other engineers also took up positions in foreign firms, some eventually returning to Germany.

As a final comment on the turbojet work of Junkers, it can be said that their engines powered not only Germany's and the world's first operational jet aircraft but also the first Soviet and French jet aircraft, and were planned to power the first jet aircraft of countries such as Italy. Today, by virtue of the number built, the 109–004 B turbojet is the most prolific of preserved examples of German turbojets, and the engine can be seen in many of the world's museums.

After his pioneering achievement in getting the first turbojet engine into mass production, Anselm Franz went on to oversee, at Avco Lycoming, the development of other successful engines. These included the T53 gas turbine helicopter engine (Cobra, Mohawk, UH-1, etc), T55 series of turboshaft engines and the T55 high-bypass turbofan engine designated the ALF-502. The 1960s saw the development of the AGT-150 gas turbine for the M1 Abrams main battle tank. Franz died in 1994, aged 94, after receiving many awards and honours.

For more details of post-war work on turbojets stemming from Junkers work see pp.251/256 of *Turbojet – History and Development 1930–1960, Vol.1* (A. L. Kay)

Fig 2.74A: Work at a very early stage on a Horten Ho IX all-wing jet fighter. Using a dummy engine, the engine fitting in the structural frame is being refined.

Bayerische Motoren Werke AG (BMW)

Brandenburgische Motorenwerke GmbH, Berlin-Spandau (Bramo), and origin of the P.3302 (TL) and the short-lived P.3304 or 109–002 (TL) — BMW Flugmotorenbau GmbH, Munich, and the short-lived P.3303 (TL) — BMW turbojet development facilities and personnel — The P.3302 (TL) V1–V10 series — The P.3302 (TL) V11–V14 series — The 109–003 A-0 (TL) — The 109–003 A-1 and A-2 (TL) — Description of the 109–003 A-1 and A-2 — The 109–003 E-1 and E-2 (TL) — Production of the 109–003 A and E — Testing the 109–003 — Utilisation of the 109–003 A-1 and 2 and E-1 and 2 — The 109–003 C (TL) — The 109–003 D (TL) — The 109–003 R (TLR) — Further 109–003 developments — The 109–018 (TL) — The 109–028 (PTL) and a twin-PTL project — The P.3306 (TL) — The expendable P.3307 (TL) — Conclusion

The experience of the Bayerische Motoren Werke GmbH (later AG) with aero-engines went back to 1917 when the Rapp-Motorenwerke GmbH of Munich was absorbed. The technical lag in aero-engine development induced by the post-First-World-War years led to the importation of American Pratt & Witney air-cooled, radial engines (Hornet and Wasp) which were built under licence and formed the basic for the new, larger engines which the RLM was demanding by 1933. At the end of 1934, a subsidiary, the BMW Flugmotorenbau GmbH of Munich, was formed to deal separately with aero-engines. Also in the same field was the Brandenburgische Motorenwerke GmbH (Bramo) of Berlin-Spandau, and this firm began pooling its development work with BMW in 1938. In 1939, Bramo was absorbed by BMW to form one of Germany's major aero-engine manufacturers of the Second World War. One of the organisation's distinctions was that it was the only one in Germany to develop all the major forms of aircraft power-plant, e.g. piston engines, turbojet engines, and rocket motors. To begin the turbojet history, a look will first be taken at Bramo up to the time when a single turbojet programme was pursued by both Bramo and BMW following the absorption of the former by the latter.

Brandenburgische Motorenwerke GmbH, Berlin-Spandau (Bramo), and origin of the P.3302 (TL) and the short-lived P.3304 or 109–002 (TL)

In 1936, the Brandenburgische Motorenwerke GmbH, or Bramo, was formed at Berlin-Spandau as a separate subsidiary of Siemens & Halske AG (later Siemens Apparate und Maschinen GmbH), of which it had previously been the aero-engine section. Bramo's head of research, Hermann Oestrich, had already made studies in 1929 of the possibility for jet propulsion, and had concluded that, for the aircraft speeds at that time envisaged, jet propulsion would prove too inefficient to be practical. Thus, when in 1938 the company was asked by Mauch and Schelp of GL/C3 to study the field of jet propulsion, Oestrich turned his attention to ducted fans driven by piston engines (ML units), which appeared more suitable for speeds then in prospect.

By the end of 1938, Bramo had built their first ML unit, which consisted of a 160 hp Bramo 314A seven-cylinder, air-cooled, radial engine driving a multi-bladed airscrew at 2,200 rpm within a short annular duct or ring. This unit was installed in a Focke-Wulf Fw 44 Steiglitz (goldfinch) two-seat biplane in place of the normal engine and airscrew. Flight test results indicated that the ducted-fan unit was worth pursuing, and so an MLS unit, with provision for afterburning to further accelerate the airflow, was designed and constructed. The ducted fan of this unit was driven at 2,600 rpm by a 1,200 hp Bramo 323 air-cooled, radial engine. Unfortunately, very poor results were obtained from this MLS unit when it was tested in a wind tunnel at the AVA, Göttingen, these results being partly caused by a low fan efficiency and low rpm.

As we have seen, very much higher rotational speeds were striven for in the development of the Mueller ducted-fan units at the Heinkel company, but these had the benefit of later experience. In any event, Bramo did not pursue their MLS project far, and abandoned it early in 1939 to concentrate on turbojet (TL) studies, which looked more promising as aircraft speeds increased. Even so, Bramo now believed that, for high-speed flight, ML and TL units were theoretically equally efficient, but that the ML unit could give the best take-off power and specific fuel consumption under partial load. Against this, the TL unit promised a less complicated construction, and this, coupled with the disappointing performance of the MLS unit tested, appears to have influenced the decision to concentrate on the turbojet (TL) studies, which were begun at the end of 1938.

In these studies the selection of the best type of compressor was considered, and with the parameters of both weight and size in mind, were put in the order of favour as contra-rotating axial, straightforward axial, and centrifugal. However, when the question of simplicity and ease of development by a company without gas turbine or turbojet experience was considered, it was decided to make a start on a turbojet engine with a straight-forward axial compressor. This engine was designed between December 1938 and April 1939, and later incorporated a single-stage turbine wheel under development by BMW at Munich, since, during this period, the two companies began pooling their development work. Other data for the turbojet design were gathered from Bramo experiments with combustion systems and so forth.

When, in the summer of 1939, Bramo were taken over by BMW, Bramo became known as the BMW Flugmotorenbau Entwicklungswerk Spandau. Following this, the Bramo turbojet design was given the project designation of P.3302. In agreement with the RLM, this engine was at that time considered as a research engine to assist in development of the more complicated turbojet engine with contra-rotating axial compressor, which it was believed could be made with a smaller diameter. The contra-rotating engine was given the project designation of P.3304 and the official designation of 109–002.

In the design of the 109–002, assistance was given by the independent engineering consultant Helmut Weinrich of Chemnitz, who already had some relevant designing experience. In 1936, Weinrich had submitted plans to the RLM for a turboprop unit (PTL) using a contra-rotating compressor and turbine, and had followed this up with experiments, but officialdom was apparently not ready to accept such radical power-plants at that time. Another effort in the field was Weinrich's patent (663,935) of August 1938 for a method of *introducing fuel to combustion turbines*, and other work of Weinrich's is discussed in Section 4 under Brückner-Kanis. By 1940, intensive work was being performed on the 109–002, although only about twenty designers and engineers were available for this development at Spandau. Individual stages of the contra-rotating compressor were built and tested, the forged aluminium alloy compressor blades being produced by WMF of Geislingen, and an annular combustion chamber was settled upon although not developed. As experience was gained, the realisation dawned that the technical difficulties in developing even a straightforward axial compressor turbojet were very great.

Thus, by early 1942, the 109–002 project was abandoned in order that all BMW resources could be concentrated on the development of the straightforward axial compressor turbojet, the P.3302. This development, which was to evolve into the BMW 109–003, is examined after the following look at the early turbojet work of BMW-Munich

BMW Flugmotorenbau GmbH, Munich, and the short-lived P.3303 (TL)

When, in 1938, BMW Munich accepted an official jet propulsion study contract, the work was put in the charge of their research head, Kurt Loehner. The decision was made that valuable experience could be gained by the actual construction and testing of an experimental turbojet engine, utilising, where possible, previous experience that BMW had with their exhaust-driven turbo-superchargers for piston engines. This supercharger work was, from 1937, largely in the hands of Dr Alfred Müller (the later activities of whom are related in Section Three), working for BMW as an independent consultant. Under Müller's name, the design of a hollow, air-cooled blade for gas turbines (especially in turbo-blowers) was filed for patent in December 1937, and the design of a *waste-gas turbine for aircraft* using hollow blades was granted a patent (682,744) in October 1939.

A typical arrangement of a BMW supercharger consisted of a single-stage axial turbine driving a single-stage centrifugal impeller or compressor. By 1938, the hollow, air-cooled turbine blades in these superchargers were allowing turbine wheels to run at working temperatures of up to about 900 °C. The elements chosen, therefore, by Loehner for the experimental turbojet engine were a single-stage axial turbine (with hollow, air-cooled blades) driving a *two-stage* centrifugal compressor and a combustion system with the high working temperature of 900 °C. While this engine, which is believed to have received the project designation of P.3303, was being designed in more detail, experiments proceeded with the turbine wheel in which the first difficulties arose. With the outbreak of war in September 1939, the impetus was given for less experimental and more developmental work, so that it was agreed that BMW-Munich should drop the work on Loehner's centrifugal turbojet and concentrate on the necessary production of piston engines for the Luftwaffe. Where possible, the experience with Loehner's engine was to be used to assist the development of the P.3302 axial turbojet at BMW-Spandau (formerly Bramo).

BMW turbojet development facilities and personnel

When the Bramo company was absorbed by BMW, the intention was that all jet-engine development should eventually be centred at Spandau. To a large extent this plan was followed, but the complicated network of the BMW organisation, partly because of the diversification in power-plant types and, later, through dispersal of plants and sub-contracting, caused various aspects of the turbojet work to be conducted in various places. However, most of the turbojet development work did centre at Spandau (with the aid, at first, of outside organisations), until the bombing of Berlin caused a large part of the work to be dispersed to caves at Wittringen (near Saarbrücken). Following the liberation of France in 1944, the work had to be removed from this border location, and finally, after short stays at a couple of other sites, finished up in a disused salt mine (Shafts 6 and 7) at Neu Stassfurt. This location was given the code name Kalag, and its twenty spacious, underground workshops were well equipped. Engine test beds were located at the head of Shaft 6.

At Spandau in 1939 only about forty engineers and designers were available for jet-engine work, while construction and experiments had to be performed in the same workshops used for piston engines, which had priority. The position improved in 1942, when not only were the 109–002 turbojet and various projects, e.g. MLS, dropped, but piston-engine work was also stopped at Spandau so that more personnel were available to work on one turbojet project, namely the P.3302. Thus, in 1942, there were available about 200 designers, scientists, test personnel and help for the same and about 700 people in the workshops. By the end of the war in 1945, these figures had increased to approximately 550 and 1,000 respectively, largely by transferring BMW personnel from piston-engine work.

The managing director of the whole organisation was Fritz Hille until about November 1944. By then, the increasing power of the Speer Ministry was taking effect, so that Hille was replaced by Dr Wilhelm Schaaf, while the director in charge of all engine production was Dr Stoffregen, both men coming from Albert Speer's staff. From the Oberwiesenfeld plant near Munich all BMW aircraft power-plant development was directed by Dr Bruno Bruckmann, who had below him Dr Biefang (later succeeded by Dr Rolf Amman) directing piston-engine development, Dr H. Oestrich directing turbojet development, and H. Zborowski directing rocket motor development. Kurt Loehner appears to have turned his attention elsewhere when the P.3303 centrifugal turbojet was dropped at Munich. From about 1942 until the end of the war, the principal members of Oestrich's turbojet staff included the following, but exact dates of appointments are unknown

Name	Job
Dipl.-Ing. H. Hagen	Basic design and project calculations. (Assistant to Oestrich.) Design and testing problems
*H. Roskopf	Chief designer
Dipl.-Ing. H. Zobl } Dipl.-Ing. Kruppe }	Combustion and fuel atomiser research
Dr W. Sawert	Metallurgical research
*H. Wolff	Experiment and research. (Moved to Heinkel-Hirth in 1943.)
*Domsgen	Turbine blade development
Dr Bosse	Designer
Dipl.-Ing. Menz	Designer. (Assistant to Bosse.)

(* Titles unknown)

In addition, the following personnel were amongst those who came under Bruckmann for general engine work, and therefore assisted to an increasing extent with turbojet development also:

Name	Job
*Wilzmann	Chief technician
Dr H. Wiegand	Chief metallurgist
Dr Scheinhost	Assistant to Wiegand. Concentrated on turbines

Name	Job
Dr Noack	Control and governor problems
Dipl.-Ing. P. Kappus	Chief of project study
Dipl.-Ing. Huber	Project studies. (Assistant to Kappus.)
Flugkapitän Staege	Chief of flight-testing
Dr Denkmeir	Responsible for jet flight-testing (from July 1944)
Dipl.-Ing. Stoekicht	Mechanical design
Dr Donath	O-series production

(* Titles unknown)

Like Junkers, BMW utilised in the first stages of turbojet development their existing piston-engine facilities. Combustion test rigs had, at first, sufficient airflow to test only a part of a combustion system, so that BMW experimented with a segment of an annular combustion chamber. At least one test blower was driven by a 900 hp Bramo 323 radial aero-engine. Gradually, the facilities were improved and became more specialised, and in this respect the special BMW high-altitude-engine test plant is worthy of particular note. Construction of this test plant, which was designed by Dr Soestmeyer, began in May 1940 on the northern edge of the BMW Oberweisenfeld factory site at Munich. Heavy bomb damage was sustained in May 1944, but by October 1944 the plant was repaired and was in regular service from then until the end of the war, testing both BMW 109–003 and Junkers 109–004 turbojets.

The test plant consisted of an altitude chamber measuring about 3.80 m (12 ft 5 in) diameter by 7.30 m (24 ft) length in which either a piston engine or turbojet engine was mounted. Its layout is shown diagrammatically, together with performance curves, in Figure 2.75. A three-stage centrifugal compressor absorbing 2,800 hp and with a compression ratio of 2.4:1 gave a maximum airflow of 25 kg/sec (55 lb/sec). The air first passed through a heat exchanger, which could be used to heat the air if required, but normally the air was then cooled by water sprays. It then passed through a Freon plant, which brought the temperature down to –15 °C, and was then expanded through a turbine doing useful work, which reduced the air temperature to –70 °C if required. As the diagram shows, control valves were connected in a bypass circuit between the compressor outlet and turbine inlet so that any desired air temperature could be obtained. From the expansion turbine, the air passed to the engine in the test cell, giving the necessary ram pressure to simulate speeds up to 900 km/h (559 mph). Altitude conditions were produced in the exhaust from the test cell by first cooling with water sprays, bringing the temperature down to 250 °C, and then through a cooler, bringing the temperature down to 40 °C. Two exhausters, or vacuum pumps, were connected by means of a valve, either in series or parallel depending on the altitude conditions required, to produce the requisite low pressure in the exhaust. If the exhaust pumps were placed in series, an intercooler was used. The pressure range at entry to the test cell was 1.5 to 0.008 atmospheres and a pressure inside the cell could be maintained equivalent to an altitude of up to 16,500 m (54,120 ft). The second curve in the graph with the diagram shows the projected increase in performance planned with additional equipment, this equipment having been manufactured but dispersed to parts of western Germany.

Since the power load necessary for operation of the plant was in the region of 10,000 kW, only night operation, under Munich off-peak conditions, was possible. A crew of ten was used for daytime maintenance and a crew of ten for night test-running, although one man could control the conditions in the test cell. A periscope in the control room enabled combustion conditions to be examined at two longitudinal positions. To measure power output, an hydraulic brake capable of absorbing up to 5,000 hp was used for piston engines, while jet thrusts were measured mechanically, and later, using load cells.

Another altitude engine test cell was built by Brown Boveri & Cie of Heidelberg for BMW at Oberürach (near Stuttgart). This tunnel provided an airflow of 20 kg/sec (44 lb/sec), but an auxiliary unit could be added to raise the airflow to 30 kg/sec (66 lb/sec). Two similar test cells were also made for E-Stelle Rechlin, but because the buildings were not ready in time, they were not erected but stored in woods near the airfield.

The P.3302 (TL) V1–V10 series

During 1939, BMW-Spandau laid down the design of the P.3302 turbojet to produce a 600 kp (1,323 lb) thrust at 900 km/h (559 mph), or about 700 kp (1,544 lb) static thrust at sea level. An axial compressor driven by a single-stage turbine was planned, together with a very low combustion temperature of 600 °C. However, when, later in the year, Loehner's P.3303 centrifugal turbojet project was stopped at Munich, it was decided to incorporate the turbine wheel being developed for that

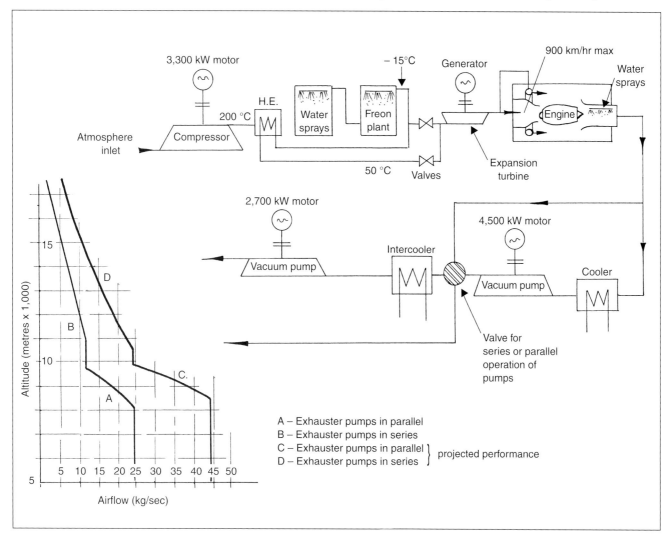

Fig 2.75 Schematic layout and performance graph of BMW high-altitude test chamber.

engine into the P.3302. With this hollow-bladed turbine wheel, it was considered that the combustion temperature could be raised to the high figure of 900 °C, while at the same time lowering the planned pressure ratio of the compressor. By this time the RLM had ordered ten experimental engines (V1 to V10) to be built.

The compressor form for the P.3302 V1 was designed by Prof. Encke of the AVA. It was a six-stage axial type with a small percentage of reaction, and gave a low pressure ratio of 2.77 to 1 at 9,000 rpm. The rotor blades had AVA, Göttingen, high-speed profiles, and test-stand results indicated an 80 per cent efficiency for the compressor.

Based on promising BMW-Spandau experiments in 1938 with annular combustion chamber segments, an annular combustion chamber was chosen for the P.3302 which resulted in an engine diameter of only 0.670 m (2 ft 2 in), the RLM having asked for the minimum possible frontal area. Experiments with the turbine wheel led to the conclusion that the combustion temperature of 900 °C was too high, and so this was lowered to 750 °C before the end of 1939. When the first engine, the P.3302 V1, first ran in August 1940 the result was disappointing, and only 150 kp (331 lb) thrust was given at 8,000 rpm on the test stand.

The causes of this poor result were several and included:

(i) Insufficient airflow for a combustion temperature of 750 °C, since 900 °C had originally been designed for.
(ii) Poor turbine efficiency since the lowering of the combustion temperature had lowered the velocity of the hot gases. This, in turn, lowered the power to the compressor and its air delivery.
(iii) Uneven temperature distribution from the combustion chamber, with as much as 370 °C temperature difference between the hottest and coldest parts. This caused distortion of the combustion chamber inner liner and turbine inlet nozzle diaphragm, which created friction in the engine bearings.
(iv) The efficiency of the annular combustion chamber was only 60–70 per cent, and a lot of fuel left the chamber unburned. There were also high pressure losses caused by the baffle plate system used.

To eradicate these faults, a new compressor, combustion system and turbine, plus innumerable detail improvements, were needed. This, of course, amounted to a complete re-design, and the development of new components was immediately begun. For this work, the P.3302 V2–V10 engines were used in addition to other research facilities. Outside assistance was obtained from Brown Boveri, Brückner-Kanis and MAN in the development of a new turbine, while Henschel and the Technischen Akademie der Luftwaffe (TAL) assisted with the new combustion system. The branch of TAL responsible for combustion work was located at Eckertal (Harz) under Dr Nagel.

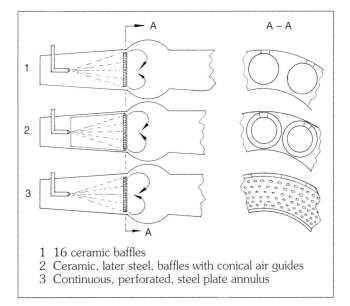

1 16 ceramic baffles
2 Ceramic, later steel, baffles with conical air guides
3 Continuous, perforated, steel plate annulus

Fig 2.76 Three main systems of baffle plate burners tried by BMW.

BMW were, however, keen to tackle as much of the work as possible themselves. Already, in 1938, combustion work was started at Spandau to establish the best way of obtaining steady, efficient combustion without blowout. This, it was decided, could be achieved by creating eddy regions in the airflow, and so work began on the development of baffle plates inside an annular combustion chamber. Three main systems were tried and are shown diagrammatically in Figure 2.76. In the first system, 16 ceramic baffle plates were mounted in the combustion chamber, and sprays from the fuel nozzles, located upstream, impinged on the forward faces of these. The total airflow passed around the baffles and was to carry the fuel into the lower-velocity eddy regions for combustion. However, not only were the blowout characteristics unsatisfactory but the airflow around the baffles was not uniform around the annular chamber, with the result that temperature and velocity distribution was not uniform downstream and operation under full airflow conditions was not possible without blowout. This system was then modified to give a second system in which conical guide tubes were positioned in front of the ceramic baffles. These guide tubes permitted operation with full airflow without blowout of the combustion, but temperature and velocity distribution still remained poor. Also poor was the combustion efficiency, and the ceramic baffles frequently broke under thermal shock. Replacing these baffles with ones of perforated steel gave an unlimited baffle life but did nothing for the system as a whole, other than bestowing a slight improvement in the blowout characteristics. In an attempt to obtain an even airflow around the baffle system, the individual circular baffles were abandoned and replaced by a continuous, perforated, steel plate annulus, but this gave no improvement, and in fact actually increased the pressure losses within the chamber.

It was probably the annular baffle plate that was used in the P.3302 V1, and a careful look at the poor results drew the conclusion that impinging the fuel sprays onto the baffle plate was wrong. Experiments whereby the fuel was injected upstream into the eddy regions were tried and immediately gave a 20 per cent improvement in combustion efficiency over the previous poor figure of 60–70 per cent. There was also a 25 per cent improvement in the temperature and velocity distribution, although this was still far from good, and the total airflow still passed around the baffle plate and straight into the combustion chamber. Nevertheless, this improved baffle combustion system was used for subsequent engines in the P.3302 series. (It was not until the next series that the baffle system was dropped and another combustion system, which was under parallel development and which used a divided airflow of primary and secondary streams, was introduced.)

By the summer of 1941, the P.3302 had been persuaded to give about 450 kp (992 lb) thrust on the test bed, mainly by the combustion improvements mentioned above, and flight tests began at Berlin-Schönefeld, using a Messerschmitt Bf 110 (twin-airscrew fighter) with the turbojet suspended below. In the autumn of that year, two of the turbojets were delivered to Messerschmitt's Augsburg plant for fitting to the Me 262 V1 (PC+UA) prototype of the twin-engined fighter. Held up for turbojet engines, this aircraft had been flying since 18 April 1941 on the power of a Jumo 210G piston engine and airscrew in the nose, and this engine was retained when, on 25 November 1941, an attempt was made to take off with two of the P.3302 engines installed under the wings. This attempt was a failure, however, since the turbines failed at take-off revolutions, although the power of these turbojets, even with the airscrew running, was most probably too low anyway. Troubles had been experienced with the turbines for some time prior to this, and failures had occurred at speeds as low as 8,000 rpm. The type of turbine construction used in these early P.3302 engines is shown by the sketch in Figure 2.77. Classical gas turbine impulse blading was used, the turbine wheel being 530 mm (1 ft 8^m in) in diameter and the rotor blades being 90 mm (3 in approx.) long. Each blade was constructed by welding together two pieces of metal, and the blades were attached to the turbine wheel by welding at the base. Cooling air entered the turbine wheel at the forward face only and passed by diagonally bored holes into the hollow blades and out through their tips. While this form of construction proved satisfactory in the smaller sizes used in BMW superchargers, the larger size of the turbojet turbine proved too much and failures occurred in the welded joints through fatigue and in the blades through heat-induced brittleness. However, all the causes of failure took time to identify, and this form of turbine construction appears to have been persevered with for most, if not all, the ten engines in this series.

Early in 1942, two of the P.3302 V2–V10 engines were again delivered to Messerschmitt and fitted to the Me 262 V1. On 25 March 1942, with the nose-mounted piston engine and the two turbojets running, Fritz Wendel cautiously prepared to take off in this aircraft. Improvements to the turbojets had raised the static thrust to about 550 kp (1,213 lb) each, but this was still some way below the planned thrust. The aircraft managed to take off, however, although greatly underpowered, but at a height of about 50 m (160 ft), the turbine blades of the port engine failed and the starboard engine immediately followed suit. With great skill, Wendel brought the stricken aircraft in to land after the shortest possible circuit on the piston engine alone. This unfortunate episode doubtless did much to sound the death knell of BMW's chances of providing engines to power future Me 262s. (The damaged Me 262 V1 was repaired and flew again on 2 March 1943, using two Junkers 109–004 B-0 turbojets, and practically all subsequent aircraft in the series were also powered with Junkers engines.) Even so, up to about April 1942, both the Junkers and BMW experimental engines were running below the designed thrust by about the same amount, and both were beset with their own difficulties. Final known data for the P.3302 V1–V10 series, without accessories or variable-area exhaust nozzles, are as below:

Thrust	550 kp (1,213 lb) static, at sea level
Speed	9,000 rpm approx.
Weight	750 kg (1,654 lb)
Pressure ratio	2.77 to 1
Specific fuel consumption	2.2 per hour (petrol)
Airflow	15 kg/sec (33 lb/sec) approx.
Specific weight	1.36
Diameter	0.670 m (2 ft 2 in), later 0.690 m (2 ft 3 in)

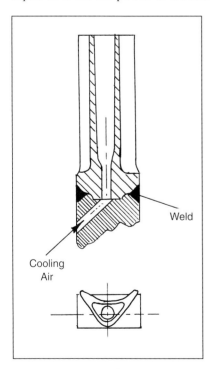

Fig 2.77 Sketch of turbine construction for BMW P.3302 V1–V10 series.

The P.3302 (TL) V11–V14 series

Following the experiences with the P.3302 V1–V10 series of engines and the testing of new components, BMW-Spandau began the construction of further experimental engines during the summer of 1942. Steps had been taken to increase the engine's airflow by designing a new compressor with a 30 per cent greater mass flow and an increased pressure ratio of 3.1 to 1 at a speed of between 9,000 and 9,500 rpm. The new compressor was designed and built at Spandau and had seven instead of the previous six stages. The profiles of the rotor blades were changed from the previous AVA Göttingen high-speed type to thicker NACA type, probably on the advice of Prof. Dr Betz of the AVA, and 30 per cent of the pressure rise now occurred in the stator blades. The lengths of the blades were decreased.

An improved turbine with greatly increased efficiency was also introduced, the rotor blades of which had a rounded leading edge and a longer, sharp trailing edge instead of the previous form with similar leading and trailing edges. To improve blade cooling, and therefore durability, more air was now brought from both sides of the turbine wheel and into the roots of the hollow blades. The previous welding method of blade attachment was also abandoned, and the new blades had longer roots, each being secured to the wheel by three axial pins which were buried half in the blade root and half in the wheel. The pins, which were arranged two on one side and one on the other side of the blade root, held the blades against centrifugal forces, while shoulders at the base of each blade countered radial or rocking forces. The previous, simple, curved metal vanes forming the turbine inlet nozzles were now replaced by hollow vanes having profiles similar to the turbine rotor blades. This not only gave greater rigidity and resistance to distortion for the inlet nozzle assembly, but allowed for air cooling. Cooling air entered the inlet nozzles at their bases and emerged through trailing edge slots. The combustion system, at this stage, was still of the baffle annulus type with upstream injection of the fuel into the airstream eddy section and without division of the airflow.

Test-bed running began with the P.3302 V11 towards the end of 1942, but no attempts were made to fly an aircraft with the new engine, and flights were confined to the Bf 110 flying test-bed. At some stage, a new intake cowling was introduced and the first tests with a variable-area exhaust nozzle began.

For the P.3302 V1–V10 series, the intake cowling and duct had conformed to the standards of the Verein Deutscher Ingenieure (VDI), and gave a slight increase in the local air velocity over the outer cowling contours for the purpose of giving uniform distribution of air velocity prior to entering the compressor. This intake proved adequate for low speeds and test-bed running, but it soon became apparent that a more accurate form was needed for flight engines. For high-speed flight, it was necessary to avoid the increase in local air velocity over the outer cowling contours in order to avoid flow separation, while the conversion of ram air into dynamic pressure in the intake duct had to be as favourable as possible. On the low-speed side, the best air intake conditions were also necessary in order to obtain full thrust and the best take-off run. Dr Küchemann of the AVA, Göttingen, worked on a revised form for the outer cowling, while wind-tunnel experiments at the LFA, Braunschweig, gave a corrected form for the inlet duct.

The type of variable-area exhaust nozzle tried with the P.3302 V11–V14 series is shown by the sketch in Figure 2.78. The device was of the so-called Ringspalt-Schubdüse (or annular-slot exhaust nozzle) type, in which a tapered ring was moved axially to increase or decrease the exhaust nozzle area. Operation was manual (possibly through rack and pinion gear), while cooling air, tapped off from the fourth compressor stage, was fed to the hollow, supporting spokes or vanes of the ring and passed out through slots in the trailing edge of the ring, thus cooling both spokes and ring.

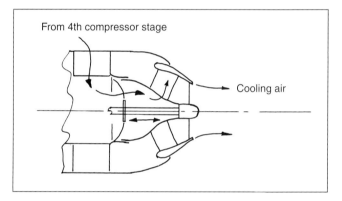

Fig 2.78 Annular-slot exhaust nozzle for BMW P.3302 V11–V14 series.

First runs with the new engine yielded at least 550 kp (1,213 lb) static thrust, but were disappointing, and a number of deficiencies were revealed, such as:

(i) Fractures due to vibration occurred in the first-stage rotor blades of the compressor after a short time. Starting characteristics of the new compressor were not as good as with the old compressor since aerodynamic defects gave very high drag during starting.

(ii) Serious difficulties remained with the baffle plate combustion chamber, and its poor efficiency and temperature distribution could not be improved.

(iii) Turbine blade failures were less but still frequent.

(iv) Distortion of the exhaust nozzle ring occurred, together with overheating of its operating mechanism.

(v) Compressor axial thrust bearing was prone to overheating because of an insufficient oil feed during starting or accelerating.

The following data for the P.3302 V11–V14 engines, which naturally varied considerably in performance, include the best figures obtained at the end of the series:

Thrust	600 kp (1,323 lb) static
Speed	9,500 rpm
Weight	650 kg (1,433 lb)
Pressure ratio	3.1 to 1
Specific fuel consumption	1.6 per hour (petrol)
Airflow	18.8 kg/sec (41.45 lb/sec) approx.
Specific weight	1.08 approx.
Diameter	0.690 m (2 ft 3 in) ex auxiliaries

With the end of 1942, solutions had been found or were within sight for the most serious difficulties, and a new engine was under construction. This new engine, which represented the final major re-design, was the responsibility of Hans Roskopf, now chief designer at Spandau. The P.3302 had by then received the official designation of 109–003, the pre-production models of which were designated 109–003 A-0.

The 109–003 A-0 (TL)

The new features incorporated in the 109–003 A-0 were designed to eliminate the above list of previous faults. Some, if not all, of these features may have been tested

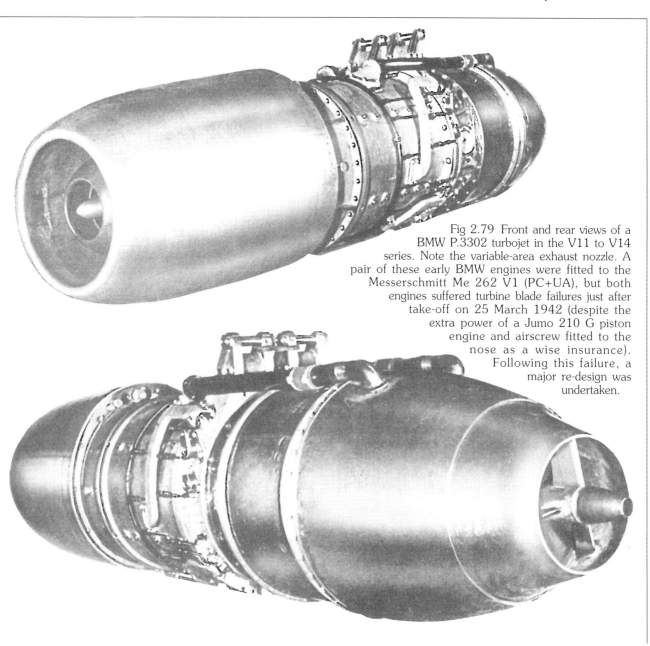

Fig 2.79 Front and rear views of a BMW P.3302 turbojet in the V11 to V14 series. Note the variable-area exhaust nozzle. A pair of these early BMW engines were fitted to the Messerschmitt Me 262 V1 (PC+UA), but both engines suffered turbine blade failures just after take-off on 25 March 1942 (despite the extra power of a Jumo 210 G piston engine and airscrew fitted to the nose as a wise insurance). Following this failure, a major re-design was undertaken.

in two experimental engines designated P.3302 V15 and V16, but improvements were still necessary at the pre-production stage, 109–003 A-0, so that true pre-production engines were later designated 109–003 A-00.

The first-stage rotor blades of the new compressor used with the previous experimental engines (P.3302 V11–V14) had failed within one hour of running, and this was obviously a serious problem requiring close scrutiny. Accordingly, a plastic window was fitted into the compressor casing of an engine facing the first-stage rotor blades. A small steel point or scriber was then fitted into the tip of one rotor blade at approximately the centre of gravity of its profile section. By this means it was found that the 75 mm (3 in) long rotor blade took up a vibrating motion on its natural frequency, the amplitude of which reached up to 2 mm ($5/64$ in) under the most critical condition at approximately the rated speed of 9,500 rpm. The blade frequency was, however, far too low to be excited by variation of aerodynamic pressures caused by the rotor blades passing through the wakes of the stationary vanes in front of the rotor, and it was found out that the trouble emanated from further forward at the profiled ribs (or spider) supporting the rotor front bearings. These ribs generated a wake sufficient to create aerodynamic disturbances which were cut by the rotor blades. Unfortunately, when running at about rated rpm, the first-stage rotor blades cut the disturbances at a rate equal to the natural frequency of the blades, so that a resonant vibration was excited which led to blade fatigue and failure. None of the rotor blades after the first stage ever gave trouble, since the aerodynamic disturbances grew less as they travelled rearwards, while the natural frequency of the blades increased as the blades in each stage progressively shortened.

Two modifications were introduced to cure the compressor trouble by creating as wide a gap as possible between the critical speed of the rotor and the rated speed of the engine. Obviously, the rated speed of the engine (9,500 rpm) had to remain unaltered if possible, so the modifications introduced were:

(i) A change in the rotor blade profiles. Where the previous blades had a thickness of 9 per cent at the hub and 6 per cent at the tip, the new blades were given 12 per cent at the hub and 5 per cent at the tip. This greatly increased their natural frequency and decreased centrifugal and bending stresses in the blades.

(ii) A new support spider for the rotor front bearing. The number of ribs was decreased, which reduced the frequency of the excitation transmitted in the rib wakes. Also, to break the sequence of the impulses, the ribs were positioned with unequal spacings.

Apparently, these modifications were entirely successful and no further compressor blade failures were suffered, but the compressor efficiency was reduced from 80 per cent to 78 per cent. It is interesting to note that similar conclusions on the cause of compressor failure were arrived at by Dr Bentele at Heinkel-Hirth in October 1942 in connection with their 109–011 V1 engine. However, whereas Heinkel-Hirth had redesigned an offending support spider, they had resorted to other means of reducing blade root bending stresses, such as the introduction of freely rocking blades. BMW also tried, at one stage, compressor blades with a degree of looseness in the rotor. They found that running with such blades was quite possible, centrifugal force keeping the blades in a stable position, but the critical vibrations were worse than with the normal blade fittings so that firmly fastened compressor blades were preferred.

Despite the previous considerable improvement of the combustion chamber, results had remained poor and it was decided that the baffle plate combustion system was basically wrong. A second type of combustion system was therefore put under development and was ready for introduction into the 109–003 A-0 (see Figure 2.80). A basic difference of the new system was the division of the airflow into primary and secondary airstreams. The primary stream (60–70 per cent of the total) passed to burner cones to support combustion, while the secondary stream (30–40 per cent of the total) was introduced downstream to mix with the hot gases. Each of the 16 burner cones had its own primary air supply, the downstream edge of the cone producing toroidal eddies into which the fuel was sprayed directly. Further downstream there were 80 hollow fingers which protruded into the annular combustion chamber from the outer and inner walls, and secondary air flowed from these fingers into the gas stream. A later improvement admitted a small amount of air around the actual fuel nozzles, which reduced the temperature at the centres of the flames and cooled the nozzles. The new combustion system was a success. Combustion efficiency was raised to 90 per cent or more, pressure losses were greatly reduced, and the temperature and velocity distribution was greatly improved. At this stage petrol was still being used as fuel.

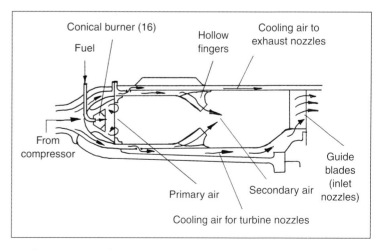

Fig 2.80 Sketch of combustion chamber for BMW 109–003

Improvements to the turbine section of the engine concentrated on improving the cooling and extending the life of the rotor blades. To replace the previous turbine blades made from two welded halves, blades forged from tubular stock by the Leistvitz company were tried, but although these gave a longer life, they proved unsuitable for mass production in the envisaged quantities. BMW then developed a new blade made by rolling sheet stock, folding and welding down the trailing edge. Inside the hollow blade was a sheet-metal insert which restricted the cooling-air passage to a small volume around the inner wall of the turbine blade. Adequate cooling was thus achieved using the minimum of air, the cooling air amounting to about 2 per cent of the total engine airflow in the 109–003 A-0. Not only did the cooling-air insert (Kühleinsatz) afford outstanding economy in cooling the turbine blades, but it contributed largely to the subsequent freedom from blade failures; friction between the insert and the blade occurred every time the blade flexed, and thereby gave a strong damping action. Since the new blade had a less solid foot than before, a new method of attachment to the wheel was needed. Figure 2.81 gives a sketch of the new method, whereby a pin was welded to the blade and cooling insert feet, and the whole assembly was fitted with two side pins into an appropriately shaped cut-out in the turbine wheel rim. A shaped circular plate on each side of the wheel prevented longitudinal movement of the blades and pins, and also served to duct the cooling air to the blades. This cooling air was tapped off from the fourth stage of the compressor.

The material chosen for the blades was FBD (17.0 Cr, 15.0 Ni, 2.0 Mo, 1.15 Ta-Nb, 1.0 Si, 0.1 C, blce Fe) made by Gebrueder Boehler of Kapfenberg Steuermark, Austria, while the cooling insert was made in another heat-resisting steel known as Remanit 1880 S (18.0 Cr, 8.0 Ni, 0.1 C, blce Fe) made by DEW. Mass-production of the new blades was undertaken by the Württembergische Metallwarenfabrik (WMF) of Geislingen/Steige, where Prof. Dr Arthur Burkhardt headed development work.

Results with the new turbine wheel were good. The life was considerably increased and the amount of heat-resisting steel used was reduced. Furthermore, the 109–003 A-0 was designed so that the turbine wheel could easily be replaced without removing the engine from the aircraft, an advantage not enjoyed by the Junkers 109–004. When turbine blades did fail (initially after about twenty hours), the fracture occurred at the blade neck, just above the foot, and was caused by hardening of the metal.

Since the annular-slot exhaust nozzle had proved deficient, a new method of varying the exhaust nozzle exit area was devised. The new device was known as the Pilz-Schubdüse (mushroom exhaust nozzle), and in principle was the same as the movable bullet type of device already described for certain Heinkel-Hirth and Junkers engines. The power to move the bullet axially,

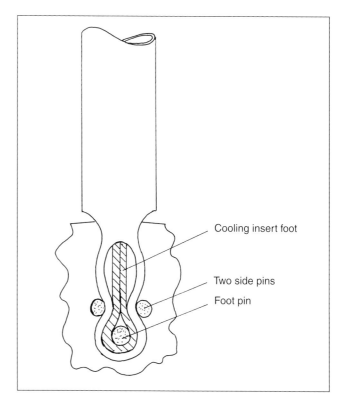

Fig 2.81 Sketch of method of attaching turbine blades in the BMW 109–003 A-0.

and thereby alter the exit area, was derived from an electric motor and transmitted via a shaft to bevel gears, a radial shaft to the engine axis and other gears which rotated a nut. This nut rotated on a threaded shaft connected to the bullet. Air was used to cool part of the drive mechanism, and the sheet-metal exhaust structure and the new system was less susceptible to, and less affected by, distortion.

The final principal modifications introduced with the 109–003 A-0 were a new and extensive lubrication system and the use of anti-friction (ball and roller) bearings at all the main points. Previously, some bearings had been of the plain, bushed type. All auxiliaries were fitted, including a Riedel starter motor.

With the appearance of the 109–003 A-0, the BMW flight-testing section began using a twin piston-engined Junkers Ju 88 as a test bed. The turbojet being tested was mounted beneath the aircraft's fuselage, with its fuel tank in the bomb bay, and the first flights with the Ju 88 so equipped were made in October 1943. The new features introduced with the 109–003 A-0 proved to be on the correct lines, and with systematic development the engine reached a static thrust of 800 kp (1,764 lb), which was the new rated thrust aimed for, and trouble-free runs of at least twenty hours were obtained.

During 1943, the time came once more to try out BMW turbojet engines as the sole power-plants of an aircraft. By this time, Heinkel-Hirth had been forced to abandon their 109–001 (HeS 8) and 109–006 (HeS 30) turbojets and were turning to other firms' engines to power the prototypes of the Heinkel He 280 twin-jet fighter. Thus, the He 280 V4 first flew on 15 August 1943, using two BMW 109–003 A-0 engines, while the similarly powered V9 (final He 280 prototype) first flew on the 31st of the same month. In October 1943 both the He 280 V5 and V6 first flew on 109–003 A-0 engines, the V5 having flown two months previously on 109–001 engines. As far as is known, the BMW engines functioned satisfactorily while powering the He 280s, the flights apparently being as much to test the engines as the aircraft. Maximum and cruising speeds at altitude were in the region of 900 km/h (559 mph) and 820 km/h (510 mph) respectively. Despite its good performance, the year of 1943 saw the end of competition for a production order between the He 280 and the Me 262 fighter prototypes, the former being abandoned largely for the reason of its shorter range.

The 109–003 A-0 engine was next called upon to power experimental prototypes of the Arado Ar 234 A (see Figure 2.82). Earlier, prototypes of this aircraft had flown on twin installations of the Junkers 109–004 engine, but the Ar 234 V6 and V8 were each proposed for installations of four of the BMW turbojets. The Ar 234 V8 (GK+IY) flew first on 1 February 1944 with the engines in paired nacelles beneath the wings. Severe shock waves were generated between these nacelles and the fuselage, so that the Ar 234 V6 (GK+IW), which first flew on 8 April 1944, had the four engines in four separate nacelles spaced beneath the wings. This arrangement proved a little more successful, but by the time the Ar 234 V13 first flew in August 1944 (with four 109–003 A-1 engines), the paired-nacelles arrangement was improved and was reverted to as the most efficient.

The foregoing has been confined to a résumé of the new features introduced with the 109–003 A-0 without giving a detailed description of the engine, since in the main essentials it was similar to the production A-1 and A-2 engines described later. Data for the 109–003 A-0 follow:

Thrust	800 kp (1,764 lb) static
Speed	9,500 rpm
Weight	570 kg (1,257 lb)
Pressure ratio	3.1 to 1
Specific fuel consumption	1.35 to 1.4 per hour (petrol)
Airflow	19.3 kg/sec (42.55 lb/sec)
Specific weight	0.712
Diameter	0.690 m (2 ft 3 in) basic
Length	3.565 m (11 ft 8 in) with exhaust bullet extended, 3.460 m (11 ft 4 in) with bullet withdrawn

By August 1944, 100 109–003 A-0s had been built, 60 of which were intended for flight, and over 4,000 hours of running had been performed on test stands for developmental and experimental purposes alone. At this time, however, the engine was lagging behind the Junkers 109–004 in performance, and further delays were to come in adapting the engine to run on J2 fuel instead of petrol. Even so, the official opinion was that the BMW

Fig 2.82 The Arado Ar 234 V8 prototype was fitted with four BMW 109–003 A-0 turbojet engines as part of the trials to ascertain the best arrangement of engines for the more powerful Ar 234 C. *(Phillip Jarrett)*

engine had a great development potential (its annular combustion chamber, superior accessibility and smaller frontal area and weight were looked upon with favour, for example), and it was hoped to produce it more rapidly and economically than the Junkers engine.

The 109-003 A-1 and A-2 (TL)

The first examples of the mass-production model, the 109–003 A-1, began coming off the assembly lines around October 1944, though a few examples were released months earlier for testing. The differences between the A-1 and A-0 model were mainly in detail design to suit mass production, and whereas performance was hardly altered (except for engine life), three serious limitations had to be taken into account. These limitations, officially requested, were that the engine had to be produced for about 500 man-hours per unit, that each engine should use no more than about 0.6 kg (1.323 lb) of nickel (far less than for the Junkers 109–004 B) and that the engine should be broken down into the maximum number of easily made components. Another requirement, easy accessibility, was already largely met in the A-0 model.

Thus, before the engine was committed to mass production, the design was given to a Dr Fattler and staff to ruthlessly simplify wherever possible without falling below the performance of the A-0. Dr Fattler had experience with mass-production techniques in the American automotive industry before the war, and substituted many cast and machined parts with sheet-metal parts. The enthusiastic substitution of such parts would have been taken further than it was, but long and acrimonious conferences between the staffs of Dr Fattler and Dr Oestrich finally thrashed out a compromise between the purely mass-produced and the technically good engine. The engine produced did not meet the exact requirements of production time and material limitations but came reasonably close.

To increase the life of the engine, most attention was given to improving the turbine section. An improvement to the turbine blade fixing is shown in Figure 2.83, whereby a wedge was positioned on each side of a blade. These wedges were forced tight by centrifugal action and prevented the tangential blade movement which wear or bad fitting had previously allowed. This modification was actually tried on the last few A-0 models, and Figure 2.84 illustrates a turbine wheel from the 90th A-0 engine.

Fig 2.84 Turbine wheel for the 90th BMW 109–003 A-0 pre-production turbojet engine, illustrating the method of attaching the hollow blades. Some blades were removed for blade heat transfer experiments in England after the war.
(Cranfield Institute of Technology, Bedford)

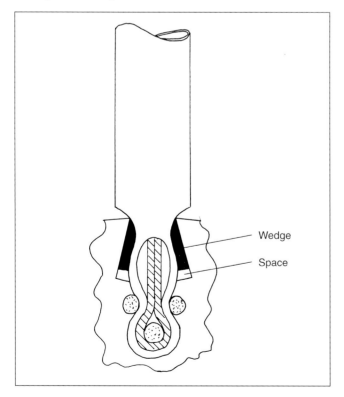

Fig 2.83 Improved turbine blade attachment with the BMW 109–003 A-1.

The illustration clearly shows the pins at each blade root, while, where the blades have been removed, the extra, upper cut-outs in the wheel rim for the wedges can be seen. Other small but effective improvements eventually allowed the 109–003 A-1 to run up to 50 hours or more without trouble, the final limit still being set by the turbine wheel. By careful metering, a reduction in cooling air was also found possible, and air for the turbine was tapped off after the last compressor stage instead of at the fourth stage as in the A-0 model.

Changes made in the 109–003 A-2 engine were almost exclusively to ease production, and not all the latest modifications to improve the performance had been incorporated on the production lines by the end of the war. Thus, while experimental versions of the A-2 model are reputed to have given thrusts of up to 1,200 kp (2,646 lb) static, the general performance of the production models remained similar to the A-0 and A-1 models. The weight, however, increased by 30 to 40 kg (66 to 88 lb) because of the simplification and strengthening of various parts. While the number of turbine rotor blades was decreased to 66 (compared with 77 for the A-1), the thickness of the blade metal was increased from the previous metric gauge of 2.15 up to 2.65. Two small fins, which were fitted to the exhaust cone of the A-2 and late examples of the A-1, reduced vibration.

A set-back for the production engines came in 1944 when an official decree required that all turbojets should run on the more-available J2 fuel, the use of which had been fully proved by Junkers turbojets. A BMW J2 fuel system was not ready, so that a system similar to that for the Junkers 109–004 B was adapted, but this still required considerable time to accomplish. The most obvious change was the replacement of the Henschel centrifugal pump with a Barmag gear pump in order to reach the pressure of 60 atmospheres required at sea level. Changing to J2 fuel initially gave some sooting problems, but these were overcome by correction of the fuel nozzle spray angles, and when the burners were finally matched to the new fuel system, the heat distribution was found better than with the petrol system. When, occasionally, hot spots did develop in the combustion system, they were sometimes visible on the exhaust cone. The following description of the 109–003 centres on the A-1 as the principal production model.

Description of the 109–003 A-1 and A-2

A view of the complete engine is shown in Figure 2.85; a section through the complete engine is given in Figure 2.88; and an engine dismantled into its four main sections of intake, compressor, combustion chamber/turbine and exhaust nozzle is shown in Figure 2.86. Reference can also be made to Figure 2.87, which illustrates sheet-metal components only. The following describes the 109–003 A-1, with references to the A-2 also, from nose to tail.

Intake section

A streamlined aluminium alloy nose cowling was designed to fair into the cowling for the whole engine when used in an exposed mounting, e.g. underwing. The air intake duct which led from the nose cowling was also of light alloy sheet (reinforced by external channels riveted on), and its internal diameter increased towards the rear, the smallest diameter at the front being 0.375 m (1 ft 2^i in).

Around the intake duct, behind the nose cowling, were mounted an oil tank and an oil cooler. The oil tank occupied the top position and had a capacity of about 16.0 litres (3.52 gallons), while immediately behind this tank was a smaller tank of about 1.0 litre (0.22 gallons) capacity containing the petrol/oil mixture for the Riedel starter motor. For late A-1 models and all A-2 models, another small tank was added to carry petrol for initiating turbojet combustion prior to switching to the main J2 fuel supply. The oil cooler was of the tube type and occupied the lower, port side of the intake duct. Air for this cooler was tapped off through curved scoops about halfway along the intake duct. This air then moved forwards (in a reverse direction to the intake air), through the cooler and then rejoined the mainstream by emerging from an annular slot formed between the nose cowling and intake duct.

The central cowling, positioned well back inside the intake duct, housed the Riedel two-stroke starter motor, which was attached by its flange to the central gear casing for the auxiliary drives. Intended for either hand-starting or electrical-starting, this Riedel starter was of the same type already described and illustrated for the Junkers 109–004 B-1.

Compressor casing and central casting:

The compressor casing was cast in magnesium alloy (Elektron), and represented one of the largest single castings in the engine. This casing (which was not split longitudinally and thus differed from the Junkers 109–004 compressor casing) carried most of the auxiliaries and a panel for most of the leads to the aircraft. At its forward end was bolted the flange of a cast ring containing an assembly of a streamlined support spider and a central casing for the auxiliary drives, gears and front compressor bearings. The intake duct was attached to the forward end of the cast ring. At the rear end of the compressor casing was a cast disc supporting the compressor rear bearing and having eight separated circumferential slots for the exit of the compressor air. Being the principal structural member of the engine, the compressor casing was provided at its top rear end with two of the three ball-jointed attachment points.

The spider-supported casing, situated between the Riedel starter and front of the compressor shaft,

GAS TURBINES FOR AVIATION

Fig 2.85 BMW 109–003 A turbojet.

Fig 2.86 A BMW 109–003 A in its four major sections.

Fig 2.87 A high proportion of sheet metal was used in the components of the BMW 109–003 turbojet. In addition to those shown, sheet metal was also used for the starter motor fairing and the hollow turbine rotor blades.

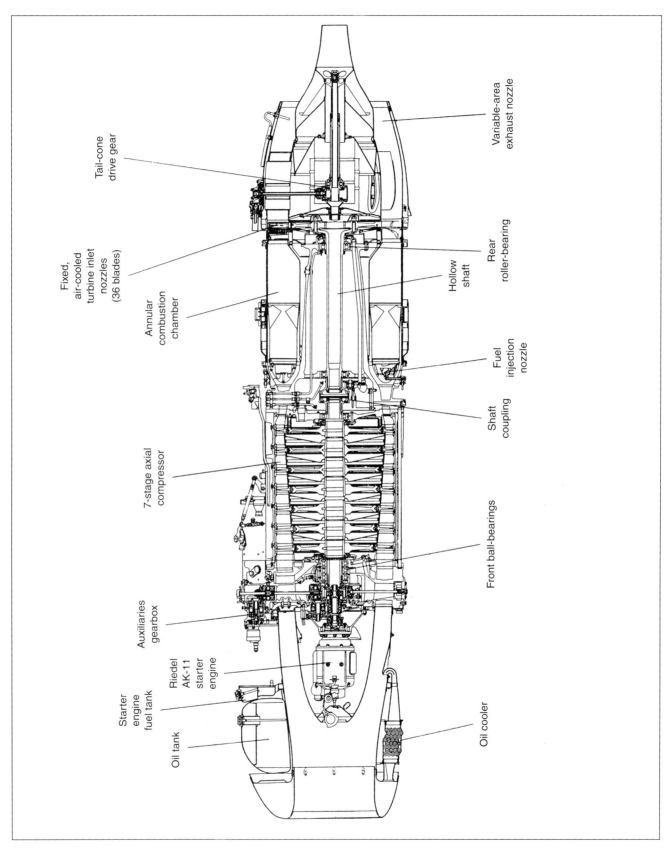

Fig 2.88 BMW 109–003 A-0 turbojet.

contained a complicated drive system between the starter and the compressor shaft and between the compressor shaft and auxiliaries. This complication was because the compressor revolved in the opposite direction to that of the Junkers 109–004 and necessitated reversing the drive while maintaining the final drive on the central axis and providing the requisite reduction. Housed in the rear of the casing was an assembly of three single-row ball races for the front of the compressor shaft, while the cast disc at the rear of the compressor casing held a single-row roller race for the rear of the compressor shaft. The assembly of three ball races took considerable axial thrust from the compressor.

Bolted to the cast disc at the rear of the compressor casing was the central casting, which was similar in shape to that illustrated for the Junkers 109–004 B-1 but without the ribs. The forward end of the central casting formed an annulus with eight separated, circumferential slots corresponding to the air exit slots from the compressor section. A large central hollow in the annulus provided a space for the coupling between the compressor and turbine shafts. At the rear of this hollow was housed a single-row ball race for the front of the turbine shaft. From the rear of the annulus, the central casting tapered sharply down (inside the annular combustion chamber) and formed at its rear extremity a housing just in front of the turbine for the turbine shaft rear bearing, which was a single-row rollar race. The central casting housed lubrication pipes for the central and rear bearings and also supported the combustion chamber, turbine and exhaust assemblies, and was indirectly connected to the rear ball-jointed attachment point.

Compressor:
The seven-stage axial compressor had a rotor blade tip diameter of 0.550 m (1 ft 9^l in), and used a small amount of reaction, the blades having almost symmetrical profiles and a constant 40 mm (1.574 in) chordal length from root to tip. The thickness of all blades was 5 per cent of chord at the tip and 12 per cent of chord at the base. At 9,500 rmp and a tip speed of 273 m/sec (895 ft/sec), a compression ratio of 3.1 to 1 at sea level was obtained on the test bed, while a compression ratio of 3.9 to 1 was expected at 900 km/h (559 mph) at sea level. Compressor efficiency, based on torque and total head measurements when driven by a steam turbine, varied between 83 per cent at the top point and 78 per cent at the rated condition. (Temperature measurements of efficiency worked out at about 2 per cent higher.) The performance of the compressor was far from unusual, but the efficiency curve was very flat, the designed Mach number was fairly high at 0.8 and no surging trouble was given.

For the first three stages, the compressor rotor blades were of magnesium alloy (Normen No. 3510) while the other higher pressure and temperature blades were of duralumin (Normen No. 3115). Each rotor blade had a

Fig. 2.89 One of the seven stages of the BMW 109–003 A-1 axial compressor.

tongue at its base which fitted into a groove machined in the periphery of its rotor disc, a single pin passing longitudinally through the disc serving to retain each blade. Figure 2.89 illustrates one rotor stage of the compressor. The duralumin compressor discs increased in diameter from front to rear and had steel bushes pinned inside their bosses, the steel bushes being a force fit on the compressor shaft. On the compressor shaft there was a collar after the third stage, the first three stages of compressor discs being forced on from the front and the others from the rear. The complete compressor shaft was supported by three ball races at the front and a single roller race at the rear end, a labyrinth gland being formed on the compressor side of each bearing set.

There were eight rows of compressor stator blades, these starting in front of the first rotor stage and finishing after the seventh rotor stage. The duralumin stator blades of each row were secured between an inner and outer shroud ring, the inner rings being a close fit between the rotor blade roots and the outer rings being secured by intermediate, overlapping rings bolted to the compressor casing.

Each compressor rotor disc was fitted with its blades and balanced statically before fitting on the compressor shaft. The complete compressor rotor was then balanced dynamically before dismantling ready for interleaving with the stator rows. Since the compressor casing was in one piece, it was then necessary to fit stator rings and rotor discs alternately, beginning with the fourth row of stator blades and working out to the front and rear. No check was made on the balance once rotor and stator blades had been interleaved. While the rotor design was simpler than that for the Junkers 109–004 B, the stator design and the process of assembling the complete compressor (requiring extensive jigging) was more complicated.

Compressor blades were forged by WMF of Geislingen, polished and then chemically treated (anodised?) against corrosion. After assembly, the compressor rotor was enamelled and stoved at a temperature of about 200 °C to give further protection. This enamel did not affect the balance and stood up to a normal 50- or 100-hour test, but soon deteriorated if sand or other abrasive was drawn into the compressor.

The seventh or last-stage rotor disc carried at its rear a collar which was finned around its circumference. This finned collar rotated with the compressor, and being a close fit inside a fixed circular shroud, formed an air seal between the compressor final stage and the inner rear bearing section.

Whereas in the previous 109–003 A-0 engines cooling air had been tapped from the compressor fourth stage and fed by pipes through the central casting to the turbine section, the production A-1 and A-2 models used a much simplified system. The hotter air being delivered from the last compressor stage was found usable and was tapped off just inside the combustion chamber secondary air space and led straight to the turbine section.

Combustion chamber:
The inner and outer sheet-metal sections of the annular combustion chamber are illustrated in Figure 2.90, while reference can again be made to the sketch in Figure 2.80. A sand-cast aluminium alloy ring supporting inside 16 main fuel burner cones was bolted to the rear face of the central casting annulus. To the cast ring were bolted the sheet-metal inner and outer sections with hollow fingers (illustrated) which formed about half the length of the combustion chamber, the final half being simple sheet-metal rings leading up to the turbine inlet nozzles.

The 16 burner cones injected the main J-2 fuel downstream and imparted a certain amount of swirl to assist atomisation. Around the top half of the combustion chamber there were two auxiliary fuel injectors for petrol and two sparking plugs for starting purposes only. (Late examples of the A-0 model had used six starting injectors.) Current for the sparking plugs was fed via a booster coil.

The hollow fingers, which directed secondary air into

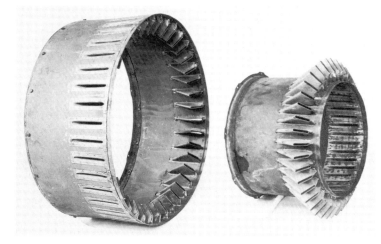

Fig 2.90 The combustion chamber of a BMW 109–003 A-1 turbojet engine. Note the hollow fingers projecting from each section for secondary air.

the hot combustion zone, numbered 40 on the outer circumference and 40 on the inner, and they were arranged so that one set pointed towards the spaces between the other set. Upon the angle and disposition of these hollow fingers depended very largely the temperature distribution of the gases at the turbine. The hollow fingers also experienced the highest temperature in the combustion chamber, the temperature at the roots of their leading edges being 800 °C but dropping to a mere 160 °C at the tips of the leading edges, i.e. at the points between the inner and outer fingers. Consequently, these fingers were deep-drawn from heat-resisting chrome steel known as Sicromal, which was used in various compositions (e.g. Sicromal 10 contained 12.0 to 14.0 Cr, 1.5 to 2.0 Al, 1.0 to 1.5 Si, 0.1 C, blce Fe.) For the rest of the combustion chamber's sheet-metal parts, aluminised mild-steel was found adequate in the face of the enforced stringent use of heat-resisting steels.

Relatively low gas velocities were noteworthy in this BMW combustion chamber. Air entered the section just before the burner cones at a velocity of 110 m/sec (360 ft/sec), and accelerated to its highest figure of 120 m/sec (393 ft/sec) inside the cones and also between the hollow fingers. The secondary air velocity was as low as 27.5 m/sec (90 ft/sec) as it flowed around the ring holding the burner cones and into the annular spaces leading to the fingers, where it accelerated up to 75 m/sec (246 ft/sec) and more. The life of the combustion chamber was in the region of 200 hours.

Turbine inlet nozzles and turbine:
A ring of fixed turbine inlet nozzles was positioned at the end of the combustion chamber and immediately in front of the turbine. In order to ease production, the profiles of both inlet nozzles (fixed blades) and turbine blades were kept uniform along the whole radial length

and had no twist. BMW had made comparative tests of both twisted and untwisted blades. They found that twisted blades gave a higher peak of efficiency by giving optimum angles of attack throughout their length, but this efficiency fell off much faster once above or below optimum conditions than was the case with untwisted blades. The untwisted blades were, therefore, a compromise in efficiency while being most suitable for mass production.

The turbine inlet nozzle assembly consisted of 36 hollow blades secured between an inner and outer shroud ring. The outer ring was attached by a flange to the rear of the combustion chamber outer wall, while the inner ring mated with the combustion chamber inner wall and then continued into a perforated drum which was bolted to the rear of the central casting. At the forward end of the combustion chamber, small holes admitted cooling air at full compressor delivery pressure to the space between the combustion chamber inner wall and the central casting. At the end of the central casting, this air entered the perforated drum and passed via holes in the inlet nozzle inner shroud ring into the fixed, hollow blades to emerge through slots in the trailing edges. The blades forming the inlet nozzles were made from Sicromal 10 heat-resisting steel, while the rings and drum were made from mild-steel sheet, sprayed with aluminium.

A hollow shaft, separate from the compressor shaft, was provided for the single-stage turbine and was carried by a single ball race in front and a single roller race at the rear, these bearings being mounted inside the central casting. The front of the turbine shaft was connected to the rear of the compressor shaft by means of a splined coupling which allowed some longitudinal movement. After passing through the rear roller bearing, the shaft opened out into a flange to which was bolted the turbine disc. The disc was of a steel alloy (0.80 Mn, 0.90 Cr, 0.80 Mo, 0.40 Si, 0.15 to 0.19 C, blce Fe) heat treated to give the requisite tensile strength. Axial location of the disc was ensured by means of a machined lip (which fitted over the edge of the shaft flange), and also by providing the eight holes in the disc with internal bevels which corresponded with the securing nuts. A complete turbine bolted to the shaft can be seen at the end of the combustion section in the illustration of a dismantled engine.

The method whereby the hollow turbine blades were secured to the rim of the disc with pins and wedges has already been described and illustrated for late examples of the 109–003 A-0 model. A shaped plate, fitted to each side of the turbine disc, served not only to prevent longitudinal movement of the blade pins, but also to duct cooling air into the blades at their roots. The cooling air, from the perforated drum already mentioned, was directed inside the shaped plate on the front face of the turbine disc; it reached the rear face by passing through 16 holes in the disc. A very low consumption of cooling air for the turbine rotor blades was achieved by means of the insert, which left only a very narrow (about 1 mm) air space inside each blade. The resulting thin layer of cooling air accordingly reached a high temperature by the time it emerged from the blade tips, but stresses at the blade tips were low in any case. Beneficial friction damping given to the turbine blades by the cooling inserts has already been mentioned. Using approximate figures, the cooling air used for the turbine wheel, expressed in percentages of the total engine airflow, was 1 to 1.5 per cent passing through the blades and out of the tips into the hot gas flow, 0.5 to 1 percent passing through openings at the blade roots and 0.5 per cent leaking between the revolving front turbine plate and the fixed, perforated drum.

Because of the settling-in of the turbine blade roots and distortion of the sheet-metal shrouding ring, it was necessary to build new engines with a blade tip clearance of about 2.75 mm (0.108 in), and the least clearance permissible was 1.50 mm (0.060 in). After about 50 hours' running, the turbine blades of the A-1 engine hardened and cracked, so that an inspection was carried out after every 10 hours' running. An increasingly longer life was being obtained from the A-2 engine (some improvements had still not been incorporated by the end of the war), but the turbine was still considered the engine's chief life-limiting component. Accordingly, production was geared to give 50 per cent spare turbine discs and 100 per cent spare turbine rotor blades. Even so, to change the turbine wheel on either the A-1 or A-2 model was a remarkably rapid operation (compared with the same job on the Junkers 109–004 B), and took only about two hours. Once the exhaust nozzle section had been detached, aircraft engine nacelles being designed with that section left exposed, the turbine wheel with its attachment bolts was completely accessible.

Various heat-resisting steels were used for the turbine rotor blades, which numbered 77 in the A-1 engine and 66 in the A-2 engine. One such steel was the chrome-nickel steel FBD (17.0 Cr, 15.0 Ni, 2.0 Mo, 1.15 Ta-Nb, 1.0 Si, 0.1 C, blce Fe), but a later steel, FCMD (13.7 Cr, 15.5 Mn, 0.42 Mo, 0.22 Nb, 0.45 Si, 0.20 V, 0.12 C, blce Fe), was introduced to dispense with the use of scarce nickel by replacing it with manganese. The chrome-nickel steel Remanit 1880 S used for the cooling inserts was doubtless also replaced, possibly with a type of FCMD (18.0 Mn, 13.0 Cr, blce Fe) or FCM (16.0 Mn, 15.0 Cr, blce Fe). All the steels mentioned here were produced by Gebrueder Boehler, except for Remanit, which DEW produced.

The process used by WMF for the manufacture of the hollow turbine blades was as follows. From 2 mm thick cold-rolled plate, strips measuring 90 mm (3 in) wide by 125 mm (4 in) long were cut. These were then taper-rolled (0.7 mm at the tip to 2.6 mm at the root), annealed and cut to size where the section thickness was just right. Slots were cut to form the cooling air inlet and the material was folded over longitudinally. With a profile core inserted, the blade was then pressed to shape in an eccentric press, and the trailing edge was welded up using an atomic hydrogen process. The cooling insert (a thin, pressed and folded metal strip) was looped over a solid pin and pushed inside the blade. Flaps on the blade root were then folded over the insert and pin and welded, the complete foot then being ground to shape. To finish, the completed blade was heated to 1,150 °C and quenched in water, the foot was shot blasted, and the blade hand-polished and treated against corrosion.

Exhaust system:
The exhaust system consisted of a nozzle with a movable bullet or tail cone to vary its exit area, and had an internal spider or supporting vanes to carry the tail cone structure and final drive mechanism. The casing of the nozzle was double-walled and provided with small scoops to admit external air for cooling purposes. This air followed a devious path between the two walls of the nozzle, through the hollow vanes of the supporting spider, around the internal drive mechanism and along and out of the rear hole in the moveable bullet, where the engine's exhaust flow produced a considerable ejector effect. The main support for the exhaust system was by attachment of the outer casing to a reinforced ring on the rear of the combustion chamber, which indirectly put the load onto the rear of the central casting.

Longitudinal movement of the exhaust bullet was by means of an electric motor which drove a shaft leading to bevel gears on the exhaust nozzle casing. These gears took the drive at right angles to another shaft, passing through one of the hollow vanes of the spider, which drove the final, internal bevel gears. The final bevel gears rotated a threaded sleeve, supported by two ball races, which screwed a shaft connected to the movable bullet backwards or forwards. Since the exhaust system was constructed from sheet metal, two vanes on the bullet were found necessary to counteract vibration. An electrical switch in the aircraft cockpit was used to select various positions of the bullet and hence the exit area, these positions being:

Position	Condition	Nozzle exit area
A	Starting and idling	Maximum
S	Take-off and climbing	Minimum
H	Horizontal flying	Intermediate

The indications are that, for a horizontal bullet travel of 105 mm (4 in), the nozzle exit area could be varied between 970 sq cm (150 sq in) minimum and 1,400 sq cm (217 sq in) maximum, but that less than this full range was used in the production engines. On earlier (109–003 A-0) engines, a fourth bullet position was used for high-altitude, high-speed flying, and this may have been introduced again on 109–003 E engines. As described later, an automatic control for continuously variable operation of the exhaust bullet was under development.

Although the temperature in the exhaust nozzle reached about 600 °C, it was necessary to make only the hollow vanes of the supporting spider in Sicromal 10 heat-resisting steel, while all the other sheet-metal parts were made from aluminised mild steel. This aluminising method of superficial protection, which has been frequently mentioned in connection with German turbojet engines, was carried out by one of two processes for BMW at least. The component was either immersed in molten aluminium (Aluminitieren process) or painted with an aluminium lacquer, developed by Zarges-Weilheim, and then baked at 400 °C (Alumakieren process). The latter process was considered the best.

Auxiliaries:
The main starting and auxiliary drive gears were situated in a complicated assembly in a casing between the starter engine and the front of the compressor shaft. A Riedel type AK 11 (9–7034) air-cooled, two-stroke, two-cylinder, petrol engine (already described and illustrated for the Junkers 109–004 B-1 turbojet) was fitted to the front of the gear casing inside the intake central fairing. To the drive shaft of this starter engine was fitted a starter claw which, when running, connected to a short shaft in front of the compressor shaft. The drive was taken by the short shaft to another, offset, stub shaft and from there to the compressor shaft. Once the turbojet was running, the starter claw shaft revolved independently of the starter motor and drove three auxiliary shafts via bevel gearing. These three auxiliary shafts were set at right angles to the engine axis and passed out through the hollow vanes of the support spider.

The top vertical auxiliary shaft drove the main auxiliary gearbox situated on the top of the engine, just forward of the compressor casing. A fuel pump, governor and fuel regulator were mounted on this gearbox. The lower vertical auxiliary shaft drove the front oil scavenge pump, while the third auxiliary shaft drove the oil pressure pump located on the starboard side just forward of the compressor casing. A spur and bevel gear drive was also taken from the front of the turbine shaft to drive a second oil scavenge pump situated inside the engine behind the compressor.

Oil, specified as Flugöl S3, was delivered at the rate of about 180 litres/hr (39.7 gallons/hr) by the starboard-mounted pressure pump to the main and auxiliary gears and to the front and rear bearings of the compressor and turbine shafts. For the two central and one rear main bearings, a manifold was situated on the outside centre of the engine, from which led three oil pipes to the three bearings. The oil was sprayed by jets onto these bearings, ran down inside the central casting and was drawn out by the second scavenge pump. Oil also drained to the bottom of the central gear casing and was drawn out by the front scavenge pumps. Both scavenge pumps used a common pipe to deliver the return oil to the cooler, which was provided with a relief valve to avoid overloading when the oil was cold. The bypass relief valve opened at a pressure of about 10.55 kg/sq cm (150 psi).

A small centrifugal pump, for delivering petrol or B4 starting fuel only, was driven by its own electric motor, while the main J2 fuel pump was of the Barmag gear type and was driven from the main gearbox. Other fittings on the outside of the engine included a panel for making electrical, hydraulic and fuel connections to the aircraft, and a thermocouple for reading the jet exhaust temperature (EGT).

Fuel system and operation:
Main fuel from the aircraft tanks passed via a filter on the turbojet engine to the Barmag fuel pump, which delivered it to the all-speed governor, fuel regulator and the 16 burner cones. The full rmp capacity of the pump was about 2,275 litres (501 gallons) per hour at 70 kg/sq cm (996 psi) pressure, and was close to engine requirements at sea level on a cold day, thereby giving some protection against overspeeding in the event of a governor failure. Because BMW had to rapidly convert their engine for J2 fuel operation, the fuel system used was derived from that used on the Junkers 109–004 B, and was therefore basically similar.

The centrifugal all-speed governor, which regulated the fuel flow and engine speed, had its servo mechanism supplied from the aircraft hydraulic tank. Engine speed was regulated by the pilot's throttle acting on the flyweight springs of the governor, which in turn operated a spill valve. This spill valve diverted fuel in excess of engine requirements back to the fuel pump.

To limit the speed of engine acceleration, a special accelerator valve opened a valve to bypass any excess fuel until a predetermined engine speed was reached. This valve was controlled by an aneroid capsule responsive to the pressure difference across the compressor, and was developed by the firm of Kinsler at Donau-Eschlingen. The system allowed for compressor/turbine inertia during initial acceleration, and prevented full fuel flow until sufficient air was flowing through the engine. By thus avoiding overheating, an exhaust nozzle temperature limitation of between 540 and 620 °C was to be maintained. Although development work was in progress on methods of automatically controlling the exhaust nozzle exit area, this did not reach fruition in time, so that the chief means of setting the nozzle was by the pilot-operated switch already described.

For starting, the engine was run by the Riedel starter up to 1,200 rpm, when the electric petrol pump and ignition was switched on. The starter was left engaged to assist acceleration until a speed of 2,000 rpm was reached, when the main J2 fuel system could be cut in. Since the centrifugal part of the governor did not become effective until the engine reached 6,000 rpm, the pilot's lever operated directly on a throttle valve below this speed.

The idling speed on the ground or test stand could be brought down to 3,500 rpm when a thrust of 55 kp (121 lb) was produced, but in flight the lowest permissible idling speed was 6,500 rpm, which prevented a flame-out up to an altitude of about 10,000 m (32,800 ft). Acceleration from 6,500 rpm to the full speed of 9,500 rpm took between five and seven seconds. Overspeeding on a quick acceleration amounted to about 150 rpm.

The main fuel injectors shown in Figure 2.91 (which also shows a starting fuel injector) were designed for altitudes up to 12,000 m (39,360 ft), but operation at 13,000 m (42,640 ft) was possible at full throttle, although a flame-out would occur if power was reduced at that altitude. The possibility of using the engine at the high altitude of 13,000 m was first demonstrated in September 1944, when Hauptmann Bispink of the BMW's flight test section flew the Ar 234 V13, using two early 109–003 A-1 engines. For operation at greater altitudes and more efficient operation at all altitudes, a duplex fuel injector was under development, but was not ready in time.

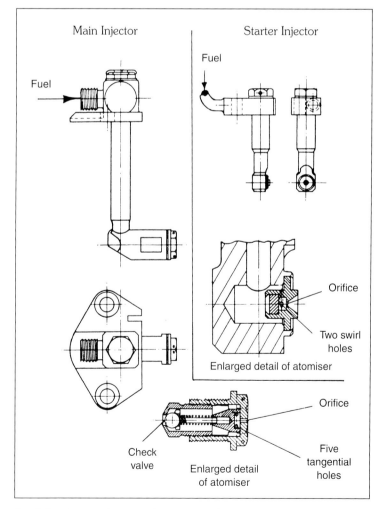

Fig 2.91 Main and starting fuel injectors for BMW 109–003 A.

If a flame-out did occur, restarting in flight was only possible at or below 3,500 m (11,480 ft) altitude, when it was necessary to first windmill the engine at an airspeed of 400 km/h (248 mph). At 2,000 m (6,650 ft) altitude, restarting could just be obtained at an airspeed as low as 250 km/h (155 mph) when the engine windmilled at 2,000 rpm. However, at this altitude, the minimum airspeed preferred for restarting was 550 km/h (342 mph), which gave an engine windmilling speed of 2,500 rpm.

Typical data for the BMW 109–003 A-1 and A-2 follow, differences between the two models being indicated where they occur:

Thrust	800 kp (1,764 lb) static
Thrusts at 900 km/h (559 mph)	705 kp (1,554 lb) at sea level 565 kp (1,246 lb) at 4,000 m (13,120 ft) 315 kp (695 lb) at 10,000 m (32,800 ft) Speed: 9,500 rpm
Weight:	A-1 model: 570 kg (1,257 lb) A-2 model: 600 to 610 kg (1,323 to 1,345 lb)
Pressure ratio	3.1 to 1
Specific fuel consumption at static thrust	A-1 model: 1.33 to 1.35 per hour (petrol) A-2 model: 1.40 to 1.47 per hour (J2)
Airflow	19.3 kg/sec (42.55 lb/sec)
Specific weight	A-1 model: 0.711 A-2 model: 0.756
Gas temperature	800 °C max.
Diameter Length	Basically as for 109–003 A-0

The 109–003 E-1 and E-2 (TL)

The 109–003 A-1 and A-2 engines were produced in modified forms as the E-1 and E-2 models respectively to permit installation above a wing or fuselage instead of below a wing. Chiefly, this meant providing three attachment points at the bottom of the engine, rerouteing cable and pipe connections to the aircraft, providing cooling air scoops around the top part of the exhaust nozzle and providing a new cowling. This cowling gave the engine its characteristic hump-back appearance, when mounted above an aircraft, owing to the auxiliary equipment being positioned on the top forward section of the engine. A 109–003 E-1, with the main cowling removed, is shown in Figure 2.93.

On the test stand, the performance of the E models was the same as for the corresponding A models, while differences in weight were negligible. There were some aerodynamic differences between the two models because of different cowl shapes and interferences between cowling and airframe. The requirement for the 109–003 E

Fig 2.92 A BMW 109–003 A on a test stand. The mounting cradle is connected to electrical load cells on each side to measure the thrust. *(Photo Deutsches Museum Munchen)*

was raised in September 1944, when various companies began designing single-engined jet fighters in answer to an official requirement for a cheap Volksjäger (people's fighter). Heinkel Ho 162s were the only production fighters to fly with the 109–003 E engine (the prototype He 162 V1 Spatz (sparrow) making its maiden flight on 6 December 1944), while the sole completed example of the Henschel Hs 132 dive-bomber did not fly before the war ended. Both types were similar in layout.

Fig 2.93 A BMW 109-003 E-1 turbojet (c/n: 394646) on a hoist, showing part of the cowling opened.

Fig 2.94 The BMW 109–003 E-1 turbojet engine. Modified from the A-1 model to allow attachment above, instead of below, a wing or fuselage. Most E-1 and E-2 models were earmarked for the Heinkel He 162 Salamander fighter.

Production of the 109–003 A and E

Manufacture of the 100 109–003 A-0 and A-00 pre-production engines was carried out in BMW plants at Spandau and Basdorff, with the final assembly and testing at Eisenach (50 km west of Erfurt). In March 1944, before production of the 109–003 A-1 began, the decision was taken to move production underground, or at least to dispersed sites. This decision followed the first real damage to BMW plants the previous month, although BMW officials had requested as early as 1937 that production plants should be built underground.

For the Eisenach plant (code-named Town-Werke), the Speer Ministry assigned the following dispersals:

Heiligenroda: This was a salt mine near Dorndorf (30 km SW of Eisenach) with 40,000 sq m (430,000 sq ft) of floor space for production. It was to produce complete engines, but was only in partial operation by the end of the war.

Abterroda: A much smaller salt mine near Heiligenroda, where parts such as the oil cooler were to be made.

Bad Salzungen: A salt mine (25 km SSW of Eisenach) for assembly but not used since the engine components did not materialise.

Springen: This was the largest salt mine in the Eisenach area, and had 176,800 sq m (1,900,000 sq ft) of floor space assigned to BMW. Although 500 machine tools were installed, full operation was not achieved.

An integrated production was planned for the plants in the Eisenach area and plants dispersed in the Magdeburg area, chief of the latter being:

Ploemintz: A salt mine near Baalberge, Bernburg (35 km SW of Magdeburg), principally for the machining of large parts.

Egeln: Near here (about 20 km SW of Magdeburg) was a series of potassium mines, some of which Junkers from Dessau were using. Shaft 8 had extensive facilities for the assembly of Heinkel He 162 jet fighters, while Shaft 7 was a dispersal for some of the BMW Spandau plant. There, 109–003 engines were made, while test stands were erected in a small wood just north of Shaft 7. In the same area was the BMW 'Kalag' dispersal development plant.

Production at Spandau (Berlin) was transferred chiefly to Zühlsdorf, near Oranienburg (30 km NW of Berlin), while another independent production line was set up in Main Tunnel B (near the V1 flying bomb assembly line – see Figure 8.11) of the Mittelwerke (Nordhausen). At least for the Mittelwerke operation, most components were manufactured outside. One of the major sub-contractors was the VDM airscrew manufacturers using their factories at Frankfurt and Marklissa. In addition to the three major manufacturing and assembly complexes of the Eisenach-Magdeburg areas, Zühlsdorf and Mittelwerke, production of components was also planned in a great number of small towns and villages where small workshops, garages and so forth made components for shipment to the main plants. Hence, the reason for designing the engine with a maximum number of easily made components.

On 11 September 1944, the first official production schedule for the 109–003 was issued calling for a monthly production of 6,000 engines by January 1946 (compared with a planned peak of 5,000 per month for the Junkers 109–004). By the combined efforts of all the sources mentioned above, this production rate was to be worked up to as follows:

1944		1945					
Nov.	Dec.	Jan.	Feb.	Mar.	Apr.	May	Jun.	Jul.
75	100	180	400	800	1,400	1,900	2,365	2,800

1945				1946
Aug.	Sep.	Oct.	Nov.	Dec.	Jan.
3,350	4,100	4,700	5,300	5,750	6,000

Needless to say, these production plans were extremely optimistic, and were set back by the effects of delayed development, bombing, inaccessibility of underground plants and so forth. The first production 109–003 A-1 engines were delivered from Zühlsdorf in October 1944, and in fact this plant produced the majority of the engines completed by the end of the war. Excluding the 100 109–003 A-0 and A-00 pre-production engines, the production figures actually achieved by all sources was as follows:

1944			1945		
Oct.	Nov.	Dec.	Jan.	Feb.	Mar.
30	65	94	100	120	150

This gives a total of 559 engines, mostly of the A-1 and E-1 type, the latter being for the Heinkel He 162 fighter. The peak of 150 in March included about sixty of the A-1 model. The poor production figures, together with increasing Allied bombing raids and a swing in priority to the Junkers 109–004 B (for the Me 262 fighter), led to a revision of production schedules for the 109–003 E-2 model. On 8 March 1945, the revised schedule (Lieferprogram für gerat 003) was issued and is summarised thus:

	1945					
	Mar.	Apr.	May	Jun.	Jul.	Aug.	Sep.
Zühlsdorf	180	150	200	300	450	500	500
Eisenach & Magdeburg complex	20	100	250	400	550	700	700
'Village' production	–	100	200	300	500	500	500
Mittelwerke (Nordhausen)	–	–	20	80	200	400	800
Total planned production per month	200	350	670	1,080	1,700	2,100	2,500

Although each engine was planned by Dr Fattler to be produced in 500 man-hours, the time taken was actually longer. Nevertheless, the figure was realistic and was made up as follows:

Work	Planned man-hours	Achieved man-hours (approx.)
Machining	220	220
Sheet-metal work	160	160
Starter, governor, etc.	80	60
Miscellaneous		100
Assembly	40	60
Total man-hours	500	600

As a further breakdown, we have figures for the production of the sheet-metal parts:

Components	Man-hours
Intake section	21.5
Starter fairing	3.0
Oil cooler inserts	1.75
Oil cooler complete	25.0
Compressor stator blade rings	36.5
Combustion chamber	20.75
Turbine nozzles	14.5
Turbine rotor blades	10.0
Exhaust nozzle	27.0
Total	160.00

With the exception of the starter fairing and turbine rotor blades, the sheet-metal parts listed above are illustrated in Figure 2.87, which gives an impression of the high proportion of sheet metal in the engine.

Some of the reasons for the failure to achieve the planned production of the 109–003 turbojet have already been mentioned briefly, but one of the most serious problems was caused by the large reliance on production in disused mines. Their main disadvantage was that the small, old lifts limited machinery size and reduced efficiency, and more modern lifts were only slowly being built. In the salt mines, humidity was low and the walls would dissolve if too much moisture was allowed to escape from concrete pouring; in the lifts and vertical shafts, humidity was much higher and corrosion presented a problem with lifts and other equipment there. Working hours in the salt mine factories were eight hours per day per person as a practical limit, but this was increased to 12 hours towards the end of the war. A large percentage (up to 55 per cent) of the workers often comprised impressed and slave workers. To mention a final production problem, there were the constant changes in materials, one raw material for the 109–003 being changed seven times, for example, without any obvious advantage in performance or economy. Such was the confusion by 1945.

Testing the 109–003

In addition to testing by BMW, functional and performance tests on both production and experimental engines were officially performed by E-Stelle Rechlin in a similar manner to that described already for the Junkers 109–004. However, whereas the Junkers engine passed both bomber and fighter official type tests (see Introduction), the BMW 109–003 had only passed the fighter official type test by the end of the war.

For measuring turbojet thrust, BMW found a pendulum-type balance (as used for commercial scales) the best device after rejecting hydraulic and electric devices as unsatisfactory. The best ground-testing installation BMW had was, of course, the previously described high-altitude chamber at Oberweisenfeld, Munich. This and other elaborate test cells were supplemented with many test stands which were dispersed and of a simple brick and concrete construction.

A great deal of BMW's test work was performed by their flight test section, which at one time operated from Berlin-Schönefeld under Flugkapitän Staege. In July 1944, Dr Denkmeir was made responsible for the sub-section devoted to turbojet flight tests, but towards the end of the war, when long-range research became pointless, Dipl. Ing. Peter G. Kappus was put in charge of flight testing and was primarily concerned with readying the 109–003 for the He 162 fighter. Previously, Kappus had headed project studies (department EZD) but was far from being chair-bound since extensive flight training with the DVL before the war enabled him later to fly jet aircraft.

The first aircraft used for 109–003 flight tests by BMW was a twin piston-engined Ju 88, which began flying with an underslung A-0 turbojet in October 1943. Further flight test installations for turbojets were considered at a meeting held by BMW on 2 May 1944, and requests for more aircraft were made. This resulted, in the latter half of 1944, in two more aircraft being taken on as flying test beds, these being the Arado Ar 234 V15 and V17. They were originally Ar 234 B Blitz bomber prototypes, and each was fitted with two 109–003 A-1 engines, the V15 having made its maiden flight on 10 July 1944.

The main work involved in the flight-testing was exploring the altitude flame-out and restarting limits with the fixed orifice fuel injectors, testing the variable-area exhaust nozzle and testing the J2 fuel system which had been introduced at a late date. Another flight programme covered the testing of a new surface oil cooler which was to be introduced on late 109–003 A-2 engines; this cooler was integral with the intake cowling to combat compressor icing and was easier to make than the tubular cooler. However, work on the fuel system became the most urgent, so that by 1945 the BMW flight test section had acquired a further Ar 234 B prototype (possibly the V18), an Ar 234 C prototype (four engined), another Ju 88 and two Messerschmitt Me 262s to facilitate the work. Nevertheless, because of the acute fuel shortage, flying time was progressively and drastically reduced as each main problem was solved.

Also, towards the end of the war, the BMW aircraft were moved about to different bases for safety from the advancing Allies. The first such move in 1944 was to Oranienburg, which was followed by a move to Burg (near Magdeburg). Orders then came to evacuate Burg, where it is believed the two 109–003-powered Me 262s were abandoned since they were not ready for testing. BMW flights were to continue from the Autobahn south of Munich, but it is doubtful if much work was possible there at that late stage. In any event, Kappus himself flew one of the Ar 234 B prototypes from Burg to Lechfeld (south of Augsburg) and then on to Neubiberg (south of Munich), where BMW finally blew up their equipment as the US 42nd Division approached Munich. Unfortunately, the only figure we have for total 109–003 flight tests is a very vague one of 'between 500 and 1,000 hours'.

While on the subject of 109–003 flying test beds, mention should be made of the Heinkel He 219 night-fighter that the Heinkel Company had at Vienna, a known photograph of which appears to be a fake. This twin-piston-engined aircraft was fitted with a single 109–003 turbojet beneath the fuselage, but the purpose was not to test the turbojet. The aim was to gather data on the spreading of the exhaust stream in the neighbourhood of a fuselage, such data assisting in the aerodynamic development of aircraft with fuselage-mounted turbojets (e.g. He 162). The conditions under which the exhaust stream will adhere to a fuselage wall vary according to the angle of incidence of the fuselage in flight, but optimum conditions can be obtained by the correct location of the engine and by suitably shaping the fuselage behind the engine. By this means, a cushioning airflow can be interposed between the fuselage and the exhaust stream to prevent adhesion under most normal conditions. It is not known how far the He 219 tests proceeded, but earlier, water-tunnel tests by N. Kunze (1944) showed that the exhaust stream could be expected to adhere to the fuselage at an incidence of –10 ° but would not do so at +10 ° down to at least 0 °.

Utilisation of the 109–003 A-1 and 2 and E-1 and 2

Whereas the Junkers 109–004 B turbojet saw considerable war-time service, the BMW 109–003 was only just entering service by the end of the war, since its development had fallen behind the Junkers engine by about nine months. The chief use for the BMW engine was in powering the He 162 and particular versions of the Ar 234, and according to one source, the engine was known unofficially as the Sturm (storm).

While, in the summer of 1944, experimental reconnaissance missions were being made with Me 262 and Ar 234 B (2 × Junkers 109–004) aircraft, development proceeded on the proposed Ar 234 C (4 × BMW 109–003). Already, in the previous year, experiments had been made with the Ar 234 V6, V8 and V13 to ascertain the best arrangement of the four engines beneath the wing of the aircraft, the final arrangement chosen being that of pairing the engines in two nacelles. To develop various versions of the Ar 234 C reconnaissance and bomber aircraft, 12 prototypes were planned, these being the Ar 234 V19 to V30 inclusive (each with four 109–003 A-1, A-2 or improved model engines).

The Ar 234 V19 made its maiden flight on 30 September 1944, and was followed the next month by the V20. First production aircraft were to be the Ar 234 C-0 pre-production machines, Ar 234 C-1 reconnaissance machines and Ar 234 C-2 bombers. Other versions planned were the C-3 multi-purpose aircraft (V21 acting as prototype), the C-4 special reconnaissance machine, the C-5 two-seat bomber (V28 acting as prototype), the C-6 two-seat reconnaissance aircraft (V29 acting as prototype) and the C-7 night-fighter. Enlarged versions of the C-7 were also planned as the Ar 234 P-1 and P-2, still using the four-engine arrangement. Ten of the prototypes, together with 14 Ar 234 C-0 pre-production and C-1 reconnaissance aircraft, were completed by the end of the war, but delays in receiving the 109–003 A-1 or A-2 engines resulted in none of the machines being delivered to the Luftwaffe.

While the development of the Ar 234 C aircraft was proceeding (albeit too slowly because of disruption of the Arado plant concerned and the number of sub-variants involved), events were taking place which were to bring another aircraft into being, powered by the 109–003. On 8 September 1944, the specification of the Volksjäger (people's fighter) was issued as the outcome of a conference of industrial leaders and the Speer Ministry in the late summer of 1944. The specification, which arose largely from the ideas of Karl Otto Saur, now chief of the Technisches Amt, called for 'A design to make use of existing aircraft components, only the barest essentials to be carried in the way of equipment. The power to be supplied by a BMW 003 turbojet rated at 800 kp static thrust. Top speed to be 750 km/h, endurance of not less than 20 minutes at sea level, gross weight not more than 2,000 kg, wing loading not more than 200 kg/sq m. The aircraft to be regarded as a "piece of consumer goods" and to be ready by 1 January 1945.'

Great pressure was put on the five aircraft companies willing to submit designs for this specification, and Heinkel's P.1073 design was finally selected on 30 September and given the designation He 162 Spatz. A month later, day and night work had produced detailed drawings, and the first prototype flew on 6 December with Flugkapitän Peter at the controls. Although this aircraft later crashed, work went ahead at high speed to solve the major structural and aerodynamic defects and to pursue general development with thirty or so prototypes. Simultaneously, production went ahead with the He 162 A-1 and then A-2, the aircraft by then being known as the Salamander instead of the Spatz.

Modifying the 109–003 A engine into the E model for over-fuselage attachment presented little difficulty, but the same was not true of engine production, and considerable writing-down of the original schedules was necessary. Since the He 162 was regarded as a 'piece of consumer goods', only about 20 per cent engine spares were planned compared with 40–50 per cent spares for the 109–003 A-1 (for the Ar 234).

Whatever the difficulties, though, the Nazis, who largely conceived the He 162 and were by then controlling almost everything, were determined to get the jet fighter into service at a time when the Me 262 was not being produced in sufficient quantity to effectively combat the Allied air onslaught on Germany. Every possible facility was pressed into service in a well-planned production programme (from the airframe point of view at least), and by January 1945, the operational proving of the He 162 was begun. This work was performed by Erprobungskommando 162 (alias Einsatzkommando Bär) based at E-Stelle Rechlin, and later at München-Riem. On 6 February 1945, the unit I./JG 1 transferred from the Eastern Front for He 162 training at Parchim, and II./JG 1 moved to Warnemünde the following month for similar training. Encroaching Allied forces disrupted the numerically extensive training programme planned so that, by 4 May 1945, units under training had been merged into one unit (designated I./(Einsatz) Gruppe/JG 1) at Leck in Schleswig-Holstein with a nominal strength of about fifty He 162s.

Combat with the enemy was, however, forbidden since more training was needed to reach operational readiness, and on 8 May 1945 the unit at Leck was obliged to surrender with all its aircraft intact to British forces. Five days previously, Adolf Galland's élite Me 262 unit (JV 44), which had been joined by the He 162s of Einsatzkommando Bär, had also been forced to surrender. Thus, while the He 162 was the aircraft which most nearly took the BMW 109–003 turbojet into full operational service, all the great effort and ingenuity spent on the Volksjäger programme came to nothing in military terms, since time ran out. By the end of the war, about 275 He 162s had been completed and about 800 nearly completed. A large proportion never received engines. Typical performance expected from an He 162 A-2 Salamander with a 109–003 E-1 or E-2 engine was a maximum level speed of 835 km/h (522 mph) at 6,000 m (19,680 ft) altitude, an initial climb rate of 1,290 m/min (4,231 ft/min) and a maximum range of 1,000 km (620 miles).

Although by September 1944 the bulk of 109–003 engines were earmarked for the He 162 programme, other aircraft, apart from the Ar 234 C, continued to be planned for the BMW engine. March 1945 saw construction under way on the Henschel He 132 V1 prototype of a dive-bomber with very similar layout to the He 162 but with a transparent nose for a prone-lying pilot. A single 109–003 E-2 engine was to provide power, and the first flight was expected in June 1945, but the V1, together with two other prototypes (partially built) was captured by Soviet forces at Schönefeld before flight tests began.

Above: Fig 2.95 Two views of the Heinkel He 162 A-2 Salamander (Werk Nr. 120 222) formerly of JG 1. This aircraft was tested in the USA as T-3-504.

Top: View on the port side, tied down. *(Deutsches Museum München)*

Below: A View on the starboard side. *(Smithsonian Institution)*

Fig 2.96 The Heinkel He 162 A-2 Salamander in flight. *(Phillip Jarrett)*

As we have seen, the Ju 287 V1 low-speed test aircraft for a bomber with swept-forward wings was flown in August 1944 on the power of four Junkers 109–004 B-1 turbojets. Although official backing for this development was soon dropped because of the demand for fighters, backing was again given in March 1945. Thus, work went ahead on the first true prototypes and pre-production examples of the Ju 287, each of which was to be powered by six 109–003 A-1 turbojets. These engines were mounted one on each side of the fuselage nose and two in a paired nacelle forward of each wing main spar, the latter four engines giving mass balance to the wings. Such engine mounting was aerodynamically unsound, and consideration was being given to the use at a later date of fewer, but more powerful, engines. However, the Ju 287 V1 and unfinished V2 were captured by Soviet forces, and the programme was continued in the USSR after the war. It is said that the Ju 287 V2, albeit modified and with swept-back wings, reached a speed of about 1,000 km/h (621 mph), presumably with BMW engines.

Fig 2.97 Some of the Heinkel He 162 fighters were flight-tested by the Allies after the war while others were dissected. Here, an He 162 A (possibly Werk Nr. 120 017 evaluated as T-2-494 at Wright-Patterson Air Force base) is being dismantled in the USA. Its BMW 109–003 E turbojet is in the foreground. *(Richard T. Eger)*

GERMAN JET ENGINES

RIGHT:
Fig 2.98 Heinkel He 162 V20 (Werk Nr. 220 003) war-damaged at München-Riem in May 1945. This machine was built at the Heinkel Hinterbrühl plant.

BELOW:
Fig 2.99 A captured Heinkel He 162 A Salamander (Werk Nr. 120 086) on display, possibly at Hyde Park. *(Phillip Jarrett)*

Another aircraft which the Germans planned to fly with BMW engines was the Horten Ho IX V1 flying wing. As the illustration in the Junkers section shows, this design necessitated a very compact installation of the principal equipment in the wing centre section. Thus, when the two BMW 109–003 A-1 engines turned out to have a larger diameter than the aircraft designers at Göttingen anticipated, they could not be fitted and the Ho IX V1 was completed as a test glider. Subsequent prototypes were designed for Junkers engines, which, although larger, were more easily obtained, and so BMW missed another chance.

It now remains to mention but a few of the German aircraft projects designed around the 109–003 engine. One of the most interesting was the Junkers Ju EF 130 bomber, which was slightly larger than the Ju 287 and with a completely different layout. The EF 130 was another all-wing design with virtually no vertical surfaces. Four 109–003 E turbojets were grouped together above the rear of the wing centre-section, and the crew were accommodated in a leading-edge gondola. A maximum speed of 990 km/h (615 mph) at 6,000 m (19,680 ft) was expected, together with a range of 5,900 km (3,665 miles), but the project was shelved.

There were, of course, quite a few fighters projected in response to the Volksjäger requirement, for the fulfilment of which the He 162 was selected. At the time, many considered the Blohm und Voss P.211 a better design which was allegedly bypassed for political reasons. The P.211 was a rather bolder design in that the 109–003 turbojet was mounted inside instead of outside the fuselage, and worries over intake duct losses may even have influenced its rejection. Simple, constant-chord, swept-back wings and a boom-mounted tail unit were other features. A top speed of 1,035 km/h (643 mph) at 15,000 m (49,200 ft) altitude was, perhaps optimistically, estimated for the P.211. Another Volksjäger competitor was the Arado Ar E.580, which had a very similar layout and proportions to the He 162, with the single 109–003 E engine mounted on top of the fuselage and a similar estimated performance.

An interesting variation on external engine mounting was shown in the projected Gotha Go P.60. This was designed as a semi-delta night-fighter with small vertical surfaces near the wing tips and a nose section extending forward of the wing apex for the crew and equipment. Two 109–003 engines (one A model and one E model) were mounted one above and one below the rear centre section of the wing.

There were many other aircraft projects for the BMW turbojet (though not as many as for Junkers turbojets), mostly for fighter aircraft and largely from the Focke-Wulf and Blohm und Voss companies. Various projects planned the use of a combination of piston and turbojet engines, such as the Fw project P.222–004 day-and-night-fighter which would cruise on a Jumo 222 piston engine and boost its speed with two 109–003 engines. In view of the fact that, for project designing purposes, Junkers or BMW turbojets could be specified, it hardly seems necessary to mention more than the foregoing examples of aircraft designed to use the 109–003 A and E engines. One 109–003 B was built, its static thrust being 900 kp (1,985 lb). We must now return to the main object by looking at the further development of the 109–003 engine.

The 109–003 C

Following a suggestion of the GL/C-E2 section of the Technisches Amt in 1941, the industrial turbine firm of Brown Boveri & Cie (BBC) at Mannheim were given the job of designing and developing a new axial compressor for the 109–003 A-0. At that time, this engine was being designed with the aim of eliminating faults in the previous P.3302 experimental engines, major troubles being experienced with the compressor. Although BMW were themselves working on resolving their compressor problems, the BBC compressor was, if successful, to be an interchangeable alternative.

The work, which was known as the Hermso project, was the responsibility of Dipl.-Ing. Hermann Reuter in the department known as TLUK/VE, and his personnel included Hermann Schneider, Joseph Krauss and Karl Waldmann. A major difference in the BBC compressor design, compared with the AVA design, was that 50 per cent reaction blading was used which shared the pressure rise equally between rotor and stator blades. The object was to achieve greater efficiency, air mass flow and pressure ratio while still being interchangeable with the AVA compressor. This object was achieved, and the Hermso I seven-stage compressor reached an efficiency of 84 per cent and gave an air mass flow of 20 kg/sec (44.1 lb/sec) with a pressure ratio of 3.4 to 1. These results were gained in runs with the compressor only, which was housed in a specially built casing with multiple pressure-reading points, a vertical outlet duct and a rear extension shaft for an external drive.

Unfortunately, the work took a considerable time, so that the BMW engine intended to receive the Hermso I compressor, the 109–003 C, was not in full production by the end of the war*. BBC also built another compressor intended for the 109–003 engine. This was the Hermso II 10-stage axial compressor (see Figure 2.100), which achieved an efficiency of 90 per cent, higher than any other German compressor. Data for the projected 109–003 C engine with the Hermso I compressor follow:

Thrust	900 kp (1,985 lb) static
	800 kp (1,764 lb) at 900 km/h (559 mph) at 8,000 m (26,240 ft) altitude.
Speed	9,800 rpm
Weight	610 kg (1,345 lb)
Pressure ratio	3.4 to 1
Specific fuel consumption	1.27 to 1.3 per hour (with J2 fuel)
Airflow	20 kg/sec (44.1 lb/sec)
Specific weight	0.677
Diameter	0.690 m (2 ft 3 in) basic
Length	3.415 m (11 ft 2 in) with exhaust bullet extended

The increase in overall efficiency which the BBC compressor was expected to confer on the 109–003 C meant that less heat needed to be released in the combustion chamber, even though the thrust was greater than for the A-2 model. This meant a lower specific fuel consumption with consequent increase in the range of a given aircraft. However, no specific aircraft projects to utilise the 109–003 C are known, but this is hardly surprising when not even a prototype of the engine was built.

The 109–003 D

During 1944, BMW were requested to redesign the 109–003 to give greater thrust and fuel economy while still retaining the same installed volume. The thrust required was 1,100 kp (2,425 lb) static, while the particular applications in mind were long-range reconnaissance versions of the Arado Ar 234 which would operate at ceilings up to 17,000 m (55,760 ft) and benefit by using engines with a decreased specific fuel consumption. The 109–003 D was to be a Class II engine, and while Heinkel-Hirth were already developing a slightly larger engine (109–011 A) in this class, it was thought that the BMW engine (with its less devious flow path) would be more efficient and more suitable for long-range work, whereas the Heinkel-Hirth engine would have the merit of robustness for combat work.

* One 109-003 C was found hidden in Austria at the war's end and was taken to Sir Roy Fedden's London office. One source states that the 003 C was used in the Arado Ar 234 C-3 to C-7 inclusive.

While based as far as possible on the 109–003 A for such items as combustion system, variable-area exhaust nozzle and structural design, the 109–003 D was in fact a new engine. It had a new eight-stage axial compressor designed to give 30 per cent greater air mass flow, a higher pressure ratio and improved efficiency, while the turbine was of the two-stage type.

Development of the new compressor was begun by BMW and Brückner-Kanis, but if the 10-stage axial compressor (Hermso II) being developed by Brown Boveri & Cie turned out to be more efficient, that was to be used instead. At the Brückner-Kanis company, the compressor work for BMW was in the hands of Rudolph Friedrich, who, years earlier, had designed the compressor for the advanced Heinkel HeS 30 (109–006) turbojet. As with the 109–006 compressor, Friedrich chose 50 per cent reaction blading for the new BMW engine.

By the end of the war, no prototype of the 109–003 D had been built, though some documents state 'in construction'. Considerable headway was made with the initial compressor development, however. A test compressor, without the full number of stages, was running at an efficiency of 89 per cent to give a pressure ratio of 3.2 to 1, and the indications were that the requisite 4.95 to 1 pressure ratio could be achieved with an efficiency of 85 per cent, but a compressor with the full number of stages was not completed in time. The performance and data laid down for the 109–003 D were:

Thrust	1,100 to 1,150 kp (2,425 to 2,536 lb) static
Speed	10,000 rpm
Weight	620 kg (1,367 lb)
Pressure ratio	4.95 to 1
Specific fuel consumption	1.1 per hour
Airflow	25 kg/sec (55 lb/sec)
Specific weight	0.56
Diameter	0.70 m (2 ft 3 in) basic
Length	3.656 m (11 ft 8 in) with exhaust bullet extended

The 109–003 R (TLR)

During 1943, when the development of the 109–003 A-0 was beginning to yield results, BMW's project study

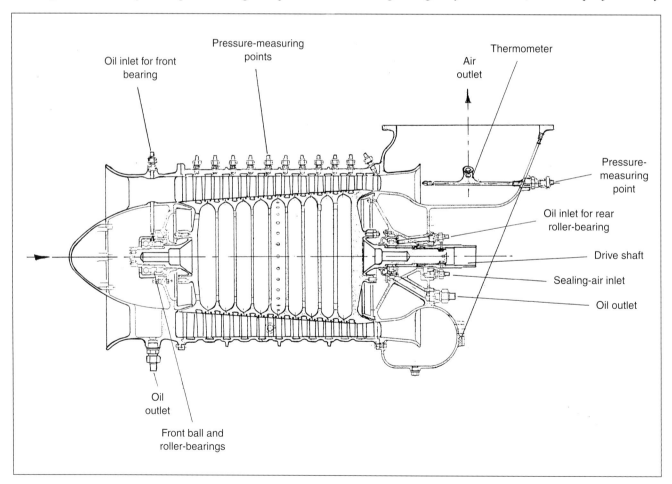

Fig 2.100 The Hermso II experimental 10-stage axial compressor built by Brown Boveri & Cie (BBC) of Mannheim for the BMW 109–003 D turbojet had a higher efficiency than any other German compressor.

department (EZS) under Dipl.-Ing. Kappus identified the combination turbojet/rocket unit (TLR) as a superior concept. With such a unit it was hoped to obtain the advantages of cruising and fast flight with reasonable fuel economy on the turbojet, while having in reserve the rocket unit for rapid acceleration and fast climbing without much penalty in weight. The rocket could also be used for short-time increases in altitude. Since BMW were already working on rocket engines for missiles and aircraft, the Technisches Amt accepted their TLR proposal, and a development contract was issued in September 1943. Thus, work began on the world's first combination jet power unit, which received the official designation of 109–003 R. This unit was to consist of a 109–003 turbojet modified to drive the fuel pumps of a liquid-propellant rocket motor mounted on top of the engine.

The initial objective, therefore, was the development of the rocket engine, which received the BMW project number of P.3395 and the official designation of 109–718. Previously, BMW rocket work had begun in 1938 to supplement similar work being performed by other German organisations. In the autumn of 1939, Dipl.-Ing. Helmut Philip von Zborowski was asked to act as technical consultant during the planning and construction of a BMW rocket research and testing facility at their Zühlsdorf factory. At the time, Zborowski was working on rocket problems at Brunswick, first under Prof. Busemann and then under Eugen Sänger, and had previously worked at BMW-Munich in 1934. His experience and technical ability therefore led him to return to BMW as the Zühlsdorf facility was nearing completion, and he directed rocket development there until the end of the war. As the scope of BMW rocket work increased, the Allach plant (near Munich) was brought into being, while personnel for the later Bruckmühl plant (Upper Bavaria) were drawn from the main Zühlsdorf plant after this latter was bombed in 1943.

When BMW first embarked on rocket development, a careful study was made to select the oxidant which would be used to support the burning of the fuel in the combustion chamber. Liquid oxygen, as used in the A-4 (V2) long-range rocket, was rejected because its low boiling point, necessitating special storage, made it unsuitable where missiles and aircraft must be permanently ready for instant action. Hydrogen peroxide, being used with success by the Walter Werke, was rejected because its instability gave a poor safety factor whilst its sensitivity to low temperatures required special storage precautions and set unrealistic limits on vehicle altitudes. A survey of all other chemical possibilities brought a decision in favour of adopting nitric acid as an oxidant despite the fact that this dangerous (though untried) liquid was viewed with considerable disfavour by rocket engineers. This suspicion of nitric acid, prominent in official circles, did much to delay its introduction into general use, but its selection by BMW was based on the fact that it had a higher density than the accepted oxidants, which compensated for differences in specific impulse and also gave a missile, for example, a higher ballistic cross-sectional loading (section density). Also, the firm of I.G. Farben confirmed that economic production of unlimited quantities of concentrated nitric acid (SV-Stoff or Salbei) could be achieved.

For fuel, methanol (M-Stoff or Mell) was originally used because it gave safe ignition with the nitric acid when a pyrotechnic or electrical ignition system was used (as in the 109–510 and –511 motors). Later, methanol was supplanted by liquid hydrocarbon fuels because of their higher density, and finally the standard aviation fuels were employed to simplify the supply in service. The desire to simplify and increase the safety of rocket engines, by dispensing with auxiliary ignition systems, led to the search for fuels which would spontaneously ignite when brought into contact with the nitric acid oxidant. At first, organic metal compounds, such as zinc ethyl, were tried, but these had the disadvantage of ageing in atmospheric oxygen, which degraded their properties. Initial tests with turpentine fuel, which was accidentally found to work when copper was added, led to a search for a self-igniting fuel among the amines which finally resulted in the range of fuels collectively termed R-Stoff or Tonka by BMW. Tonka compositions were many and complicated but were finally reduced to three, namely Tonka 93, 250 and 500, all of which contained raw xylidine.

By the time the 109–718 rocket motor for the TLR unit began development, Tonka self-igniting fuel and nitric acid oxidant were well-established BMW propellants. A thrust of 1,250 kp (2,756 lb) for a maximum three-minute period was aimed for, the thrust to be turned on and off at will. From the start of the development, high combustion temperatures generated by the self-igniting agents gave trouble, burning and buckling even the externally cooled SAS-2 (chrome-nickel steel) combustion chamber liner used in the first 15 test units. A stock of 80 duralumin liners, which had been ordered in error, was then used, and although these melted in about ten seconds, they served to develop an improved injection and mixing system for the propellants. It was still necessary, however, to supplement the external cooling system to prevent a constant temperature increase in the walls of the combustion chamber. Working along the lines originally suggested by the rocket pioneer Dr-Ing. Oberth, a solution was found. A proportion of the Tonka fuel flow was diverted direct from the head of the injection nozzle in such a manner as to form a film of cooling liquid around the inner wall of the combustion chamber. Thus, the chamber then had two cooling systems, namely a nitric acid external cooling jacket and a Tonka fuel internal cooling film.

Results with the new combustion chamber were gratifying, and the immediate improvement was such that an hour's continuous running was made without trouble. Further improvements reduced fuel consumption, raised the thrust, raised the exhaust velocity to 2,040 m/sec (6,691 ft/sec) and reduced the combustion chamber weight. Previously, a 109–718 combustion chamber had weighed 24.92 kg (54.97 lb), with 37 per cent of the parts in duralumin or other light alloy. Largely because of the success of the new cooling system, it became possible to redesign the combustion chamber using almost exclusively (93 per cent) light alloy, which reduced its weight to 12.77 kg (28.16 lb). Thus, expensive heat-resisting steel components were almost dispensed with and the chamber was made very much easier to machine. The next step was the adaptation of pumps and control gear (from the previous 109–510 rocket unit) and the combination of the rocket with the 109–003 turbojet, for which purpose the rocket and turbojet teams at Zühlsdorf worked together.

A complete 109–003 R is illustrated in Figure 2.101. For the rocket combustion chamber, two mounting brackets, spaced 0.360 m (1 ft 2 in) apart, were attached to the turbojet combustion chamber section. The rocket propellant pumps were mounted on a gearbox at the forward end of the rocket, the 200 hp required to drive them being taken from the turbojet main auxiliary gearbox to the rocket pump gearbox via an extension shaft (with two universal joints) and an electrohydraulic coupling. The pumps were of the centrifugal type and were only started up when the turbojet was running at its full speed of 9,500 rpm when the rocket fuel pump ran at 16,300 rpm and the oxidant pump at 20,000 rpm. Other equipment included the necessary piping and electrohydraulic shut-off valves, while the rocket propellants were carried inside the aircraft.

The mixture ratio, by weight, of the propellants was 1 part Tonka fuel to 3.5 parts oxidant. Later, it was intended to substitute normal J2 turbojet fuel and an igniting agent in place of the Tonka fuel, which was more expensive and took more time to produce. When filling up the fuel tanks with spontaneously igniting propellants, it was emphasised that personnel must wear protective clothing, rubber gloves and shoes with rubber soles. Furthermore, it was stressed that one person should fill the rocket fuel tank and another person the rocket oxidant tank. The planned total weight of rocket propellant to be carried by a given aircraft is difficult to assess owing to the early stage of the development of such aircraft and the variance in propellant mixture ratios, but for a twin 109–003 R installation in an Me 262, propellant weight was about 1,000 kg (2,205 lb) for 140 seconds' burning time.

Operation of the 109–003 R was as follows. When the pilot operated a tumbler switch, an electrohydraulic coupling or clutch engaged the shaft taking the drive to the rocket pumps. Once the necessary pressure had built up for both propellants, pressure valves opened and admitted the propellants to the rocket combustion chamber, where spontaneous ignition occurred. If the requisite propellant pressure was not reached within three seconds of switching on, the drive coupling disengaged automatically, while for twin engine

Fig 2.101 The BMW 109–003 R (TLR) turbojet engine with its BMW 109–718 liquid propellant rocket attached. This was a way to boost combat power before the advent of the afterburner (reheat).

installations, one rocket unit was automatically cut out if the other one failed. The propellant feed pressure into the combustion chamber increased gradually in order to give an increasing thrust up to the fixed maximum, but throttling of the thrust was not possible.

Once the development of the 109–003 R was completed, the time to consider flight tests arrived. Official type-testing and passing of the rocket unit presented little difficulty since it used a combustion chamber, pumps and valves based on those used for the already proven BMW 109–510 rocket unit built for rocket aircraft use. The 109–718 rocket was put through about fifty runs, each of three minutes' duration, as a type test. A meeting was held on 19 November 1943 to discuss the installation of two 109–003 Rs in an Me 262 test aircraft. (Further development of the new power unit was also discussed at this meeting.) The order was then given to proceed with the manufacture of an experimental batch of 109–003 Rs and to begin the installation in a test aircraft at Messerschmitt's Augsburg factory. However, while development of the engine had proceeded remarkably rapidly, experimental manufacture of the 109–718 rocket components at BMW's Allach and Bruckmühl plants fell a good year behind schedule through bombing and dispersal.

By November 1944, installation of two 109–003 Rs into a prototype of the Me 262 C-2b was begun at Augsburg, but during the first ground test of the aircraft at Lechfeld, a rocket unit exploded, apparently because of careless handling. The extent of damage to the aircraft is unknown, but it was not until early in the following year that the same or another aircraft was ready for testing, the aircraft having the works number of 170 078. On 28 March 1945, Flugkapitän Karl Baur flew the aircraft from Lechfeld on its maiden flight, the rocket units being started once the aircraft was in the air. Only two flights were made before a fault developed in one of the turbojets, and an engine change could not be effected before the war ended. Although these flights were reported as spectacular, the performance obtained is unknown. However, by turning the rockets on at 700 km/h (435 mph) at sea level, the Me 262 C-2b was expected to achieve an initial climb rate of 5,100 m/min (16,728 ft/min), reach an altitude of 10,000 m (32,800 ft) in one minute 55 seconds and an altitude of 13,000 m (42,640 ft) in two minutes 20 seconds. Maximum level speed expected was 900 km/h (559 mph) at 9,000 m (29,520 ft) altitude. The best altitude could be obtained by switching on the rockets at the normal service ceiling, when, if enough propellant was carried, a theoretical ceiling of 18,000 m (59,040 ft) was possible. On the other hand, the best range could be obtained by switching on the rockets immediately after take-off to push rapidly through the denser atmosphere and then fly relatively lighter at the most efficient altitude for the turbojets. By this method, an Me 262 with a drop tank faired into the fuselage could achieve a range of 1,700 km (1,056 miles).

There were other versions of the rocket-boosted Me 262 under development which offered a simpler solution, though a less spectacular performance. The Me 262 C-1a prototype (No. 130 186) had a Walter 109–509 A-2 hydrogen peroxide rocket unit fitted to exhaust from a modified tail, standard Junkers turbojets being retained. Karl Baur flew this aircraft on 27 February 1945, and with the extra 1,700 kp (3,740 lb) thrust from the rocket, a climb to 11,700 m (38,400 ft) was achieved in four minutes 30 seconds from a standing start. An even simpler arrangement was to be found in the Me 262 C-3a, where the Walter rocket unit and propellant tanks were slung beneath the fuselage in a jettisonable arrangement, but the sole prototype of this aircraft was not completed in time. Yet another scheme for boosting climb rate and ceiling was to fit Sänger-type ramjets* above the Me 262's turbojets. These examples will serve to illustrate the competition that BMW's TLR scheme was up against, and it is believed that the simplest of the Walter schemes (as in the Me 262 C-3a) would have been chosen for the Me 262 at least, had the war continued. Doubtless, the scheme for fitting two 109–003 A and two 109–003 R engines to the Arado Ar 234 (which the BMW project department presented in Report No. 61 dated 27 January 1945) would also have lost ground to a Walter scheme.

Under the exigencies of the time, the 109–003R unit appeared more suitable for single-engined aircraft. Again, the project department issued a report (No. 62 dated 9 February 1945) studying the performance of the He 162 with a single 109–003 R. This scheme was discussed at a meeting on 27 March 1945, but how much action was taken in the short time left is uncertain. From a standing start the TLR-powered He 162 was expected to climb to 10,000 m (32,800 ft) in two minutes 47 seconds. Another project, this one designed from the outset to use a single 109–003 R unit, was the Horten Ho XIII B semi-delta-wing fighter with the pilot housed in the base of a long vertical fin. This aircraft was intended, if built, to make its first flight in mid-1946, and with rocket and turbojet operating, was aimed at reaching a speed of 1,800 km/h (1,118 mph) at 12,000 m (39,360 ft) altitude, or almost mach 1.7.

Data for the 109–003 R follow:

Turbojet data (thrust, speed, etc.) as given for the 109–003 A or E production engines.	
Length	3.735 m (12 ft 3 in)
Data for the ROCKET component only:	
Thrust	1,250 kp (2,756 lb)
Exhaust velocity	2,040 m/sec (6,691 ft/sec)
Weight, including fittings	80 kg (176 lb) approx.
Propellant consumption at full thrust	6.5 kg/sec (14.33 lb/sec)
Rocket section approx.	1.270 m (4 ft 2 in) long × 0.292 m (11 in) high.

*Sänger's work is described in Section 9.

Further 109-003 developments

Despite the fact that the development of the basic 109–003 turbojet engine up to production status fell a good way behind schedule, BMW later found they had sufficient staff capacity to consider and work on various improvements and advances for the basic engine. Some of these developments reached the testing stage, while others remained at the project stage, depending on their priority and time of conception.

Much work was directed towards making the engine both capable of and suitable for operation at the higher altitudes being demanded as a tactical necessity. (Many Allied piston-engined aircraft had operational ceilings greater than those imposed on German turbojets.) Problems brought with high-altitude flight included intake and compressor icing, combustion flame-out and the need for pilot's cabin pressurisation. Although a new type of surface oil cooler was under test and planned for late 109–003 A-2/E-2 engines, other means of combating icing were being investigated, such as electrical heating of the first-stage compressor stator blades. Work was also about to commence on the use of the compressor to supply air for pilot's cabin pressurisation.

While the existing fixed-orifice fuel injectors were designed for operation at altitudes up to 12,000 m (39,360 ft), an engine flame-out could easily occur at altitudes above 10,000 m (32,800 ft) if the throttle was touched. In any case, even the designed altitude had to be improved upon, so investigation into a duplex type of fuel injector, with a separate spray nozzle for high altitudes, was proceeding. There was also an interest shown in a Bosch fuel system which used injection pumps.

Redesigns to increase the turbojet thrust have already been described (see 109–003 C and D), but experiments were conducted in applying water injection and afterburning (reheat) to the basic engine as a means of giving temporary increases in thrust. Methods of water injection tried were:

(i) Injection into the combustion chamber at rates up to 1,200 kg/h (2,646 lb/h). This method gave up to a 20 per cent increase in thrust and was still being experimented with at the end of the war.

(ii) Injection after the combustion chamber and before the turbine, which gave very poor results.

(iii) Injection through the hollow turbine blades at rates up to 400 kg/h (882 lb/h). Results unknown.

(iv) Injection into the compressor intake, which method was abandoned because of an adverse effect on the compressor efficiency.

The general aim of these experiments was to increase thrust by increasing the possible cycle temperature or energy level of the engine. The aim of afterburning, on the other hand, was to use up the oxygen still present in the engine's exhaust gases (diluted by secondary air) by injecting and burning more fuel after the turbine and thereby to obtain an increased thrust. In effect, a ramjet was attached to the rear of the turbojet (TLS). BMW made tests injecting fuel into the exhaust section on similar lines to the Junkers tests for the 109–004 E, but the results did not compare favourably, and stability of combustion was not achieved. Partly, this was because the 109–003 had a shorter exhaust unit and higher gas velocity than the Junkers 109–004. Naturally BMW considered their 109–003 R the ideal system for temporary thrust boosting, especially for fighters, so that the experiments in water injection and afterburning had a low priority. Nevertheless, the immediate programme for such experiments was still under discussion at a meeting held as late as 19 March 1945.

At Neubiberg, south of Munich, final base for the BMW flight-testing section, an automatic control for continuously variable operation of the exhaust nozzle exit area was ready for testing, using an Ar 234. This control equipment was designed and built at Stassfurt under the code designation T.5, but time ran out before its first test flight. Its principle of operation was that a pitot tube outside the engine nacelle, a bi-metal (heat-sensing) strip just upstream of the compressor and an electric motor for the exhaust bullet each operated a potentiometer. Those three potentiometers were connected in a balanced bridge circuit to a relay box which controlled the rotation of the exhaust bullet electric motor, the system having a follow-up or feedback nature. At full engine speed, the T.5 equipment could maintain the exhaust temperature steady within 10 °C either way, and a mechanical linkage between the throttle lever and the relay box adjusted the system for another engine speeds. A more complicated control system (one which was to be built on a lower priority than T.5) was intended to give a constant flight Mach number for a constant throttle setting, but this was only in the design stage under Dipl.-Ing. Hagen who originated the idea. Also in the design stage was a hydraulic mechanism for operating the exhaust nozzle bullet as an alternative to the electric motor system.

Having looked at some of the improvements being planned for the 109–003 turbojet, the story of the development of that engine up to the time of the German collapse is complete, and we can now turn our attention to BMW work on other designs of turbojet engine. Before doing so, however, a curious sideline should be mentioned that happened in Japan, Germany's ally. Having built the Ne-12 turbojet, the Japanese were unable to coax it up to its modest designed thrust of 340 kp (749 lb). There then came into the hands of Imperial Japanese Navy engineer Eichi Iwaya some photographs of the BMW 109–003 turbojet engine. From these, the resourceful Japanese engineers were able to design and build the Ne-20 axial turbojet, which was designed for a thrust of 475 kp (1,047 lb). Two of these engines then powered the Nakajima Kikka (orange blossom) fast attack bomber prototype on its only full flight on 7 August 1945. The Ne-20 was also scheduled to power the Ohka (cherry blossom) Model 33 piloted suicide aircraft.

Fig 2.102 Two examples of the Japanese Ne-20 turbojet which was inspired by the BMW 109–003.

The 109–018 (TL)

In accordance with the broad turbojet development programme envisaged by Schelp of the Techniches Amt, once the Class I turbojets (109–004 and 003) were developed and/or in production, more powerful engines were to begin development. As we have seen, Heinkel-Hirth worked on a Class II turbojet (109–011), and Junkers worked on a Class III turbojet (109–012). BMW were selected to work on a large turbojet, in Class IV, and were given the order to proceed with such an engine in 1942, despite the fact that their 109–003 turbojet was far from perfected and certainly not ready for production. The new engine, designated 109–018, was aimed at giving a static thrust of 3,500 kp (7,717 lb), and was intended for operation at altitudes up to between 15,000 m (49,200 ft and 18,000 m (59,040 ft) and for bringing short-range bombers up to the critical speed of Mach 0.82.

Basically, the main features and layout of the 109–018 were similar to the 109–003, but with appropriate scaling up and additional compressor and turbine stages. The chief design difficulty was that the size of the engine necessitated the use of fabricated instead of cast structural parts. Although such parts were to be normalised and stress-relieved, the distortion under running conditions was likely to give rise to some initial difficulties. (The same concern was held for the Junkers 109–012.) Owing to difficulties with the development of the 109–003, which did not go into full production until the latter part of 1944, the design of the 109–018 was not completed until the end of 1943 (or later), when work began on the construction of three developmental prototypes. Further delays were caused by the experimental group having to move four times because of bombing and the liberation of France.

Development of the 109–018 was officially stopped at some time during November/December 1944 (about the time when the Junkers 109–012 was stopped), since it was regarded as a long-term development for which no airframe was available anyway. However, work continued beyond this time on BMW's own initiative. One complete compressor was built, but to avoid its capture, it was destroyed at Stassfurt before it could be tested. The tests were to be done at either Dresden or Oberhausen. By the end of the war, combustion tests were about to commence to determine the best angles for the secondary air sandwich mixers, and three engines were partially complete at Stassfurt. The plan was that the engines should be completed at the Kolbermoor plant, followed by testing and development at Oberwiesenfeld, this plan being decided upon in March 1945. Most of the special tools, including those for blade manufacture, were moved to Kolbermoor, but with defeat imminent all engine components appear to have been destroyed.

A section through the 109–018 is shown in Figure 2.103 while one of the partially completed engines is illustrated in Figure 2.105. The compressor was of the 12-stage axial type, designed for an efficiency of 79 per cent, and was the responsibility of two engineers named Loffler and Fickert, who were presumably employees of

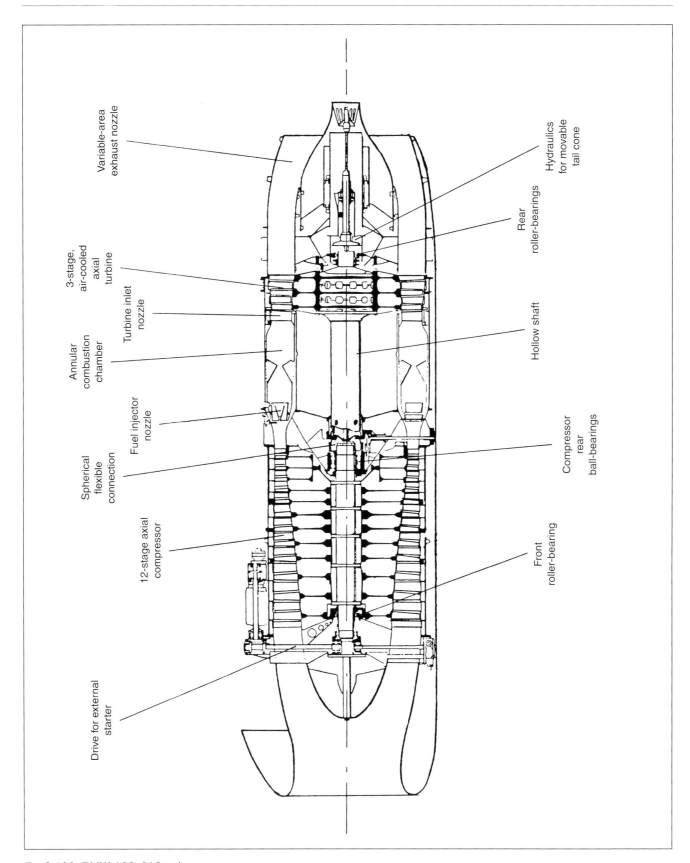

Fig 2.103 BMW 109–018 turbojet.

BMW. A light construction was used for the compressor, the first five discs being of dural and the remainder of steel. For the first seven stages, the rotor blades were of dural and attached to the discs by hollow rivets, while the remaining high-pressure rotor blades were of steel. Much of the strength of the sheet-metal compressor casing was derived from the substantial webs, or channels, formed on the outside of the compressor stator blade supporting rings, which in turn were covered by a steel skin. The compressor shaft was hollow (though not a drum type) and was supported by two large ball races at the rear and a smaller, roller race at the front.

The annular combustion chamber was of the same design as that for the 109–003, though, naturally, larger, and probably had even the same number of inner and outer hollow fingers for directing secondary air. However, there were 24 main burner cones and eight auxiliary fuel injectors, the latter being for starting purposes. A combustion efficiency of up to 95 per cent was expected. Unfortunately, more details of the combustion system are lacking, but duplex fuel injectors must have been planned in view of the very high altitudes at which this engine was to operate. The altitude aims for this turbojet were, in fact, the most ambitious planned by the Germans.

A three-stage, air-cooled, axial turbine was employed, the discs of which were held together by long bolts, these bolts being the only items in the engine which were subject to selective assembly. Departing from previous practice, the turbine had only one bearing, this being of the roller type and at the rear of the turbine. For the forward support of the turbine, a large, hollow shaft was provided which was located at its forward end inside a spherical seating attached to the compressor shaft, an offset stud transmitting power from the turbine shaft to the compressor shaft. The spherical seating arrangement permitted a certain amount of axial misalignment which structural distortions could produce, and thereby avoided serious stresses from this cause. In front of the spherical seating, a long tie-rod was attached to the turbine shaft and transmitted the thrust load on the turbine to the forward end of the compressor. This tie-rod was screw-threaded to assist in the expulsion of the turbine during dismantling of the engine. The turbine was expected to have an efficiency of 75 per cent and develop the considerable power of 40,000 hp (most of which was absorbed by the compressor, of course), the gas temperature at the turbine being in the region of 800 °C.

Cooling air for the turbine was tapped off after the compressor fifth stage and was fed to the blades via the hollow turbine shaft and hollow drum formed by the turbine discs. For cooling turbine nozzles or inlet guide vanes, some of the secondary air flowed past the combustion chamber into the guide vanes and out through their trailing edges. Of the total engine air mass flow, about 1 per cent was used for cooling the turbine and about 1.5 per cent for cooling the turbine nozzles. For the cooling of the exhaust nozzle and tail cone

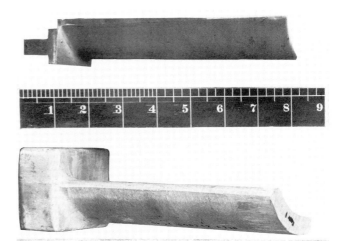

Fig 2.104 A compressor blade forging and turbine blade for the BMW 109–018 turbojet engine.

assembly, air scooped from the atmosphere was used in a similar manner to that for the 109–003, but the design of the movable bullet mechanism was different. To vary the nozzle exit area, the bullet or cone moved longitudinally on rollers fitted around a central tube. Partly inside this tube was a hydraulic cylinder to move the exhaust bullet, the internal actuating valve being operated by means of a Bowden-type cable. Initially, however, the bullet was to be operated by electromechanical means pending development of the hydraulic system.

For starting the engine, a 50 hp Riedel two-stroke petrol engine was to be used, since this was already available, but later a small gas turbine unit was to be substituted. The gas turbine was favoured by Schelp's department for starting large turbojets, and Heinkel-Hirth were given the task of developing such a starter. Notwithstanding the type used, the starter was mounted on the compressor casing and was connected to the turbojet rotor by means of bevel gears and a shaft which passed through one of the hollow arms of the front support spider, the arrangement being very much simpler than that for the 109–003. Other auxiliaries were grouped around the front of the engine but had their own gearbox driven by a separate auxiliary shaft.

The main lubrication pump had a capacity of 4,000 kg/h (8,820 lb/h) and pumped a 50/50 mixture of Flugöl S3 and hydraulic oil. The total oil circulation was 1,200 kg/h (2,646 lb/h), of which 100 kg/h was for the turbine bearing, 150 kg/h for the compressor front bearing, 300 kg/h for compressor rear bearings, 300 kg/h for cooling the starter mechanism and for the exhaust bullet hydraulics, while the remainder was for the gearcase behind the intake fairing. An oil scavenge pump was driven from the rear of the turbine and another from the rear of the compressor shaft, while a third, composite, pump was provided at the front for the central gearcase. De-aerators were proposed for the lubrication system.

The performance and data laid down for the 109–018 were:

Thrust	3,400 kp (7,497 lb) static
Thrusts at 900 km/h (559 mph)	3,000 kp (6,615 lb) at sea level
	1,460 kp (3,219 lb) at 10,000 m (32,800 ft)
Speed	5,000 rpm
Weight	2,500 kg (5,512 lb) for experimental engines but less than 2,200 kg (4,851 lb) planned for production engines
Pressure ratio:	7.0 to 1 approx.
Specific fuel (J2) consumption	1.1 to 1.15 at static thrust
Airflow	83 kg/sec (183 lb/sec)
Specific weight	0.735 (experimental engines) down to 0.647 (production engines)
Mean gas temperature	800 °C
Diameter	1.250 m (4 ft 1 in)
Length	4.950 m (16 ft 2^m in)

Pursuing the 109–018 design further, a design study was made of a TLR version (109–018 R) on similar lines to the 109–003 R layout. It is believed that the rocket unit was to be a development of the 109–718 rocket, giving a thrust of 2,000 kp (4,410 lb) for between 60 and 90 seconds. BMW considered the rocket best designed as a tri-fuel unit in the sense that it would run on J2 fuel and Salbei (nitric acid, or SV-Stoff) oxidiser, having been started with a small amount of an igniting agent such as Tonka (R-Stoff). Not only did this propellant system, desirable for all TLR designs, give increased safety, but if the rocket was not needed on a mission, its allocation of J2 fuel could be used by the turbojet engine for an increased range. There was, however, no plan to build a 109–018 R since the TLR type of engine was considered suitable for fighters only, and a large enough fighter was not then officially envisaged, although at least one high-altitude fighter was projected for the engine without the rocket unit.

An adaptation of the 109–018 for use in a twin turbo-prop layout is described at the end of the 109–028 section.

As for specific aircraft proposed to use the large 109–018, bombers were naturally the most numerous, and the following examples are given. The Junkers Ju 287 B-2 was projected with two 109–018s (or two Junkers 109–012s), as was the projected Henschel bomber P.122. The P.122 was an all-wing aircraft in the sense that there was no tailplane, but there was a cigar-shaped fuselage which supported a vertical fin and rudder at its end. Directly beneath and halfway along each low-mounted, swept-back wing was mounted one of the two turbojets. Despite its large size, the Henschel bomber was expected to reach a maximum speed of 935 km/h (581 mph) at 10,000 m (32,800 ft) altitude and to have a range of 2,000 km (1,242 miles) and a ceiling of 17,000 m (55,760 ft). Another bomber was the P.1107/II projected by the Messerschmitt company, this jet aircraft being one of the most powerful put forward by the Germans, since it was to use four 109–018 engines, giving a total static thrust equivalent to about 45 per cent of the bomber's 30,700 kg (67,693 lb) loaded weight. The engines were buried in the roots of the swept-back wings, which had leading-edge air intakes and were mounted on the centre-line of the cigar-shaped fuselage. Glazing the nose of the fuselage gave vision from the cockpit, and a butterfly tail was provided. A proposed maximum speed

Fig 2.105 A BMW 109–018 Class II turbojet engine, partially completed. The largest turbojet engine worked on by the Germans by 1945, it was remarkable for its time, being designed to operate above 15,000 m (49,200 ft) and power bombers up to Mach 0.82.

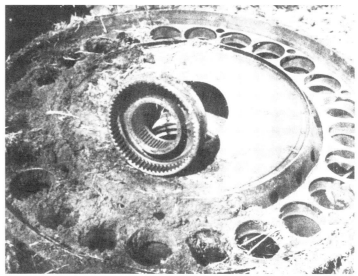

Fig 2.106 A compressor was completed for the BMW 109–018 turbojet at Stassfurt near the war's end but was destroyed to prevent use to the Allies when captured.

ABOVE: An Allied technician examining the compressor (apparently exploded with an internal charge which split open the casing and stator rings). *(Richard T. Eger)*

ABOVE: The compressor front roller-bearing race balanced (for photographing) on the sheet-metal support ring for the front of the 24 main burner cones. *(Richard T. Eger)*

of 880 km/h (547 mph) is the only performance figure known for the P.1107/II.

Apart from the airframe industry, BMW put forward aircraft projects of their own designed around their 109–018 turbojet. A BMW group was set up to make such aircraft studies in collaboration with the aircraft industry and other interested parties, and thus, along with similar work by Daimler-Benz, represents an early attempt at 'systems engineering', whereby airframe, engine and equipment were considered and integrated from the beginning. One bomber, the BMW Schnellbomber I, was of conventional layout except for the wing, which was swept forward at the roots and swept back on the main, outer panels. At the point where the wing sweep angle changed at each side was to be mounted an underslung 109–018 turbojet plus a 109–028 turboprop (described below). This very large bomber was expected to carry a 15,000 kg (33,075 lb) bomb load and have a maximum speed of 850 km/h (528 mph). Another BMW bomber project, Strahlbomber II, proposed a pod-like fuselage with swept-back wings and no vertical or tail surfaces. Two 109–018 turbojets were to be mounted in the rear of the fuselage, which had a nose air intake. This tailless bomber was to carry a 5,000 kg (11,025 lb) bomb load and have a maximum speed of 950 km/h (590 mph).

These examples of proposed utilisation of the 109–018 conclude the discussion of that turbojet, which was the last BMW design to get past the drawing-board stage before the end of the war. It now remains to discuss other BMW designs, which, because time ran out, were still in the project stage when hostilities ceased.

The 109–028 (PTL) and a twin-PTL project

As early as 1940, BMW's project department made studies for a large turboprop engine (PTL), and by early 1941 the development of such an engine had the backing of the Technisches Amt. Although the design was never finalised, the engine resulting from the development was to be designated 109–028, and by July 1942 was planned to develop 8,000 equivalent shaft hp (a combined allowance for jet and airscrew power) at 800 km/h (497 mph) at 8,000 m (26,240 ft) altitude. At about this time, the decision to develop the 109–018 turbojet was taken, so that, for a given engine size, BMW's turboprop work preceded their turbojet work instead of vice versa as in the case of other companies, e.g. Junkers 022 PTL from the 012 TL.

Nevertheless, the first designs for the 109–028 bore many features seen in the 109–018 turbojet, such as a 12-stage axial compressor of disc construction, similar combustion and exhaust nozzle assembly and the same method of supporting and connecting the compressor and turbine shafts. Although the turbine was also of the same construction as for the 109–018, another stage was added to develop power for the airscrews, so that the 109–028 had a four-stage, axial turbine using about 30 per cent reaction. The turbine stage added for airscrew power was not independent of the other three stages, but all four were bolted together. Power was transmitted to the contra-rotating airscrews by a shaft from the front of the compressor driving through planetary gearing within the airscrew boss, or hub.

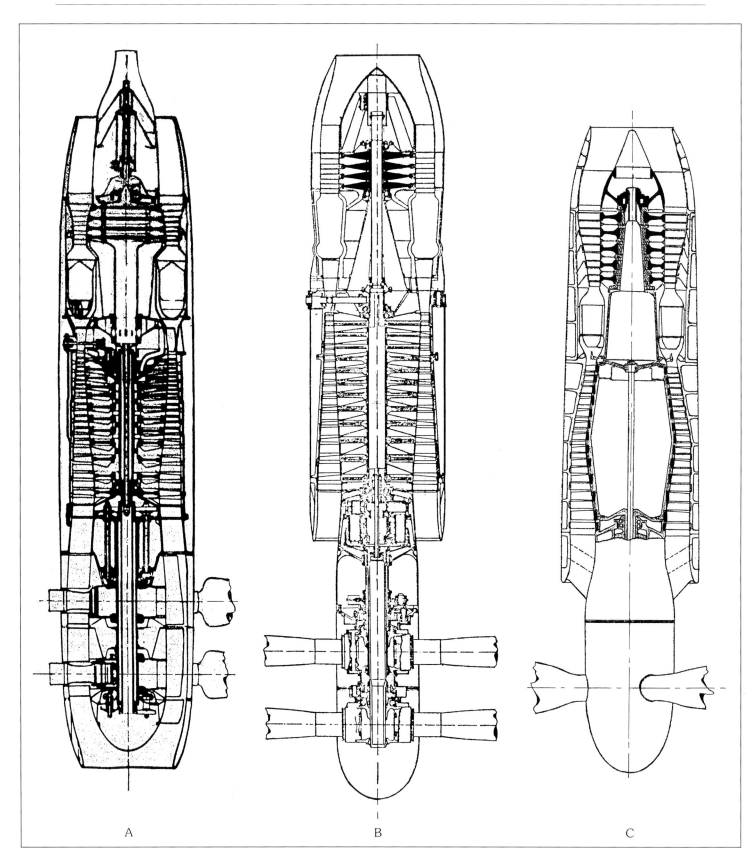

Fig 2.107 Three layouts for the projected BMW 109–028 (PTL) turboprop engine.

KEY TO DIAGRAMS OPPOSITE:

A Layout with contra-rotating propellers, ducted spinner housing, planetary gearing, 12-stage compressor and 4-stage turbine.

B Layout with contra-rotating propellers and gearing moved aft, shortened intake, 12-stage compressor and 4-stage turbine.

C Layout with single propeller, short-intake, 12-stage compressor and 5-stage turbine.

For this design of the turboprop engine (see Figure 2.107A), the airscrew hub was enclosed in a duct or shroud which began in front of the hub nose and led up flush to the engine air intake. The airscrew drive and intake section were chiefly the responsibility of Dipl.-Ing. Stoekicht of BMW and Ebert of the VDM airscrew company, and this part of the engine posed the biggest problems. Not only was the mechanical design made difficult by space and weight limitations but the aerodynamic problem of ensuring a good airflow into a compressor which was preceded by airscrews was considerable. Unless a suitable design was arrived at, the compressor would suffer a loss of intake (ram) air owing to the aerodynamic disturbances caused by the airscrews and their centrifugal effect on the airflow near the hub. The latter effect was to be minimised by using two, large airscrews revolving at the minimum speed without sacrificing efficiency, but other intake disturbances were not so easily avoided. The reasoning behind the ducted airscrew hub was that air would be drawn in before the major airscrew disturbance, but a disadvantage here was that the annular space between the duct and the hub was considerably cluttered with guide and support vanes in addition to the airscrew blade roots.

Although no airflow tests were made for this or any other intake design, another one was proposed without a duct or shroud around the airscrew boss. In this second design the aim was to position the airscrews as far forward as possible from the engine intake. A long airscrew boss therefore resulted, together with the shortest possible compressor intake, which had sharp, slightly outward-turned lips. Also, in order to minimise the cantilever load on the projecting airscrew section, its main gear drives were moved back from inside the boss to a position just in front of the compressor. Further refinements were also introduced in this design, such as a modified combustion chamber and a new exhaust nozzle bullet which did not project beyond the end of the exhaust nozzle. The nozzle area was to be adjusted in accordance with altitude and air speed, the associated control cam profile being designed so that the airscrews would not receive more than the 7,000 hp for which their reduction gears were designed.

An outline of this second major layout for the 109–028, which seems to be the layout most likely to have been adopted, is given in Figure 2.107B. Other details known for the engine, which was, after all, in a fluid state of design, are as follows. The contra-rotating airscrews were each to have three or four blades (depending on the diameter selected) with variable negative-pitch range, permitting reverse-thrust braking upon landing. For the hydraulic operation of the airscrew pitch, an all-speed governor was designed which was set by a cam, the profile of which was calculated to give good fuel consumption at part load. Engine accessories were to be driven by a radial shaft between the compressor and the combustion chamber, up to 300 hp at full speed being available for this purpose. Hydraulic operation was planned for the variable area exhaust bullet, and the turbine and exhaust sections were to be air cooled. Efficiencies aimed at were 70 per cent for the turbine and 80 per cent for the compressor, together with an average combustion temperature of 770 °C before the turbine. The exhaust temperature was to be limited by a valve which measured air mass flow (in terms of temperature and pressure upstream of the compressor) and suitably metered the fuel flow.

Finally, a complete redesign of the projected 109–028 turboprop engine was undertaken to introduce simplifications and thereby enhance the chances of early success in development. The contra-rotating propellers were replaced with a single propeller to eliminate complex gearing, and there was a more refined short intake. An all-new 12-stage axial compressor, together with a 5-stage axial turbine, was designed. This third version of the engine is shown in Figure 2.107C.

A plan existed to conduct the first flight tests by fitting two 109–028s into a Heinkel He 177 Greif bomber, which normally used two large airscrews driven by four piston engines coupled in pairs. However, since the detail design of the turboprop engine was not completed, not even the building of a prototype was begun. Data for the proposed 109–028, necessarily approximate, are as follows:

Power rating at sea level, static	
Jet thrust	2,200 kp (4,851 lb)
Shaft power	4,700 hp
	(6,570 equivalent shaft hp
Power rating at sea level and 800 km/h (497 mph)	
Jet thrust	1,425 kp (3,142 lb)
Shaft power	7,000 hp
	(12,600 equivalent shaft hp)
Power rating at 10,000 m (32,800 ft) and 800 km/h (497 mph)	
Jet thrust	790 kp (1,742 lb)
Shaft power	3,280 hp
	(6,800 equivalent shaft hp)
Power rating at 12,000 m (39,360 ft) and 800 km/h	5,500 equivalent shaft hp

Airscrew speed	925 rpm
Weight	3,500 kg (7,717 lb)
Specific fuel consumption at 800 km/h (kg per ehp per hour)	1.46 at sea level, reducing to 1.06 at 12,000 m
Airflow	100 kg/sec (220.5 lb/sec) at 800 km/h at sea level
Diameter	1.250 m (4 ft 1 in)
Length	6.0 m (19 ft 8 in)
Diameter of airscrews	3.5 to 4.0 m (11 ft 5¹ in to 13 ft 1 in)

Apart from the plan to convert an He 177 for flight tests, there were a few schemes for the utilisation of the 109–028. Calculations were made to fit two of these turboprops to the Messerschmitt Me 264, which was originally designed to fly non-stop to the USA and back, though with a small bomb load because of the enormous fuel load needed. A prototype of the aircraft began flying in December 1942 on the power of four Jumo 211 piston engines. Subsequent prototypes, intended to be precursors of a long-range reconnaissance aircraft, were held up and finally never flew. Although the performance of the Me 264 with two 109–028s was expected to be good, it was eventually decided that too much work was involved in strengthening the undercarriage (which already had to carry a normal loaded weight of about 46 tonnes) for the scheme to be useful. In the meantime, BMW put forward a bomber project, the Schnellbomber II (Schnellbomber I has been mentioned under the 109–018 turbojet), which was to use two 109–028s mounted on pylons above the forward fuselage section. Flying surfaces consisted of swept-forward wings and a delta tailplane surmounted by a large vertical surface. No performance data are known for this projected bomber, but its size would have been similar to that of the Me 264.

Because of the difficulties expected in trying to obtain a good air intake with the 109–028 turboprop, a fresh look was taken at the whole question of the layout of large PTL units. Until then, the basic idea was to take a turbojet engine and add one or two extra turbine stages to divert a major part of the power to an airscrew at the front (thus providing a larger mass of slower air suitable for larger, slower aircraft). On the advice of the independent consultant Dr Alfred Müller, however, the Technisches Amt directed BMW to make a study of a new PTL layout. This, in essence, comprised a 109–018 turbojet with an *enlarged compressor* from which air could be tapped off to supply two independent power units in the aircraft wings. Each power unit consisted of a combustion chamber supplying hot gases to a turbine driving an airscrew, the turbine operating at the then high temperature of between 900 and 1,000 °C. Assuming this interesting project (which we might designate ZPTL) was pursued, no headway could be made with it until the basic 109–018 turbojet had undergone some development. Still, the scheme makes an interesting comparison with the one looked into by Heinkel whereby two smaller turbojets were to supply hot gases to an independent turbine driving an airscrew. For both schemes, the companies concerned gained the investigation incentive from officialdom.

The P.3306 (TL)

During a conference in March 1945, the Technisches Amt sought BMW's opinion on the Heinkel-Hirth 109–011, which was supposed to be in production by then but was still beset with various technical deficiencies. Although plans were being made for BMW to give assistance to Heinkel-Hirth for the production of the 109–011, Dr Bruckmann (director of BMW power-plant development) thought that an official order was about to be issued to his company for the development or a new turbojet engine, to be interchangeable with the 109–011, but with a 1,700 kp (3,748.5 lb) instead of a 1,300 kp (2,866.5 lb) static thrust. An operational ceiling of 15,000 m (49,200 ft) was also chosen. In view of the obvious impending collapse of Germany and the fact that there had been considerable official pruning of projects and developments, Bruckmann must have had some concrete evidence that an order was imminent, since he ordered planning work to begin on a new engine, which received the BMW project number of P.3306. The project was more than an exercise to keep busy the considerable design staff which BMW had by this time, and it seems apparent that there was every intention of pushing ahead with the full development of the turbojet after making the first runs in the summer of 1946. It was thought that, once a prototype was running, full official support would be forthcoming.

The layout of the P.3306 was simplified and showed no remarkably novel features but largely followed the layout and practices used for the 109–003 while incorporating all the improvements and lessons learned during the development of that engine. A seven-stage axial compressor employing disc construction was used, driven by a single-stage axial turbine. Sheet-metal construction was used to a greater extent than in the 109–003, and castings were either dispensed with or reduced to a minimum size. The usual combustion chamber with secondary air fingers was used but with the average combustion temperature raised to 750 °C, and there was a mechanically driven bullet (open-ended) to vary the exhaust nozzle exit area. A Riedel starter motor was mounted inside the intake central fairing, and the usual auxiliaries, oil pumps, etc., were mounted externally and driven from the rotor by radial shafts. For the mounting of the main auxiliaries (on top of the compressor casing) and for the attachment of the engine, a reinforced section was fitted to the sheet metal-work. Data for the P.3306 are:

Thrust	1,700 kp (3,748.5 lb) static
	1,820 kp (4,013 lb) static allowable for a short period
Speed	8,700 rpm
Weight	900 kg (1,984.5 lb)
Pressure ratio	4.2 to 1
Specific fuel consumption	1.2 per hour
Airflow	40 kg/sec (88.2 lb/sec)
Specific weight	0.528
Diameter	0.850 m (2 ft 9 in)
Length	3.20 m (10 ft 6 in)

The expendable P. 3307 (TL)

As an expendable power-plant for missiles, the projected P.3307 turbojet was designed to use a maximum of sheet metal and to have a production time of only 100 man-hours. Unfortunately, nothing is known of the background to this project or if it was to go forward. It seems unlikely that it was a direct competitor to the Porsche 109–005 (described further on), which, although planned for the same thrust and duty, bore little other resemblance in its specification. From the weight and size point of view (and, no doubt, from the general layout point of view also), the P.3307 was similar to the 109–003, which is difficult to understand in view of the smaller thrust of the former. The compressor was of the axial type, but different sources give the number of stages as five, six or seven, which possibly indicates that the design was at a very early stage. The only other details available are that the turbine was a single-stage axial type, and the following data:

Thrust	500 kp (1,102.5 lb) static
Weight	650 kg (1,433 lb)
Specific fuel consumption	1.5 per hour
Specific weight	1.3
Diameter	0.690 m (2 ft 3 in)
Length	2.850 m (9 ft 4 in)

Conclusion

In terms of turbojet engine production, BMW's success came second to that of Junkers. However, there is every reason to believe that this gap would soon have been closed had there been time. More interesting today is the fact that, in terms of technical progress, BMW's gas turbine work was ahead of Junkers and other companies, and this lead showed every sign of increasing. Ample staff and excellent facilities had been accumulated for gas turbine development by the end of the war, and plans for future development appeared generally on the right lines, though still tempered by the German need for military utility and by a shortage of the best materials. This state of affairs was only reached after a slow start from about 1938 and protracted difficulties with the company's only developed turbojet engine, the 109–003. As we have seen, this engine finally had a longer and more reliable service life than the corresponding Junkers development, the 109–004.

At the end of the war, BMW were most assiduous in carrying out orders to destroy all evidence of their work as Allied forces approached their various plants. This effort was in vain, however, since amongst the vast quantity of material, enough was overlooked to give a good picture of the engines worked on. Also, complete sets of drawings were discovered buried in steel boxes, and there were complete engines of the 109–003 E type found fitted to He 162 aircraft. British and American interest was largely confined to testing and evaluating BMW's work, with little emphasis on further development. For this purpose, mainly existing engines and documents were sufficient, but the order was given for three 109–003 R engines to be assembled for the Allies. Two were destined for Wright Field and the US Navy in the USA, and one for Farnborough in England.

One of the most valuable plants, the BMW high-altitude engine test chamber at Munich, fell into American hands since it was within the American occupation zone. The test chamber was found serviceable, despite heavy bombing nearby, and the Oberweisenfeld airfield was very handy for flying in Allied turbojets for testing. Agreement was reached for British companies to use the BMW plant, and by July 1945 plans had been made to begin testing the British de Havilland Goblin I turbojet. After three weeks to adapt mountings and two weeks to install the engine, a series of calibrations were begun on the Goblin I up to 13,000 m (42,640 ft). After the first period of running, the German engineers assisting with the work asked if they should open the test chamber for inspection of the engine. When told this was unnecessary, they were greatly surprised, since they had no experience of a turbojet running for more than five hours without some attention. Despite the fact that the test chamber inhaled much dust from the surrounding bomb damage, the Goblin I was run for a total of 42 hours before being examined. A Goblin II was later put in the chamber and gave 71 hours of trouble-free test running.

While these tests and BMW material proved valuable, the Soviet and French authorities derived most benefit from BMW gas-turbine work in terms of subsequent development, since their countries had fallen most behind in the field. The Soviet programme set up at Kuibyshev on the Volga to examine, manufacture and learn from

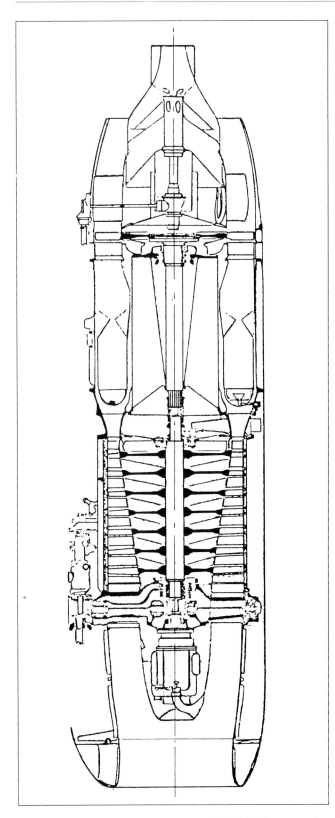

Fig 2.108 The projected BMW P.3306 (TL) was to be interchangeable with the Heinkel-Hirth 109–011 and introduce simplifications for more rapid production. It was hoped to have it running in the summer of 1946.

German turbojet work has already been described in the conclusion to the Junkers section, and this programme also included BMW developments. Machinery was transported from BMW plants in the Eisenach area, and copies of the 109–003 were soon being made at Kuibyshev under the designation RD-20. This work seems to have been kept in the training part of the programme, since only copies of Junkers and British turbojets are known to have powered Soviet production aircraft. After 1950, the German engineers were used to work on the development of a twin turboprop engine, and then a 12,000 bhp turboprop engine, but we cannot say these developments benefited from any particular German stable. Probably their origins lay in an amalgamation of both BMW and Junkers turboprop ideas. In any event, by 1955 the West became aware of the large Tupolev Tu 20 'Bear' bomber, powered by four 12,000 ehp Kuznetsov NK-012 turboprop engines which owed their existence to the Germans at Kuibyshev. In terms of power and dimensions, this Soviet turboprop compared with the BMW 109–028, but its specific fuel consumption was only about half, and its weight only about 65 per cent, that of the German engine.

The French were fortunate also. They first acquired the services of Dr H. Oestrich (former director of BMW turbojet development), and gave him the job of continuing the development of the 109–003 with the principal nationalised French engine company, SNECMA (Société Nationale d'Etudes et de Construction de Moteurs d'Aviation). A complete redesign of the BMW engine was undertaken, with the aim of first achieving a static thrust of 1,700 kp (3,748 lb), the new engine being known as the ATAR 100 (ATAR being derived from Atelier Technique Aéronautique Richenbach, a BMW plant near Lake Constance during the German occupation of France). The first ATAR experimental engine was on the test stand in May 1948, and was soon giving the designed thrust for a combustion temperature of below 700 °C. Its chief features of a seven-stage axial compressor (with 50 per cent reaction), annular combustion chamber, air-cooled turbine and a variable-area exhaust nozzle, were similar to those of the BMW 109–003. Test flights followed using a B-26 Marauder as a test bed, and by June 1948 the combustion temperature had been raised to 800 °C and the thrust to 2,200 kp (4,851 lb) static. Continuous improvement brought the static thrust up to 3,500 kp (7,717 lb) for a weight of 820 kg (1,808 lb) in the ATAR 101E which was performing endurance runs of 150 hours in the summer of 1954. Still more improvements followed, including BMW ideas on water injection and afterburning, but, without straying too far from our timescale, suffice it to say that the ATAR series of engines (eventually bearing no resemblance to the original BMW engine) were a great success and have powered many French jet aircraft, notably of Dassault design, e.g. the early Mystère fighters. Finally, a few comparative figures may prove of interest to the reader and demonstrate what development can do:

	1945	1948	1954
	BMW 109–003 A-2	ATAR 101A	ATAR 101E
Static thrust (kp)	800	2,200	3,500
Weight (kg)	600	830	1,730
Specific weight	0.75	0.377	0.494
Specific fuel consumption	1.4	1.3	1.05
Diameter (m)	0.690	0.886	0.920

For the ATAR 101, the remarkable improvement in the specific weight is noteworthy, and there was also a 2.5 times increase in the thrust per unit frontal area compared with the 109–003 A. While relating this successful post-war development story, however, one must not assume that *all* BMW development plans were on the correct lines. This remark is prompted by the fact that, despite the company's successful development of the classical constant-pressure turbojet and the correct thinking behind the selection of axial compressor, annular combustion chamber and other components, initial project work was actually started near the end of the war on a *constant-volume* turbojet engine. It was proposed to use a rotating combustion chamber with valves (presumably in order to even out the impulses on the turbine) and in this respect at least, the scheme differed from Heinkel's HeS 40. While a constant-volume turbojet engine would not be considered today (and many would not have considered it in 1945 or earlier), the post-war years were ones in which a great deal was still to be learned concerning aero-engines. The TLR unit, pioneered by BMW, and other arrangements of rocket boosting, were in great vogue from East to West for many years, and a number of experimental aircraft were built, but eventually the full development of the turbojet afterburner virtually supplanted the use of an auxiliary rocket for short periods of increased power.

Meanwhile, Dr Bruno Bruckmann (previously director of all BMW aircraft power-plant developments), his colleague Gerhard Neumann and Dipl.-Ing. Peter Kappus moved to the USA to manage the engine programme for General Electric. It was Neumann who led GE through many successful engine programmes (including the J79) and eventually became its greatest aero-engine leader.

While BMW were actively engaged in work on all types of aircraft power-plant, e.g. piston, rocket, turbojet, turboprop, they do not appear to have had over-extended themselves and did a workmanlike job on all their developments. During the war, their failure to have their turbojets (and rockets) in full operational service was due to several times missing the boat, which was caused partly by their slow start and partly by the more difficult development path usually followed. Today examples of their principal turbojet, the 109–003, are preserved in a number of museums, and the name of BMW is again associated largely with automobiles. There is, however, the associated MAN-Turbo company which works in the gas turbine field.

Fig 2.108A This Heinkel He 162 A-2 Salamander (120076) was restored after the war in the markings of 'Yellow 4' of I./JG1 for the National Aviation Museum, in Ottawa, Canada.

Daimler-Benz AG

Evolution of the Daimler-Benz turbojet programme — Description of the DB 6001 or 109–007 (ZTL) — Planned application of the 109–007 — The 109–016 (TL) — The 109–021 (PTL) — Planned application of the 109–021 — Conclusion

In 1885, following two years of experimentation, Gottlieb Daimler took out his historic patent for a *Benzin-Explosionsmotor*, or petrol engine, which he had developed from Otto's improved four-stroke gas engine. Following trials in boats and vehicles, a 2 hp Daimler engine was equipped to drive an airscrew beneath a small balloon built by Dr Karl Woelfert, and on 12 August 1888 this balloon became the world's first petrol-engined aircraft to fly when it lifted off from Daimler's Seelberg factory in calm air. Similar experiments followed, and Daimler engines powered the first airships (Zeppelin, 1900, and the first really practical airship of Henri Julliot in November 1902), but until the First World War, the main emphasis was on vehicle engines.

After the economic difficulties following the First World War, the Daimler engine company merged with the Benz vehicle company to form the Daimler-Benz AG (abbr. DB) in the summer of 1926. In the aero-engine field, DB concentrated on liquid-cooled engines under the direction of Dipl.-Ing. Bergar and Fritz Nallinger. Their first aero-engine of appreciable size appeared in 1932 as the 800 hp DB 600, subsequent engines being designated DB 601, 602, etc. The Second World War years saw DB as one of Germany's major aero-engine companies, with eleven main factories (eight of which were run by such firms as Henschel, Steyr, Avia, Fiat, etc.) making their aero-engines alone. The plant at Stuttgart-Untertürkheim was the main research, design and development centre for all DB products (including vehicles), while pre-production engine work (O-series) was performed at the Berlin-Marienefeld plant (Werk 90). As a further illustration of the company's size, there were, at the end of 1944 for example, 63,500 personnel (including just over 40 per cent foreigners) employed by DB in all its branches and factories.

Add to this a reputation for pioneering and first-class engineering products and one would have expected DB to be in the forefront of turbojet or gas turbine development, but as we shall see, such was not the case, thanks to an initial conservatism. However, before the Second World War and the Daimler-Benz merger, the Daimler Motoren Gesellschaft of Berlin did make some studies of the gas turbine. This is illustrated by their patent (426,009) granted on 3 March 1926 for a *combustion-turbine with auxiliary liquid*, whereby a petrol-air mixture was exploded in a chamber to compress a volume of water which formed a jet to impinge upon the buckets of a power turbine. Very little heat was passed to the turbine, and the curious, though interesting, proposal was a far cry from the pure gas turbine.

Evolution of the Daimler-Benz turbojet programme

When, in the autumn of 1938, Daimler-Benz were approached by Mauch and Schelp of the Technisches Amt in connection with turbojet development, the approach proved fruitless. Largely this was because of the scepticism of Fritz Nallinger, head of development at DB, who refused altogether to be involved with turbojets or jet engines. As we have seen, by 1939 all the other major German aero-engine and some airframe companies were either officially or privately involving themselves in jet engine work, so that DB then thought they might devote a small team to a turbojet project and were persuaded to accept a contract for this purpose. However, the official view was that the DB effort would be best directed along a new line, different from those of other companies, and the suggestion of a turbine-driven ducted-fan engine was taken up.

Fortunately, a first-class engineer had arrived at DB in 1939 with five previous years' experience at the DVL working on exhaust-driven turbo-superchargers and studying the general theories of jet propulsion by turbojet and ramjet engines. This man was Dr-Ing. Karl Leist, and under his direction the DB turbojet project was begun at Stuttgart-Untertürkheim. Paradoxically, in view of the early DB reluctance to enter the turbojet field, their chosen turbojet project involved what was to become one of Germany's most complicated turbojets, since not only was a ducted fan involved, but also a contra-rotating central compressor, both driven from the same turbine. It is not known if this ambitious scheme originated with Schelp and his department alone, or if Leist also held a brief for the scheme. As related earlier, other firms considered or tried turbojets with either a ducted fan or a contra-rotating axial compressor, but not both together. (The Heinkel HeS 10 (109–010), a projected ducted-fan version of the HeS 8 turbojet, was abandoned when the HeS 8 was overtaken. The contra-rotating compressor was considered for the Junkers 109–004 turbojet but not adopted, while Bramo's 109–002, which did use a contra-rotating compressor, had to be abandoned by early 1942 because of great developmental difficulties.)

Yet another novel feature marked out the DB turbojet from other German engines, and this concerned the method of cooling the turbine. Whereas the general German practice eventually involved the use of hollow turbine blades through which cooling air was passed, Leist proposed that his turbine should be only partly immersed in the hot gas stream, leaving about one third of its circumference available for immersion in a cooling air stream. Of course, this meant that less of the turbine would be available for producing work, but this was to be offset by using an abnormally high gas temperature, as high as 1,100 °C being thought possible.

Leist's ideas on such a turbine went back at least to his days with the DVL. From Aachen on 23 February 1934, he filed a patent for his *partially-impinged combustion or gas turbine* (767,078), the specification also listing earlier patents on a similar theme. Figure 2.109 gives a sketch to illustrate the scheme of patent 767,078. In this scheme (which could later have been called a scheme for a PTL or turboprop unit), a large combustion chamber supplied hot gases to drive an axial turbine which drove a compressor and an airscrew, the latter via gear wheels. Air from the compressor passed to the combustion chamber via a pre-heater fitted inside the exhaust tube and nozzle, carrying the hot gases away from the turbine. (This pre-heater would appear to foreshadow later attempts to improve the thermal efficiency of turbojet and gas turbine engines by developing suitable heat exchangers, but no work along these lines at Daimler-Benz is known.) The most important aspect of Leist's scheme for us is the fact that the hot gases were to be admitted to the turbine over only about 36 per cent of its circumference. For the remaining 64 per cent, the turbine blades passed through or were immersed in a cooling air stream drawn from the atmosphere through an intake and baffles.

By the end of 1940, the layout of the Daimler-Benz turbojet was established in its initial form, and development of components began*. The engine was to be in Class II and aimed at a thrust of 1,400 kp (3,087 lb) at 900 km/h (559 mph) at sea level. However, by about June 1942, and possibly much earlier, the engine design had been revised and a less-ambitious thrust of 960 kp (2,117 lb) at 900 km/h (559 mph) at sea level was then the aim, which barely placed the engine in the Class II category. The designation of the engine was ZTL 6001 within the company and 109–007 officially. (The initial layout mentioned above was probably designated 6000.) ZTL, or Zweikreisturbinen-Luftstrahltriebwerk, literally means two-circuit turbojet engine. The two (airflow) circuits of the 109–007 were:

(i) The airflow through the main, central compressor. This provided primary and secondary air for the combustion system, supplying hot gases for the turbine.
(ii) The airflow through the secondary compressor or ducted fan. Most of this air bypassed the combustion and turbine sections and mixed with the

*Earlier, in 1939, Daimler-Benz worked on an ML unit, later designated DB 670 and ZTL 5000.

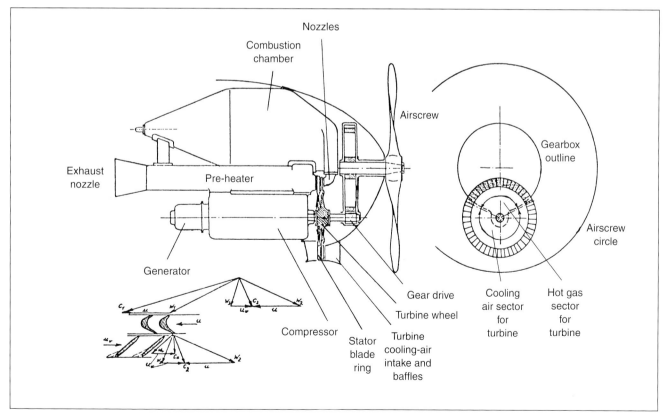

Fig 2.109 Scheme for a partially impinged combustion or gas turbine by Karl Leist (Patent 767,078).

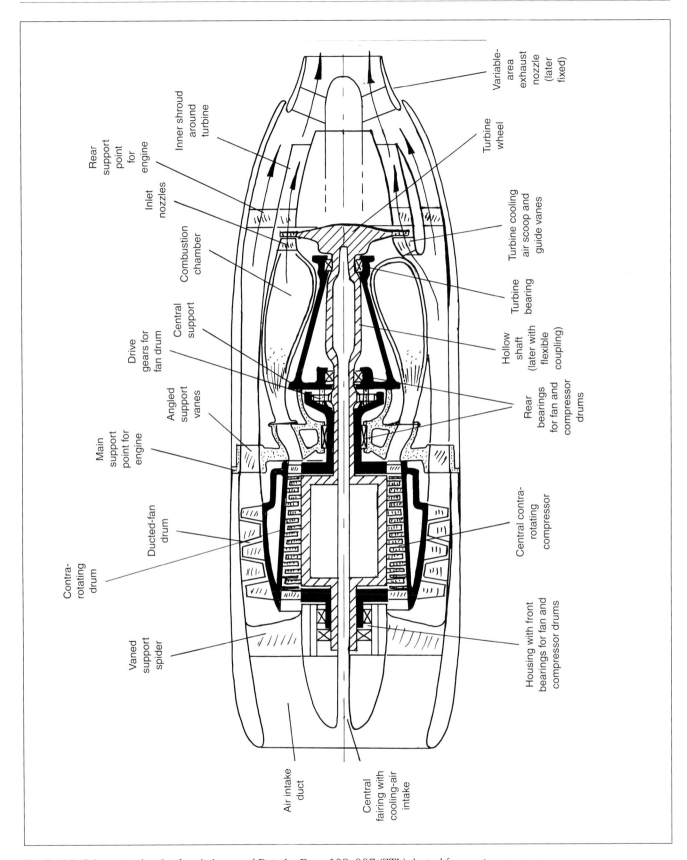

Fig 2.110 Schematic sketch of early layout of Daimler-Benz 109–007 (ZTL) ducted-fan engine. *(Author)*

hot gases only in the exhaust nozzle section. However, part of the ducted-fan airflow entered the turbine section for cooling purposes.

Complication arose from the fact that the ducted fan was built round, and not in front of, the main compressor. Thus, a rotating drum carried on its outside the moving blades of the ducted fan and on its inside what would normally have been the stator blades of the main compressor. This drum was driven at about half turbine speed via gearing from the turbine, while an inner drum, carrying the inner blades of the main compressor, was driven in the opposite direction at full speed from the turbine shaft. To obtain the necessary strength, compactness and lightness of the mechanical components was undoubtedly the major problem facing Leist and his team.

Included in this team were Drs Speiser and Laukhuff, responsible for component construction, and Drs Kamps and Stiefel, responsible for component experimentation. However, the minimum of personnel were assigned to the project, and outside assistance was sought where possible. For example, discussions were held with the AVA concerning compressor and fan blade design, while the firm of Voith was consulted on the mechanical aspects of these components. Progress on the 109–007 was extremely slow, and by the autumn of 1943 only one example of the engine was running.

Starting on 1 April 1943, this example was first used for mechanical testing of components and to tune the lubricating oil circuits. For this purpose, the engine was run without combustion and without its exhaust nozzle, and during the 21 hours' total running time, its rotational speed did not exceed 5,000 rpm – well under half the rated speed. Other important work during these first runs concerned the collecting of data on the compressor at various speeds. To run the engine without combustion, an 8,000 hp electric motor was used.

The first runs of the engine with combustion began on 27 May 1943, using watersprays to cool the turbine wheel and inlet nozzles. Subsequent development included modification to the ducted fan, but it is not known if development reached the stage whereby the water cooling could be dispensed with. Most, if not all, the engine runs were made without the exhaust nozzle attached, but a static thrust *corresponding to* a thrust of 600 kp (1,323 lb) at 900 km/h (559 mph) at 7,000 m (22,960 ft) altitude is reputed to have been reached. Such a performance is, however, doubtful since it is in line with the performance aimed at when the project started, whereas by June 1942 a lower performance was being aimed for. This lower performance would have given 520 kp (1,147 lb), and not 600 kp (1,323 lb) at 900 km/h (559 mph) at 7,000 m (22,960 ft). In any event, by the end of 1943 Daimler-Benz could plainly see the good results being achieved by other companies, and, abandoning their conservatism, they were then anxious to catch up in the turbojet field. Unfortunately, however, it was also plain that their 109–007 was still at an early stage in its development, and its complication indicated that it could not be ready for production for a couple of years. By then the war was expected to be over, one way or the other. Thus, an order from the RLM stopped the 109–007 development by about May 1944 and directed the attention of Daimler-Benz to a Heinkel-Hirth PTL project (described later). At this juncture, Karl Leist, undoubtedly a very disappointed man, left the company and went to work at the Technische Hochschule Braunschweig. Prior to this, it is known that another turbojet engine, designated 109–016, was projected or being planned by DB.

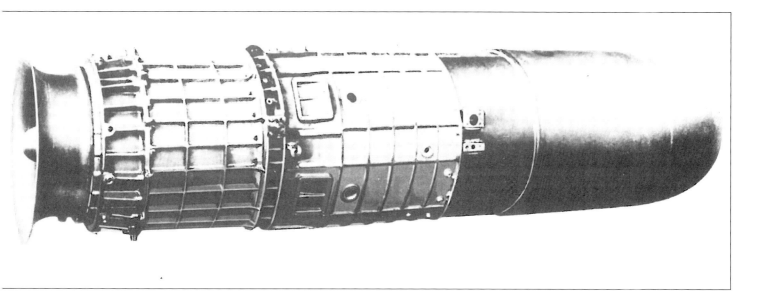

Fig 2.112 A view of the Daimler-Benz 109–007 (ZTL) ducted-fan turbojet engine.

GERMAN JET ENGINES

Fig 2.111 The prototype Daimler-Benz 109–007 (ZTL) ducted-fan turbojet engine, after a hot running test. This complex, ambitious engine is of special interest because of its ducted-fan, contra-rotating compressor and novel method of turbine cooling, but such complexities contributed towards a protracted development. The internal intake lip is clearly visible, dividing the airflow to the outer ducted fan and the inner compressor. The engine is suspended by four pendulum arms and there are electrical thrust-measuring stops on the side girders. *(Daimler-Benz AG)*

Description of the DB 6001, or 109–007 (ZTL)

Although a sectional general arrangement drawing, dated 6 May 1941, is available, much of the complicated detail of the drawing has regrettably faded away, but has nevertheless been enhanced as possible and shown in Figure 2.114. The following description is based on this drawing, but for clarity, Figure 2.110, which is based on a much earlier schematic cross-section (dated 24 February 1941), is also given as the clearest way of showing the layout of the engine. However, some of the features of the schematic, later altered by Daimler-Benz, are mentioned at the end of this description. The reader will note that the drawing and the schematic are shown different ways up. The prototype 109–007 engine under test is shown in the photograph (Figure 2.111).

Intake section:
A cast ring with a 12-vaned support spider was used to support the front bearing housing, to which was attached the central fairing inside the air intake. The light alloy air intake was attached to a lip on the front of the cast ring and, for flight engines, would have had slight rearwards divergence, but for the sole static prototype, the intake walls were turned outwards in the manner of a bell bottom. A streamlined fixed ring connected the support spider vanes at about their mid-point, this ring blending back to the front of the ducted-fan drum. Possibly at a later date a Riedel starter motor would have been fitted inside the central fairing but, on the prototype engine at least, the fairing appears to have housed a starter motor supplied from an external power source. The nose of the central fairing had a small intake which led air for cooling the bearings back by a central

tube through the engine. Inside the front bearing housing was a bearing for the compressor rotor, followed by a bearing for the ducted-fan drum, these bearings apparently being of the self-aligning roller type.

Engine casing and central support:
Following the front cast ring was a cast, externally ribbed casing which supported on its inside surface the stator blades of the ducted fan. This substantial casing surrounded and went the length of both the ducted fan and compressor, and was connected at its rear to the main structural member of the engine. The main structural member began with a channel-sectioned cast ring (which carried two engine attachment points) with internal, angled support vanes which passed through the ducted-fan airflow to connect with the central support. This central support was a casting, mainly cone-shaped, which passed behind the fan and compressor section, back between the combustion chambers and finished in front of the turbine. It carried the rear bearings for the fan and compressor drums, supported the idler gears for the fan drive, supported the combustion chambers and carried at its rear end the bearing for the turbine. The three bearings thus supported appear to have been of the self-aligning roller type. An external engine casing with substantial ribbing led back from the channel-sectioned cast ring to a point level with the turbine where a further two engine support points were fitted. Aft of this, the engine casing was of sheet metal, but most of this was omitted for the preliminary testing of the prototype engine.

Compressor:
The central, or main, compressor was of the 17-stage axial type with all blades revolving. Eight stages of compressor blading were carried on an inner drum revolving at full turbine speed, while nine stages of compressor blading were carried on the inside of a contra-rotating outer drum revolving at about half turbine speed. The compressor supplied about 29 per cent of the total engine air mass flow at a compression ratio of about 8:1. Its efficiency was calculated as 80 per cent, and the temperature of the air at the final stage was about 320 °C.

Largely of aluminium alloy, the hollow compressor drum was of almost-constant outside diameter and was constructed by mounting internally channelled rings (carrying compressor blades) onto a steel drum. Bolted to each end of the steel drum were forged flanges which extended into shafts for the front and rear bearings. The rear shaft was connected via a large flexible coupling to the turbine shaft, and also carried the first driving gear for the ducted-fan drum. The flexible coupling avoiding excessive strain (caused by misalignment, distortion and acceleration) being placed on the gear unit.

Ducted-fan and contra-rotating drum:
The ducted fan was of the three-stage axial type, supplied about 71 per cent of the total engine air mass flow and had a calculated efficiency of 84 per cent. Three stages of rotor blades were attached to the outside of the contra-rotating drum and these were preceded by three stages of wide-chord stator blades attached to the inside of the external casing. In longitudinal section, the wall of the drum was wedge-shaped (to give progressively shorter fan and compressor blades towards the rear), and was of complicated construction to facilitate compressor and fan assembly.

The contra-rotating drum was supported at the front by the first stage of the inner compressor blades, which were attached to a disc in front of the compressor drum. This disc had a hollow shaft which fitted over the compressor shaft and led forward into a bearing. A similar arrangement of compressor blades, disc, hollow shaft, and bearings supported the contra-rotating drum at its rear end, but here the hollow shaft had a bell housing with internal teeth to engage planetary gears. The planetary gears, stub shafts for which were secured inside the engine's cone-shaped central support, meshed also with a central gear on the compressor drive shaft in front of the coupling with the turbine. By this means the contra-rotating drum was driven.

Combustion system:
Four tubular, interconnected combustion chambers were used, although the engine design made provision for a fifth chamber in case it was later found possible to reduce the amount of cooling air to the turbine. The combustion system aimed at an efficiency of 95 per cent, with a maximum combustion temperature of 1,300 °C, a pressure loss of 3 per cent and a chamber pressure of 10 atm.

Curved diffuser ducts led back and outwards from the inner compressor outlet to a welded, annular chamber which supplied air to the combustion chambers. Each

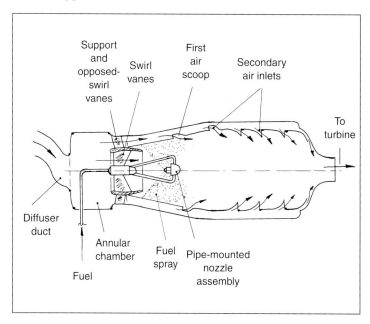

Fig 2.113 Sketch of combustion chamber for Daimler-Benz 109–007 (ZTL). *(Author)*

combustion chamber measured approximately 70 cm long × 25 cm diameter (2 ft 3 in × 9^m in) and consisted of an inner and outer tube (see Figure 2.113). Primary air for combustion entered the inner or flame tube through a ring containing angled swirl vanes. Fuel pipes led from the centre of these swirl vanes to a nozzle (developed jointly by the DB and L'Orange companies) which sprayed fuel against the swirling air, the pipe and nozzle assembly also acting as a flame-holder in a similar manner to the Junkers system. Secondary air passed around vanes, set at an opposed angle to the primary air vanes, and entered the annular space between the flame tube and the outer tube; it entered the flame tube via small scoops and fingers at a swirling angle to enhance mixing with the combustion gases.

Turbine:
The turbine was of the single-stage, axial-flow type with a tip speed of about 375 m/sec (1,230 ft/sec) and an efficiency of 64 per cent at full speed. Along 33 per cent of the circumference, the blades moved through a cooling air stream, while for the remainder of the circumference the blades worked in a gas temperature of between 1,070 and 1,100 °C. The cooling air was diverted from part of the ducted-fan airflow and entered a scoop at the beginning of the inner shroud surrounding the turbine. Inlet guide vanes at the entrance to this scoop directed the angle of the cooling air and were of broader chord than the inlet guide vanes (inlet nozzles) which preceded the turbine blades around the rest of the circumference.

A forged-steel turbine wheel was used and was bolted on its forward face to a flanged, hollow shaft. This shaft was mounted in a bearing (set inside the rear end of the cone-shaped central support) and was internally splined to fit onto a shaft flexibly coupled to the compressor shaft. The turbine blades and inlet nozzles were inserted and were of hollow, nickel-steel construction. In the design of the turbine blades, assistance from Heinkel-Hirth (and possibly other firms) was obtained.

Exhaust system:
From the turbine section to the end of the engine, the structure was of welded sheet-metal construction. From the rear diameter of the turbine wheel, a fixed cone extended rearwards to finish in a tubular section near the end of the engine. Surrounding this cone was a shorter, inner shroud which directed the exhaust gases and cooling air from the turbine into the exhaust nozzle. The main portion of the air from the ducted fan flowed through the annular space inside the engine wall and joined the hot gases just past the inner shroud, the two streams passing out through the exhaust nozzle. At one time, it was proposed to employ a variable-area exhaust nozzle of the ring type originally experimented with by BMW, but the final intention was to use a fixed-area exhaust nozzle. Although the engine was not tested with its exhaust nozzle, it seems likely that the high-speed, hot gas stream and the slower, ducted-fan stream would have only partially mixed, leaving the latter stream to surround the former as they emerged into the atmosphere. If this had been the case, the 109–007 would have been considerably quieter than other engines of comparable power in the same way that modern fan-jet engines offer a reduced noise level.

Auxiliaries:
Very few auxiliaries were fitted to the first test-stand engine, but some of the equipment for later engines was to be as follows. A Riedel 10 bhp starter engine fitted inside the intake central cowling and clutched to the inner compressor shaft would have sufficed for starting. No details are known of the auxiliary gearbox or the method of driving it. The fuel system was to consist of a Barmag gear-type pump giving a full speed delivery of 2,400 kg/h (5,292 lb/h) of J2 fuel, the engine being started on petrol and with two igniter plugs. Included in the system was a barometric fuel control, overspeed governor and throttle valve. For lubrication, a dry sump system with pressure feed to the main bearings was used.

Bearing in mind the early experimental nature of the 109–007, a combination of planned and actual data follow:

Thrust	1,275 kp (2,811.4 lb) static.
	960 kp (2,117 lb) at 900 km/h (559 mph) at sea level.
	560 kp (1,235 lb) at 900 km/h (559 mph) at 6,000 m (19,680 ft) altitude
Speed of inner compressor rotor	12,600 rpm
Speed of contra-rotating drum and ducted fan	6,200 rpm
Weight	1,300 kg (2,886.5 lb) for prototypes but 1,000 kg (2,205 lb) planned for later engines.
Pressure ratio	8.0 to 1
Specific fuel consumption (planned)	Lowest value: 1.05 per hour at 520 km/h (323 mph) at 12,000 m (39,360 ft)
	Highest value: 1.45 per hour at 810 km/h (503 mph) at sea level
Airflow through compressor	8.2 kg/sec (18.08 lb/sec)
Airflow through ducted fan	19.9 kg/sec (43.88 lb/sec)
Specific weight	0.78 approx. (planned)
Overall diameter	1.625 m (5 ft 4 in)
Diameter of inner wall of ducted-fan circuit	0.990 m (3 ft 3 in)
Length	4.725 m (15 ft 6 in)

The preceding description of the 109–007 was based on the DB drawing W601 Nr. 1 9–670 1000, dated 6 May 1944, whereas the schematic cross-section shown in

GAS TURBINES FOR AVIATION

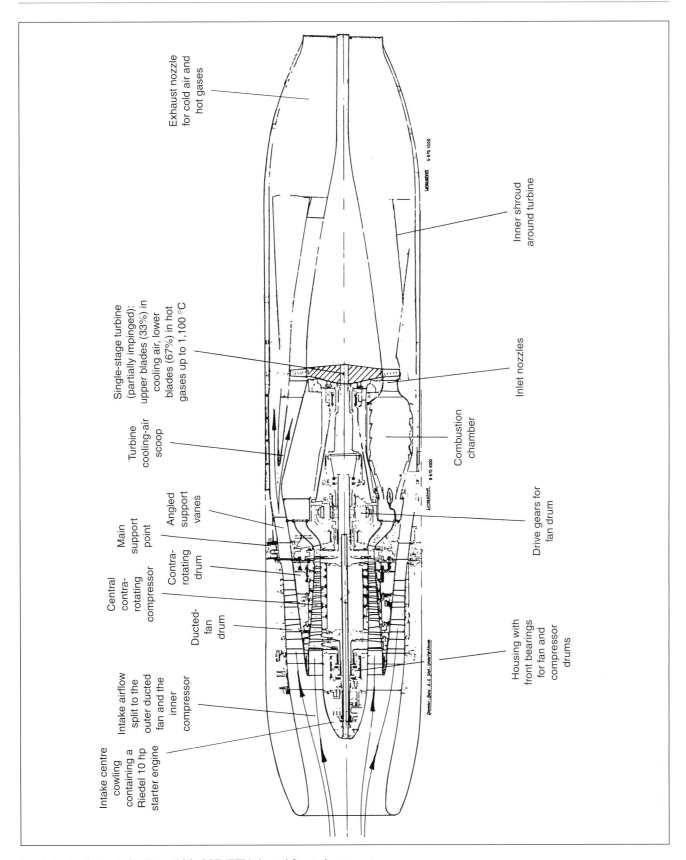

Fig 2.114 The Daimler-Benz 109–007 (ZTL) ducted-fan turbojet engine.

Figure 2.110 dates from 24 February 1941. The following notes indicate the chief points where the early schematic drawing differs from the later, final, drawing, and throw some light on the rethinking which was done on the project once the engine was actually being designed:

(i) At the intake, the streamlined, fixed ring connecting the 12 vanes of the support spider were not originally used.
(ii) The arrangement of fan and compressor blades was different.
(iii) There was no flexible coupling on the rotor shaft between the gear drive unit and the turbine.
(iv) Instead of separate combustion chambers, an annular combustion chamber may have been anticipated, blanked-off at the end where cooling air entered the turbine.
(v) A variable-area exhaust nozzle was originally anticipated instead of a fixed nozzle.

Planned application of the 109-007

Because of the experimental nature of the 109-007 turbojet, few studies were made for its application. The Daimler-Benz company itself put forward a project for a bomber with swept-back wings, a swept-back tailplane with twin end-fins and a single 109-007 engine mounted above the fuselage. This bomber was to be carried beneath a mother aircraft (having six piston engines and twin tail-booms) and released near the target zone. By this means the fast jet bomber was to have the benefit of great range, since most of its fuel was conserved for the homeward flight. Needless to say, this and other similar DB aeronautical projects remained as curiosities, but they illustrate the point that, in those days, the anticipated economy of even a ducted-fan turbojet did not recommend the turbojet engine for long-range duties, and speed was its chief attribute.

Of the regular German aircraft companies, the Arado company appears to have been one of the few (if not the only one) to have received advance details of the 109-007. There were a few projects made to fit the Ar 234 with two of the DB turbojets.

The 109-016 (TL)

By March 1944 the 109-007 had achieved 152 running hours on the test stand, but despite this, the RLM decided its development should cease, and late in 1944 asked Daimler-Benz to look at a projected turbojet designated the 109-016. This was to be an enormous engine for the time, far outside the Class IV size, and the biggest contemplated in Germany. Little more is known of the details of this turbojet other than what can be gleaned from the accompanying drawing*.

*A projected DB swept-wing jet bomber of 70 tonnes was to be powered by a single 109-016 turbojet.

Hundreds of sheets of calculations and some drawings were done. A mock-up of the axial compressor was built and this was of a non-contra-rotating type. Data for the projected 109-016 (also known as DB Project P.100 (Ü-TL-Project)) were:

Thrust	13,000 kp (28,652 lb) static.
Speed	3,500 rpm
Airflow	400 kg/sec (882 lb/sec)
Weight	6,200 kg (13,950 lb)
Specific weight	0.48
Diameter	2.0 m (6 ft 7in) excluding underslung accessories.
Length	6.70 m (21 ft 9 in).

The 109-021 (PTL)

By about May 1944, an order from the RLM stopped all DB work on their own turbojet development and ordered them to concentrate on the development of a turboprop version of the Heinkel-Hirth 109-011 turbojet engine designated 109-021. With the loss of Karl Leist, this new development appears to have been the responsibility of a Dipl.-Ing. Hertzog, and was worked on largely at Daimler-Benz's dispersal at Backnang (Württemberg). Even before May, DB had looked into the possibilities of such a development and had held discussions with the Arado company with a view to developing the 109-021 as the power-plant for a long-range reconnaissance version of the Arado Ar 234 jet bomber (twin engines).

Smaller than the Junkers 109-022 turboprop engine, the DB 109-021 was to consist basically of the Heinkel-Hirth 109-011 turbojet with an extra, third turbine stage added so that 2.50 m (8 ft 2 in) diameter contra-rotating, variable-pitch, three-bladed airscrews could be driven through reduction gearing. For multi-engine aircraft, an alternative scheme was also considered whereby a single, six-bladed airscrew could be driven through modified reduction gearing. At least 50 per cent of the engine's power was to be diverted to airscrew thrust. The reduction gearing for the airscrew or airscrews was to be of the planetary type, giving an airscrew speed ratio of 1 to 5.82 of the engine speed. Control equipment for the engine had to cater for adjustments to the airscrew pitch in addition to governing the fuel flow and setting the variable-area exhaust nozzle to one of three positions.

In other main essentials, the 109-021 was to follow the 109-011, so that the compressor system consisted of a diagonal-flow compressor followed by a three-stage axial compressor, but it is uncertain if an axial inducer was to be retained at the air intake. The annular combustion chamber was to be used with sandwich mixers and 16 of the latest L'Orange duplex-type fuel nozzles, while the turbine blades were to be air-cooled.

More specific details of the DB turboprop are unknown, but one innovation in starting arrangements

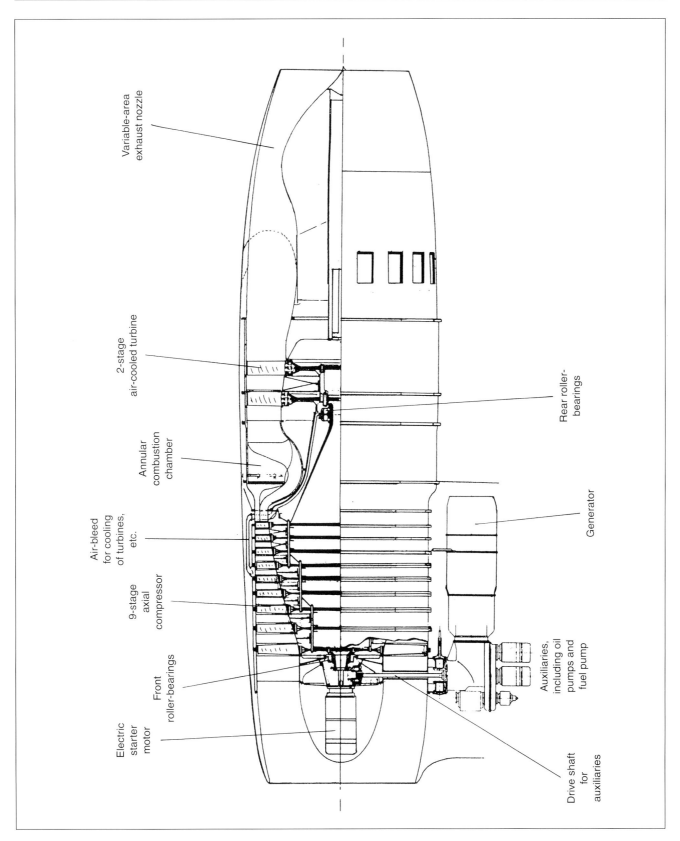

Fig 2.115 The projected Daimler-Benz 109–016 turbojet (TL) was the largest turbojet planned in Germany with the then-enormous static thrust of 13,000 kp (28,652 lb).

was under consideration. In August 1944 the Arado company advised DB that Emil Waldmann (a key man in Helmut Schelp's department in the Technisches Amt) favoured dispensing with an individual starter motor for the 109–021 and to have instead a central, electrical power source in or under the fuselage of the aircraft. From this source would be supplied smaller electric starter motors fitted to the aircraft's engines. However, notwithstanding such refinements, DB had not succeeded in building a single 109–021 turboprop engine by the end of the war. There was possibly also a lack of official faith in the engine ever being developed, since during 1944 the BMW company were asked to develop a 109–003 D turbojet to fulfil the same role as the DB turboprop engine. Estimated data for the 109–021 follow:

Power rating at sea level, static	Jet thrust 500 kp (1,103 lb) Shaft power 1,950 hp (2,400 equivalent shaft hp)
Power rating at sea level and 800 km/h (497 mph)	Jet thrust 585 kp (1,290 lb) Shaft power 2,400 hp (4,750 equivalent shaft hp)
Turbine speed	10,500 rpm
Airscrew speed	1,800 rpm
Weight, without cowling	1,266 kg (2,792 lb)
Weight, with cowling	1,306 kg (2,880 lb)
Diameter	0.910 m (2 ft 11^m in)
Length	3.696 m (12 ft 1 in)
Airscrew diameter	2.50 m (8 ft 2 in)

The above performance figures are as given in an Allied Intelligence report of September 1945, but they are curiously at variance with total power figures given on a DB graph of 12 February 1945, from which the following are examples:

Power rating at sea level and 800 km/h (497 mph)	3,300 equivalent shaft hp
Power rating at 10,000 m (32,800 ft) and 800 km/h (497 mph)	1,620 equivalent shaft hp
Power rating at 12,000 m (39,360 ft) and 800 km/h (497 mph)	1,300 equivalent shaft hp

Again, another Allied Intelligence report gives the estimated performance of 750 km/h (466 mph) at sea level as 790 kp (1,742 lb) jet thrust with 2,000 hp shaft power.

Planned application of the 109–021

As mentioned already, the prime reason for beginning work on the 109–021 turboprop engine was to provide power-plants for a long-range reconnaissance version of the Ar 234, but there were few other planned applications in view. One project study, made by Daimler-Benz, was for a curious transport aircraft which was to carry a fast jet bomber near to the bombing zone before releasing it, thus extending the range of the jet bomber. The transport aircraft had a huge, unswept wing spanning some 94.0 m (308 ft 5 in), and with very long, fixed, streamlined undercarriage legs to permit ground clearance with the jet bomber slung beneath the wing. A simple fuselage with a conventional tail unit was fitted to the wing, while above the wing were either four or six 109–021 turboprop units mounted on swept-forward pylons. (The jet bomber carried beneath had swept-back wings and a butterfly tail unit, and was powered by two BMW 109–018 turbojet engines.) A completely different application was envisaged by the Focke-Wulf company, which put forward a project for a fighter-bomber powered by a single 109–021.

Conclusion

Although one of Germany's finest piston engine manufacturers, Daimler-Benz failed to make much progress in the development of jet engines. After about five years of work, at a low tempo, the only concrete result was a single prototype of the complicated 109–007 turbojet engine, and even this development was officially stopped in view of the expected long development period ahead. Subsequent work does not appear to have been pursued with any vigour or concrete result.

The very slow pace of subsequent development was later alleged to be due to a lack of manpower, and yet there was no shortage of manpower in the Daimler-Benz organisation. As for facilities, these also were good. Although bombing attacks had heavily damaged the Untertürkheim works by September 1944 (70 per cent of the factory and many of the test beds were knocked out), dispersal had been under way since 1942, and most of the aero-engine section was dispersed by the spring of 1944. The longest delay (three to four months) was caused by a heavy raid in September 1944.

While all the evidence points towards a general lack of interest and effort by Daimler-Benz in the turbojet field, it is curious that their main project, the 109–007 turbojet, should have been of such an advanced concept – with its contra-rotating compressor and ducted fan, to mention but two features. How far the adoption of this concept was due to Prof. Leist, and how far to an official request, is not known. Although the 109–007 was doubtless one of the most interesting German turbojet engines, the revised performance being aimed at does not appear spectacular in comparison with German engines of similar power. In particular, the specific fuel consumption appears disappointing for a ducted-fan engine, while the weight and frontal area appear high for the engine's thrust.

Dr-Ing. h.c. F. Porsche KG

The expendable 109–005 (TL) — Description of the 109–005 — Conclusion

During 1930, Dr Ferdinand Porsche formed the independent design office bearing his name at Stuttgart-Zuffenhausen. Porsche, most famous for his designs of racing cars and the Volkswagen (people's car), had behind him a long and successful career in engineering. Born on 3 September 1875, he went to the Technische Hochschule in Vienna in 1894, and four years later became manager of the test department of the Bela Egger company (later Brown Boveri) and first assistant in the calculating section. Following work at other firms, Porsche designed his first air-cooled aircraft engine (regarded as the precursor of his Volkswagen engine) in 1912. His time was then largely used in designing cars for such firms as Daimler and Steyr in Austria until his own design office was formed.

Initially, the Porsche KG had only about thirteen staff, but by the time the war ended in 1945, this had risen to about forty staff and 120 workers, by which time the office had been dispersed from Stuttgart to Gmünd and Rheinau. From 1940, design work, and a certain amount of research, centred on armoured fighting vehicles, and Porsche's name is associated particularly with the Leopard, Tiger, Ferdinand, and Mouse tanks. But what of Porsche's activities in the gas turbine field? Most interesting was the work on gas turbines for armoured fighting vehicles, and this is discussed separately in Section 3. In addition, there was work on at least one project for a turbojet engine, which is described next.

The expendable 109–005 (TL)

As Allied forces moved across France in the second half of 1944, the Germans gradually lost their sites from which to launch the Fi 103 (V1) flying bombs against London. Forced back to central Holland, the flying bomb units could only attack targets such as Antwerp because of the limited range of the Fi 103. To use the flying bomb over ranges greater than the normal 240 km (149 miles) the unsatisfactory expedients of air launching from an He 111 bomber or increasing fuel load at the expense of warhead weight were resorted to. Another solution was to improve the fuel consumption of the Argus 109–014 pulsejet power unit (described in Section 7) or replace it altogether with a new, more efficient power unit. In this last respect, the Technisches Amt saw a case for a small turbojet unit which would not only increase the range of the Fi 103 because of better fuel consumption, but would also improve on other shortcomings of the pulsejet missile. Such shortcomings included the need for a special launching catapult, low speed and low operational ceiling, all of which contributed to the loss of a large percentage of missiles. Obviously, a prime characteristic of the missile turbojet had to be maximum simplicity and economy of manufacture, since the power unit was to be expendable, but it could never compare with the pulsejet when it came to simplicity.

Probably around October 1944, the official order was given for design studies to be made of expendable turbojets, but only the projects of BMW (P.3307) and Porsche are known. In any event, the Porsche design, which received the official designation of 109–005, appears to be the only serious effort in the field. Whereas the Argus 109–014 pulsejet had a static thrust of 350 kp (770 lb), the 109–005 turbojet was designed for a static thrust of 500 kp (1,102 lb). Although the turbojet was designed by Dr Porsche and his team, the project was given to Dr Max Adolf Mueller in the last months of the war in order that he could bring his greater gas turbine experience to bear and incorporate all the latest design practice known to him. As related elsewhere, Mueller's experience included work on jet engines for Junkers and Heinkel-Hirth, while in February 1945 he replaced Dr Alfred Müller on the armoured fighting vehicle gas turbine project (see Section 3). The following describes the 109–005 turbojet design as it stood at the end of the war, but the design work was not finalised and no parts were constructed. Some test work, however, such as with the combustion system, was performed.

Description of the 109–005

Most unfortunately, no drawings are available of this interesting turbojet, but its basic layout consisted of an axial compressor, eight individual combustion chambers, a single-stage turbine and a fixed-area exhaust nozzle. One detects in the design the considerable influence of Mueller from his experience with Heinkel-Hirth, one surprising feature perhaps being the inclusion of an axial inducer preceding the compressor. The whole engine was to be manufactured in 130 to 140 man-hours, or a little more than one fifth the time required for BMW and Junkers production engines.

Intake section:
The air intake duct diverged towards the rear to form a diffuser to reduce the velocity of incoming air from 180 m/sec (590 ft/sec) down to 140 m/sec (459 ft/sec) at the first compressor stage. Well forward of the compressor, the intake duct housed an axial flow inducer

which was geared to run at 4,500 rmp, or slightly less than one third of the compressor speed. Whilst, in an expendable engine, one might expect to dispense with refinements such as an inducer, four reasons were given in its favour, namely:

(i) It produced an initial swirl to the air in the direction of compressor rotation.
(ii) It heated the air and thus combated icing in the first row of stator blades.
(iii) The increase in inlet air temperature caused by the inducer permitted higher absolute velocities in the initial compressor stages without exceeding the critical Mach number.
(iv) The increase in pressure and temperature in the initial compressor stages produced by the inducer multiplied through the compressor and probably permitted one less compressor stage.

Compressor:
The number of stages used for the axial compressor is unknown, but it was designed for an efficiency of 78 per cent and a compression ratio of 2.8 to 1 under static thrust conditions and for 50 per cent reaction. A Mach number of 0.78, constant through all stages, was chosen, since, although a higher Mach number would have increased efficiency, the lower value permitted the use of blades pressed from sheet metal or cast with less accuracy. Both rotor and stator blades had a thickness of 12 per cent of chord, their profiles being a modified NACA section with increased curvature towards the trailing edge. The compressor blades were also twisted in order to maintain a constant angle of attack along their length, while for the rotor blades the tip velocity was the same for all stages, partly to ease manufacture.

Combustion system:
There were eight individual combustion chambers enclosed in an annular shell. Each chamber consisted of a fuel nozzle and an annular pilot baffle inside a primary air cylinder, the end of which projected inside the front of a larger secondary air cylinder. Combustion began in the vortex of the annular baffle and was completed in the primary cylinder, and the resulting hot gas joined the secondary air entering through the annular space between the two cylinders. The air velocity at the pilot flames was 10 m/sec (32.8 ft/sec), the average gas velocity in the combustion system was 60 m/sec (196.8 ft/sec), and on single chamber tests, the combustion efficiency was 94–95 per cent with a 60 atm. fuel injection pressure. Variable-area, spring-loaded fuel nozzles were planned supplied with J-2 fuel by a Barmag gear-type pump.

Turbine:
The air-cooled turbine blades were made from chrome nickel steel sheet and were similar to the blades for the BMW 109–003. However, instead of a purely mechanical method (pins, etc.) of attaching the blades to the turbine wheel, the blades were welded on after their V-shaped feet had been fitted into cut-outs in the wheel rim. The weld, it was hoped, would enable stresses to be spread over a much wider area. Also, the welding of blades enabled the use of a turbine wheel made from two sheet-metal discs which were flared out at the periphery. Such a sheet-metal turbine was very economical on material and manpower so that, in the event of a failure during testing, the entire wheel could be discarded and replaced.

From 800 °C in the combustion chamber, the gas temperature dropped to 660 °C leaving the turbine nozzles (or guide vanes), and then 610 °C leaving the turbine rotor blades. Unlike general German practice in cooling hollow turbine blades, cooling air was not to be tapped from the compressor but drawn from an independent source (presumably scooped from the atmosphere), since the suction action inside the revolving blades was considered quite adequate, especially in view of the short life required on a missile.

Exhaust nozzle:
Since the engine was to operate through only a narrow range of flight velocities and altitudes, a variable-area exhaust nozzle was not considered necessary. The use of a fixed nozzle therefore happily dispensed with the problems which the automatic operation of an adjustable nozzle would have posed.

Data known for the 109–005 are:

Thrust	500 kp (1,102 lb) static
	340 kp (750 lb) at 650 km/h
	(404 mph) at 4,000 m (13,120 ft) altitude
Speed	14,500 rpm
Weight, basic unit	180 kg (397 lb)
Complete power unit	200 kg (441 lb)
Pressure ratio	2.8 to 1
Specific fuel consumption	1.38 per hour at static thrust
	1.7 per hour at 650 km/h at 4,000 m altitude
Specific weight	0.40
Diameter	0.650 m (2 ft 1¹ in) } Fully cowled,
Length	2.850 m (9 ft 4 in) } and approximate

Comparative performance figures for the Fi 103 (V1) flying bomb with the normal Argus 109–014 pulsejet or with the projected Porsche 109–005 turbojet are as follows. As far as is known, the flying bomb's dimensions, warhead and fuel load were to remain the same in either case.

	With Argus 109–014	With Porsche 109–005
Maximum operational speed	645 km/h (400 mph)	650 km/h (404 mph)
Maximum possible speed	800 km/h (497 mph)	900 km/h (559 mph)
Ceiling	3,000 m (8,840 ft)	4,000 m (13,120 ft) or more
Range (ground launched)	240 km (149 miles)	700 km (435 miles)

Conclusion

While Ferdinand Porsche and his company enjoyed the very best reputation in engineering design, notably with cars and the Tiger tank, their experience in the gas turbine and turbojet field was begun at a late date, was limited and therefore had little chance to develop or be tested. While there is some evidence that Porsche designed various small turbojet engines, only details of the 109–005 are known, and as we have seen, this engine was given to the specialist Dr M.A. Mueller to complete. However, American investigators after the war were critical of the design and some of Mueller's ideas, and, for example, were sceptical of the fuel injector system in the 109–005. Upon his release from Camp Lager Schlatt, Mueller completed a detail assembly drawing* of the 109–005 for the Americans, but the subsequent history of the engine (if any) and of Mueller is unknown. From the figures given above, we can see that the flying bomb equipped with the small expendable turbojet would have been just as susceptible to Allied defences as it was with the pulsejet. However, the prime aim of increasing its range would have been achieved, and the launching procedure would have been simplified and made more flexible. Today the Porsche KG again operates from Stuttgart-Zuffenhausen in the field of cars, but Ferdinand Porsche died in January 1952.

*This drawing has proved elusive.

Section 3

Gas Turbines for Land Traction

During 1943, some serious thought was given in Germany concerning the possibilities of using the gas turbine for overland traction purposes. At that time, the only land vehicle which had been driven by a gas turbine was an experimental locomotive which the Swiss Federal Railways had ordered at the beginning of 1939. The power unit, built by Brown Boveri of Baden (Switzerland), delivered 2,200 hp to drive the locomotive through an electrical transmission system. Rail trials began after acceptance tests on the power unit in September 1941, but these trials were limited through a wartime shortage of fuel. Back in Germany, it was not until 1944, with the Second World War grinding towards its conclusion, that an official interest was shown in developing the gas turbine for land traction purposes. The official organ willing to sponsor such a development was the Heereswaffenamt (Army Ordnance Board), and its interest lay not in locomotives but in the military armoured fighting vehicles (AFVs). In the forefront of such vehicles were, of course, the tanks and self-propelled assault guns, and it was for these that a gas turbine power unit was required, if it could meet certain needs.

The reasons for initiating such a radical project at a time when Germany's resources were declining were various, but a prime reason was that the gas turbine could operate on a much lower grade fuel than the internal combustion engines then in service, and such a consideration was of great importance in view of Germany's critical fuel position. Most fuel was by then being made either synthetically or from very crude oil imported from Rumania, and the production of only low-grade fuel meant a saving of time and effort with a consequent increase in production. Another reason for interest in the gas turbine was that it offered more power for a given size and weight (and in any case, more power than was then available with developed, conventional tank engines), and therefore appeared suitable for the heavier, high-performance vehicles which were being sought to counteract the numerical superiority of Allied armour. More incidental attributes of the gas turbine, compared with the piston engine, were that it was simpler and cheaper, smoother-running, required no auxiliary cooling system and that the problem of air filtration was, if not eliminated, at least simplified. In connection with the last point, it is worth noting that German vehicles were too often rendered unserviceable through the abrasive effects of sand and grit, because, when it came to supplying spare parts to such theatres as the Eastern Front and Africa, the German supply system was stretched and gave poor results.

Overall responsibility for the gas turbine project was given to Dipl.-Ing. Otto Zadnik (of Porsche KG) who worked with four engineers, named Weitzel, Hermann, Koepell, and Jung, in a small office at Rheinau on the Austrian-Swiss border. Zadnik, who was primarily an electrical transmission specialist, had the task of designing the transmission gear and overlooking the entire installation in the vehicle. To ensure that the complete installation would meet the specified requirements, he had to co-ordinate the work of the various specialists concerned with the component parts and control equipment for the gas turbine. The principal specialist was Dr Alfred Müller, who was the chief designer of the gas turbine power units in association with the Kraftfahr Technische Versuchsanstalt der SS (Motor Technical Research Establishment of the SS) at St Aegyd, Niederdonau, near Vienna.

Dr Müller had behind him a wide experience of turbine and turbo supercharger development for aircraft, and had begun his studies at the Stuttgart Technisches Hochschule. There he made a special study of supercharger turbines (at the suggestion of Prof. W. Kamm) for his Doctor's degree in 1936, and he became convinced of the future of the gas turbine for all forms of travel. By 1937, Müller was working as an independent consultant for BMW on experimental, hollow turbine blades and other turbojet problems, and it was in 1943 that he unsuccessfully tried to interest the military in the possibilities of the gas turbine for powering tanks. In January 1944, he finished designing an excellent turbo supercharger, known as the BMW type 801, which was suitable for the piston engines of various companies, and his next major task was the AFV gas turbine.

The first projects

In a meeting with the Heereswaffenamt in Dresden on 30

June 1944, the administrative background was arranged for Dr Müller and his assistant, Dipl.-Ing. Kolb, who were to design a unit with a power output of 1,000 hp, an air mass flow of 8.5 kg/sec (18.74 lb/sec) and a combustion chamber temperature of 800 °C. Although it was realised that the land gas turbine was different in many ways from the aircraft turbojet, it was experience and information from work on the latter type that formed the basis for much of the work on the AFV gas turbine.

Figure 3.1 shows five basic gas turbine schemes considered in Germany for land traction purposes, and these will be enlarged upon at the appropriate points in the text.

The first scheme (see Figure 3.1A) was begun by Müller in July 1944, and comprised a diagonal compressor followed by a five-stage axial compressor coupled direct to a two-stage turbine to provide hot gases for a third turbine. The third turbine was mounted coaxially, but without mechanical connections, behind the other two, and provided the power drive to the gears of the vehicle. This design (shown in Figure 3.2) suffered from the drawback that when the load was removed (e.g. when changing gear), the third, working, turbine would dangerously overspeed. This led to two technically difficult solutions. Either the working turbine had to be heavily braked during its unloaded periods, or the continuously produced hot gases had to be diverted from the power turbine during such periods. A free working turbine had been adopted in the first instance because of its favourable torque characteristics, since it could be stalled during an overload and immediately regain full power when the load was reduced by virtue of the fact that the compressor remained running at all times. This design was submitted on 12 August and was criticised as being too expensive and unsuitable for rotating fuel burners, which were considered advantageous for the injection of low-grade fuels.

A new design was therefore called for in which features such as the diagonal-flow compressor and drum rotor were to be adopted from the Heinkel-Hirth 109–011 A turbojet engine in order to facilitate development. It was also stipulated that, for reasons of economy, only two bearing supports, instead of the six in the first design, should be used, and this, of course, meant abandoning the scheme of a separate working turbine. By 14 September 1944, the second design, shown in Figure 3.3, was ready for examination and had the prime feature of all three turbine stages being mounted on a common drum rotor, necessitating a method of maintaining torque outside the turbine (see Figure 3.1B). Although an electrical drive would have been ideal, the shortage of copper in Germany ruled this out, and the drive from the gas turbine was to be transmitted through some form of hydraulic torque converter, such as the Thoma hydrostatic or Föttinger hydrodynamic type. Since rotating fuel burners had been specified, these were incorporated in the design, although a glance at Figure 3.3 will show that fixed burners could be used as an alternative. In order to prevent centrifugal forces setting up an estimated pressure of 350 atmospheres in the fuel at the rotating burners, a device was designed which interrupted the hydraulic column of the fuel upstream of the burner and thereby reduced its effective length. The rotating burners avoided hot spots on the turbine stator blades, but did not, of course, do the same for the turbine rotor blades. However, as already mentioned, their adoption was intended to facilitate the use of the lowest-grade fuel, a prime reason for beginning the project at all. Although a Heinkel-Hirth type of diagonal-flow compressor had been designed for, a nine-stage axial-flow compressor was also planned for substitution at a later date because of its considered greater efficiency. Initially, a speed for both types of compressor could not be reconciled with the same turbine speed, but a speed of 14,000 rpm was finally settled on and the development of both types of compressors was ordered on 29 September 1944.

Another complication arose at this time when it was specified that the compressors should also be suitable, after modification, for aircraft turbojet engines, and accordingly the air mass flow of the AFV unit was increased to 10 kg/sec (22.05 lb/sec). (A considerable increase in this air mass flow would, of course, have been necessary for aircraft use.) The design team now decided to adopt a modified design of annular combustion chamber (already employed by the BMW 109–003 A turbojet), together with stationary fuel burners, and to concentrate on the axial type of compressor. A more intensive examination of the rotor design indicated the need to use a third bearing, positioned between the compressor and turbine sections, to avoid excessive bending of the rotor under shock loads (such as when running over a mine), which could result in destruction through the rotating blades touching the casing. All these fresh considerations led to a third design, which was the final design of the coaxial turbine layout (still basically as B in Figure 3.1), and which was designated GT 101. The basic design of this unit was completed by mid-November 1944, and it was the first to receive serious consideration for installation in a tank.

The original tank which was to be fitted with the GT 101 unit was the Panzerkampfwagen VI Tiger I (sd.kfz. 181), but it was soon ascertained that there was insufficient space for the new installation, which, although of less volume, was of greater length than the conventional engine. However, on 25 September 1944, the Panzerkampfwagen V Panther (sd.kfz. 171) became the chosen vehicle, since this was supposed to be the tank upon which all subsequent production was to concentrate (see Figures 3.4 and 3.5). Nevertheless, the first experimental trials were to be made with a Jagdpanzer VI Jagdtiger (sd.kfz. 186), since an experimental tank of this type was available at the Porsche works.

With the project settled thus far, a closer examination

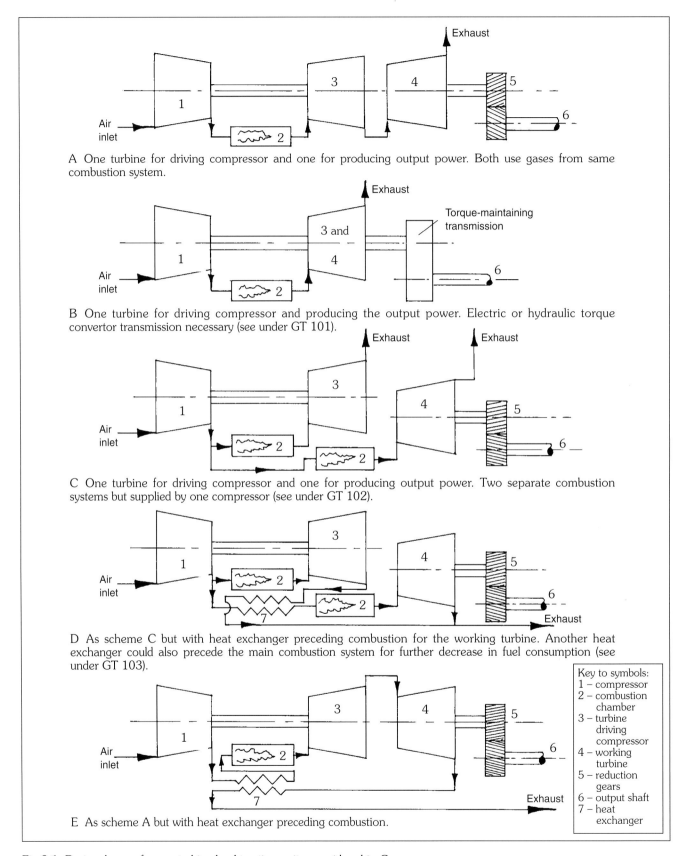

Fig 3.1 Basic schemes for gas turbine land traction units, considered in Germany.

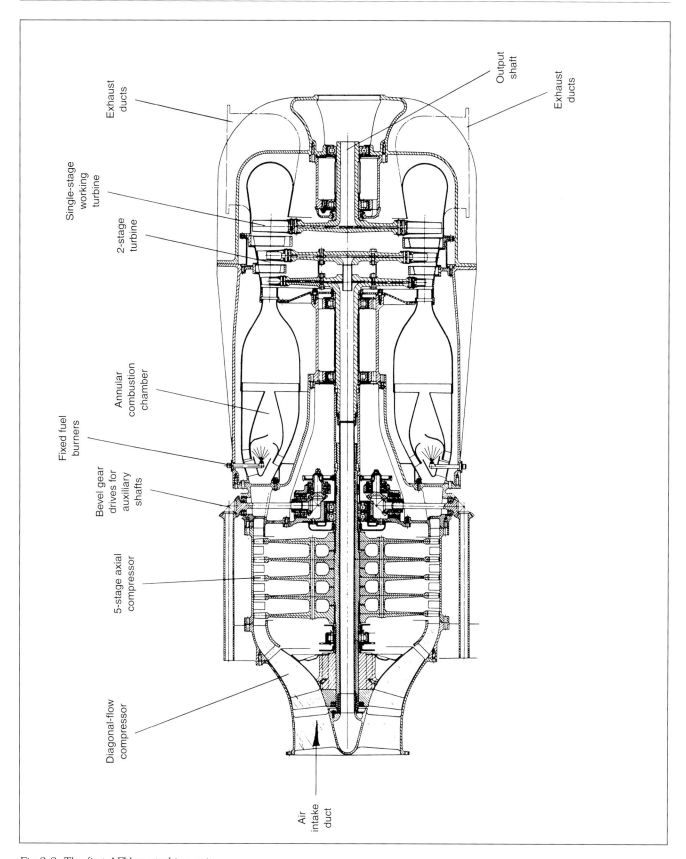

Fig 3.2 The first AFV gas turbine unit.

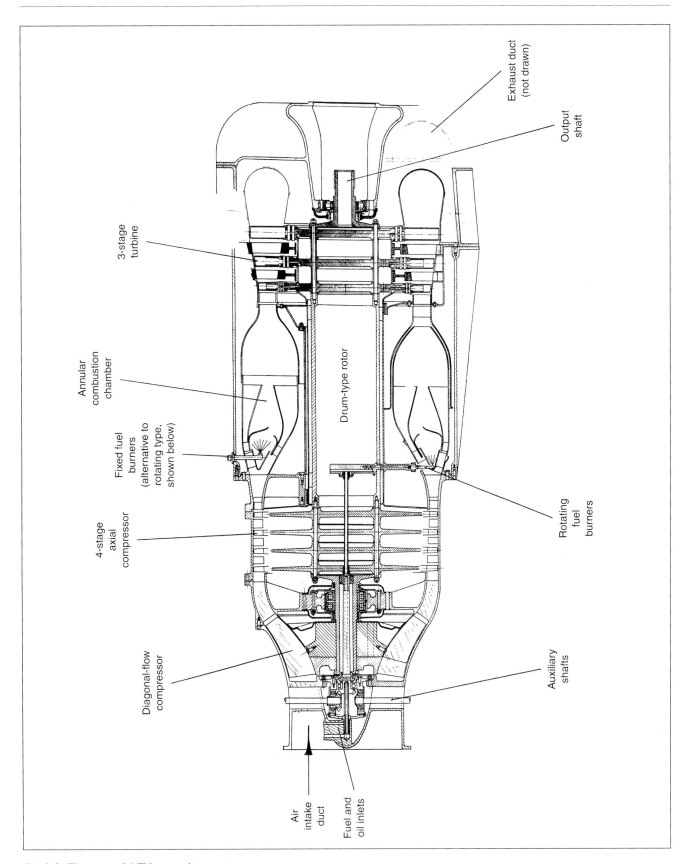

Fig 3.3 The second AFV gas turbine unit.

of its potential was possible. Of particular interest was the specific power output (or the ratio of engine power to vehicle weight), which the proposed GT Panther tank would offer, since the higher this ratio the more manoeuvrability and acceleration the vehicle would possess, with consequent tactical advantages. The following approximate figures give an indication of the specific power outputs of some principal Second World War tanks.

	Weight (tons)	Specific power Output (hp/ton)	Max. speed (km/h)
Russian T.34	26	20	53
German Panther	46	13.5	54
American Sherman	30	13	39
German Tiger	60–70	11–9	45
British Churchill	38	10	26

From these figures can be seen one of the reasons why the Russian T.34 proved so troublesome to the Germans when it was first encountered on the Eastern Front in the late autumn of 1941. Accordingly, the Panther tank was strongly influenced by the T.34, but the Germans could not achieve the requisite specific power output since the most powerful, fully developed, engine available was the 700 hp Maybach HL 230 P30 gasoline piston engine. From this developed horsepower had to be deducted some 80 hp for cooling fans and auxiliaries, which left 620 hp to give, for the Panther, about 13.5 hp/ton. To increase this figure by using a larger piston engine would mean starting the vicious circle of a larger engine compartment and larger fuel tanks, which meant more space to armour, resulting in greater weight, heavier running gear, poor running qualities and more area prone to attack. (By using petrol injection, it was hoped to increase the power of the Maybach engine to 900 hp, but this was not then available.) With the gas turbine installation, it was anticipated that the specific power output of the Panther would be doubled, and at 27 hp/ton would greatly exceed all AFVs. On the question of maximum speed, the Services would have liked, and could have got, 60 km/h (37.26 mph) on metalled surfaces, but the resultant shocks (which are proportional to the square of the speed) would have subjected the running gear and tracks to heavy wear and frequent breakdowns. These last factors were amongst those most essential to minimise, and accordingly the General der Panzerkampfwagen (Inspector-General of Tanks) limited the maximum speed to 40 km/h (24.84 mph), the continuous speed to 30 km/h (18.63 mph) and the convoy speed to 25 km/h (15.52 mph), although in practice speeds were lower than these. For designing purposes, the GT Panther was planned to have the same operating speeds as the conventionally powered Panther, e.g. 54 km/h (33.53 mph) maximum speed.

Fig 3.4 Panzerkampfwagen V G Panther (Sd.Kfz. 171) tank. In September 1944, this version of the Panther was chosen to be powered by a gas turbine unit to increase its power-to-weight ratio. With its standard piston engine, the Panther had already proved to be a formidable weapon. *(Royal Armoured Corps Centre, Bovington)*

Fig 3.5 A Panzerkampfwagen V G Panther (Sd.Kfz. 171) tank. The crew take a break but keep a watchful eye on the sky against attack. *(Bundesarchiv, Koblenz)*

The GT 101 installation

Although some difficulty was initially experienced in accommodating the GT 101 unit in the Panther engine compartment, by allowing the exhaust end of the gas turbine to project through an opening in the sloping rear wall of the tank, the installation was found to be ideal and was presented in the form shown in Figure 3.6 on 9 November 1944. In effect, GT 101 was little more than a normal aircraft turbojet engine connected to an appropriate transmission system, but nevertheless having its own peculiarities. For example, the exhaust velocity of the gases at the turbine are entirely written off as a loss, whereas this velocity, in the case of an aircraft turbojet, is the means of producing a useful thrust. For the tank unit, therefore, a divergent diffuser was arranged behind the turbine to partly convert the exhaust gas velocity into pressure and minimise exit losses. This diffuser allowed the use of a suitable blade length for the last, low-pressure stage of the turbine.

Poor torque characteristics of the GT 101 unit, caused by connecting the working turbine to the compressor, necessitated the use of a somewhat complicated transmission system which was designed by the firm of Zahnradfabrik of Friedrichshafen (ZF) and required the use of a cooling unit weighing 40 kg (88.20 lb) and having an oil capacity of 4.5 litres (0.99 gallons). This hydraulic transmission gave three self-changing torque-converter steps, and the change-speed gear had six road and six cross-country speeds with considerable overlap. Since no synchromesh was adopted, an automatic gear-change mechanism was necessary which cut in the next lowest gear when the engine speed decreased to 80 per cent because of overload, and cut in the next highest gear as soon as the engine load was reduced. This operation was initiated by a governor and electrical contacts. Further complication arose from the fact that the GT 101 unit provided no torque below a speed of 5,000 rpm. To avoid difficult vehicle starting, therefore, a small generator was driven from the engine to operate an electrical clutch only when the engine had attained sufficient speed. If necessary, the automatic control could be over-ridden by the driver's hand control.

The GT 101 unit promised to be very effective used as an engine brake, owing to some 2,600 hp being absorbed by the compressor when running at full speed. As soon as the tank tried to over-ride the transmission when travelling downhill, the engine braking effect was automatically produced by the transmission. Another attribute of the design derived from the heavy mass energy of the rotor, which amounted to 325,000 m.kg (2,350,530 ft.lb) at 14,000 rpm, and was equivalent to the entire Panther tank travelling at 43 km/h (26.7 mph). Thus, the rotor acted as a flywheel and formed an energy accumulator from which acceleration forces could be supplied when travelling in rough country.

Principal characteristics for the GT 101 gas turbine are as follows:

Turbine power	3,750 hp
Compressor power absorption	2,600 hp
Output shaft power	1,150 hp
Full load speed	14,000 rpm
No load speed	14,500 rpm
Weight	450 kg (992 lb)
Pressure ratio	4.5 to 1
Specific fuel consumption	450–500 gm/hp/h (0.992–1.1025 lb/hp/h) at full load*
Airflow	10 kg/sec (22.05 lb/sec)
Combustion gas temperature	800 °C

*or 430 gm/hp/h (0.948 lb/hp/h) at 70 per cent full load

Since the figures for fuel consumption were at least 90–100 per cent more than for a good piston engine, some improvement was considered essential, and this, it was hoped, could be obtained by the use of a heat exchanger which would reduce consumption by about 30 per cent. Notwithstanding this, the problems presented by the poor torque characteristics of the unit were considered beyond a fully satisfactory solution, and the design was superseded by the GT 102 installation. Detailed description of GT 101 is unnecessary since it was used basically unaltered to form the compressor or air-producing group of the GT 102 installation described next. Because the GT 101 unit had much in common with GT 102, it was still planned to fit it experimentally in the 71.7 ton Jagdtiger tank with almost standard transmission gear in order to obtain preliminary data.

The GT 102 installation

In December 1944, the idea of the separate working turbine which was considered in the first AFV gas turbine scheme was returned to in the design known as GT 102. Unlike the first scheme, however, the working, or power, turbine was no longer coaxially behind the compressor turbine but removed as a self-contained unit (see Figure 3.1C). GT 102, therefore, consisted of a compressor group having nine compressor stages driven by three turbine stages (essentially the complete GT 101 unit) and a separate combustion chamber for a two-stage working turbine which connected with the tank transmission. Seventy per cent of the compressor air was supplied to the combustion chamber of the primary turbine driving the compressor, whilst the remaining 30 per cent of the air was diverted via regulator valves and ducting to the combustion chamber of the secondary or separate working turbine.

It will be recalled that a major problem with the July 1944 design was that of controlling the free working turbine when the load was removed at such times as when changing gear. With GT 102, it was believed that the separate working turbine could be prevented from over-speeding fairly easily by controlling the regulating valves in the air supply ducting from the compressor to the secondary combustion chamber, and no elaborate hot gas diversion or turbine braking arrangements were called for. The new design was expected to provide a stalled torque of about 2.33 times the torque at maximum speed with the working turbine pulling like an electric motor, being excellent for climbing out of ditches and requiring only a simple gear transmission with infrequent gear changing. On the other hand, acceleration would be slower than with GT 101 because the lighter working turbine had less kinetic energy and therefore less surplus power, but it could be run right down without affecting the compressor unit and would then accelerate immediately the load was reduced. While the installation in the Panther proved more difficult than with GT 101, a preliminary solution was arrived at, which is shown in Figure 3.7.

Although the compressor had to deal with an airflow of 10 kg/sec (as with GT 101), the compressor turbine had to deal with only 7 kg/sec, since 3 kg/sec were bled off immediately after the compressor and ducted to the combustion chamber of the separate working turbine. This opened up the possibility of either shortening the main exhaust diffuser or shortening the blades of the third stage of the turbine. To produce a shorter compressor unit was most desirable in view of the tank engine compartment shape, and various means of achieving this were looked into. The idea of installing the combustion chamber above the compressor, so that the compressor unit was shortened by the length of the combustion chamber, was rejected because of bad flow conditions and because the critical war situation demanded rapid development. There was also the idea of splitting the compressor unit into the separate units of compressor, combustion chamber and turbine, interconnected with ducting and so forth, but, although this resulted in the desired installation shape, it also required too much volume. Other means of shortening the unit also presented themselves, but it was decided to develop these as a secondary project (GT 102 Ausf. 2, described later).

Description of the GT 102 unit

The compressor unit is shown in Figure 3.8, and the separate working turbine (without its combustion chamber) in Figure 3.9. For the complete installation, reference should again be made to Figure 3.7, while the following describes the GT 102 unit.

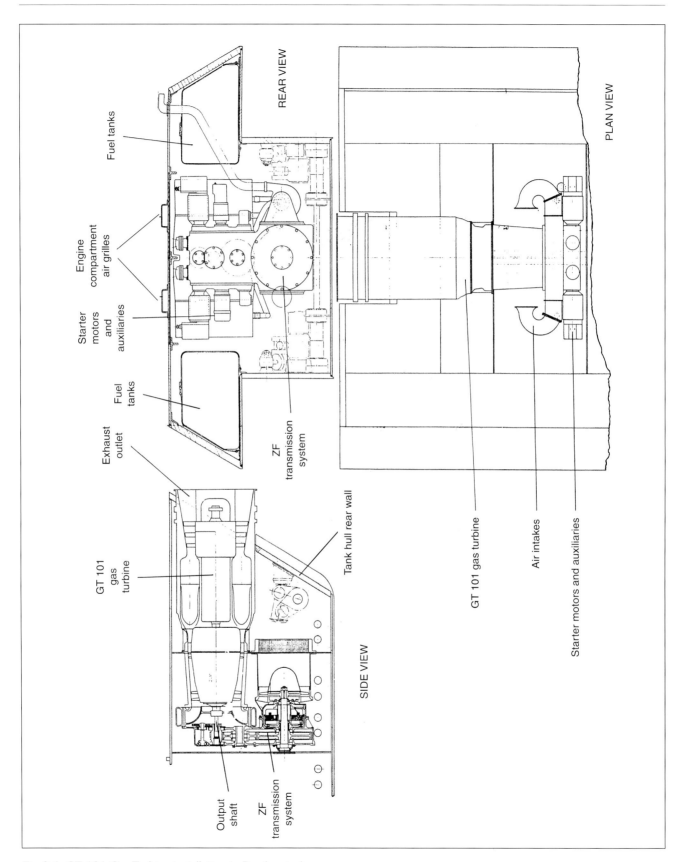

Fig 3.6 GT 101 Gas Turbine installation in Panther tank.

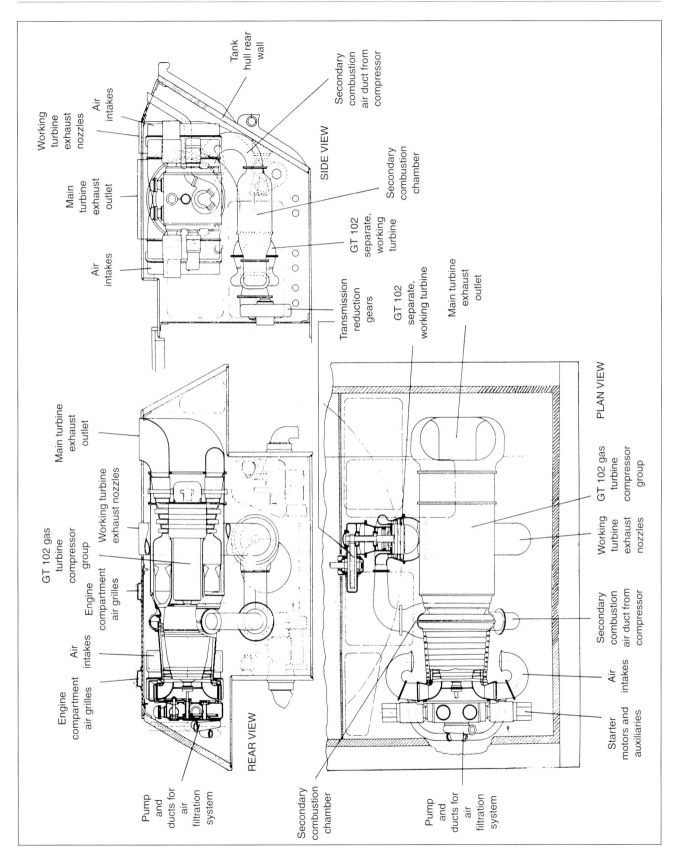

Fig 3.7 GT 102 gas turbine installation in Panther tank.

Air intake:
Air entered the engine compartment through heavy grilles in the port side of the tank hull's upper rear surface and was drawn into the front of the compressor through two curved ducts at the sides of the compressor. Attempts were being made to provide a centrifugal type of air filtration system which also incorporated a silencer, and it was hoped that this unit would reduce the noise at the air duct intakes to a satisfactory level. The filtration system used two vent pipes partitioned on the diameter to duplicate the centrifugal effect. Receptacles were provided to trap the centrifugal dust and grit which was evacuated from the traps by an independent centrifugal blower. The vent pipes were designed for the full air mass flow of 10 kg/sec (22.05 lb/sec) and an air velocity of about 80 m/sec (262 ft/sec). Chief developer of the filtration system was the Mann und Hummel company.

Compressor:
The compressor was of the nine-stage, axial type utilising 50 per cent reaction, and was designed by engineers Reuter, Waldmann, and Hryniszak of Brown Boveri & Cie, Mannheim, this firm having been chosen for the task because of the excellent results it had obtained in designing a new axial compressor* for the BMW 109–003 turbojet. Because the exit temperature from the compressor could reach 240 °C in summer, the last two rows of the rotor blades and discs were of steel, whereas for ease of production the preceding stages were of light alloy. Thus, a rotor construction resulted using metals with different moduli of elasticity and expansion coefficients. In order to maintain a positive register between the rotor discs under varying centrifugal forces and temperatures, the lip on each succeeding (warmer) disc was located inside the lips of the preceding (cooler) disc, except when changing from light alloy to steel, i.e. between the seventh and eighth stages, the light alloy disc was registered inside the steel disc.

The compressor rotor was made up partly by compressor discs and partly by rings located between the discs to lock the blades into position. The foot of each rotor blade was pressed into oblique slots in the edge of the appropriate disc, the blade foot having been first roughened to reduce the reliance on exact tolerances. A groove was then turned into both the sides of the discs and the blade feet, and into each groove fitted the lip of a locating ring. The whole assembly of discs, blades and locating rings was then pulled together with long tie-bolts, which also connected through flanges on the fore and aft bearing shafts. A similar type of assembly was used for the compressor stator (fixed) blades, although the spacing and blade rings were secured by bolts from outside the compressor housing; these gradually increased in outside diameter towards the rear so that the internal grooves in the housing could be machined in one operation. The connection between the compressor rear bearing shaft and the drum-type rotor from the turbine was not rigid but made by means of a thin membrane coupling disc. This disc easily dealt with torque and transverse stresses, but would transmit very little of the bending stresses between the turbine and compressor sections of the rotor. Axial thrust from the turbine was transmitted to the rear compressor bearing shaft by means of a long tension-bolt, which avoided putting such thrust on the above-mentioned membrane coupling disc. Thus, the rear compressor bearing (consisting of twin ball races) had to deal with the difference in thrust between the compressor and turbine, such difference being about 500 kg (1,102.5 lb). The bearing for the front shaft of the compressor was of the roller type.

Air discharged from the compressor into an annular diffuser duct leading to the main combustion chamber. About halfway along this duct, a lip in the outer wall diverted part of the airflow into a collecting annulus, from where it was ducted to the secondary combustion chamber for the working turbine.

Main combustion chamber:
Following the compressor diffuser duct, an annular combustion chamber was provided, having a similar layout to that of the BMW 109–003 A turbojet. In order to take advantage of BMW's experience, the dimensions of and between the fuel nozzle inserts, air cones, and secondary air mixing vanes or fingers remained unaltered, but the combustion chamber diameter was made smaller and the number of burners reduced from 16 to 14. A reduction in air velocity was also possible (since the main combustion chamber dealt with only 7 kg/sec of air compared with 19.3 kg/sec for the BMW engine), but was compensated for by higher air inlet temperatures and higher pressures.

Compressor unit turbine:
Particular attention had to be paid to the design of the compressor unit turbine, since upon the efficiency of this section depended very largely the efficiency of the entire installation. In addition, there was the ubiquitous problem of designing hollow, air-cooled blades made from low-quality steels because of the German shortage of heat-resisting steels. The turbine was of the three-stage type with high-pressure blading. Its rotor blades were attached to discs, the end one of which had a stub shaft, and these, together with spacer rings, were pulled together with long tie-bolts. The same bolts connected with the rear end of a hollow drum rotor which passed between the inner walls of the combustion chamber to connect with the membrane coupling disc attached to the compressor's rear shaft.

*Hermso project (see page 125).

GAS TURBINES FOR LAND TRACTION

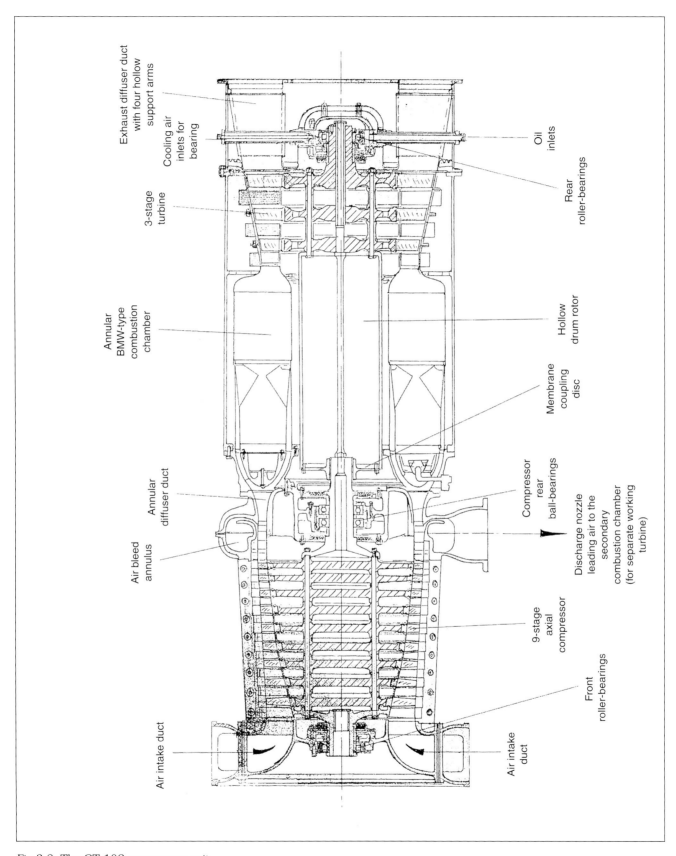

Fig 3.8 The GT 102 compressor unit.

A roller bearing was provided for the stub shaft of the turbine, and this bearing, being in a hot zone, needed cooling, which was achieved as follows. An exhaust diffuser duct after the last turbine stage caused the pressure there to be less than in the surrounding atmosphere. This pressure difference was increased by small ventilation blades provided in the last turbine stage, and the bearing cooling air was sucked through the four hollow arms welded inside the exhaust duct to support the bearing housing. After passing into the space surrounding the turbine bearing, the cooling air then passed via the ventilation blades into the exhaust duct to mix with the hot gases and assist in cooling the diffuser inner wall.

The long, hollow, rotor and stator blades of the turbine, with their consequent high stressing, could have either a welded fixing or a grooved foot fixing reminiscent of BMW turbo-supercharger practice, and both blade types could be produced from the same closed-end pressing of circular cross-section. Figure 3.10 shows a turbine rotor blade with the special cooling insert. This insert was formed from 2 mm thick sheet into a loop through which was pushed a pin, and both were then inserted into the allotted hole in the disc edge. To give the requisite cooling-air gap between the insert and the blade, the insert had a dimpled surface, the dimples beneficially increasing air turbulence and heat transfer. Similar inserts, but with longitudinal corrugation in lieu of dimples, were provided for the turbine stator blades, which were welded into inserts in the blade carrier rings.

Regarding the cooling of the turbine, Dr Müller conducted a considerable investigation. The air cooling of a gas turbine involves a certain expenditure of power, which expenditure grows with the increasing number of turbine stages. Therefore, the number of turbine stages came under close scrutiny with the aim of comparing the efficiency of numbers of turbine stages with the power loss involved in air-cooling the same. The answer arrived at was that a three-stage turbine would have an efficiency increase of 2.5 per cent, with a consequent substantial increase in power output and decrease in fuel consumption (compared with other turbine layouts). The aim in designing the cooling system was then to use the minimum of air and to ensure the air was entirely oil-free to avoid impairing cooling efficiency by carboning the inside of the blades. Cooling air was taken from the compressor diffuser and passed through the centre bearing housing and membrane coupling disc into the hollow drum-type rotor. At the turbine end of the hollow rotor, the cooling air emerged through radial holes into a hollow ring attached to the rotor. This hollow, rotating ring formed the inner limit of the first row of stator blades (or inlet guide nozzles), and the air passed from the ring into small bores in the rotor discs and from these into the blade feet. Cooling air to the feet of the turbine stator blades travelled from the compressor, around the outside of the combustion chamber and into the cylindrical housing enclosing the stator blade support rings. Exhaust gases from the turbine emerged through an opening in the starboard side of the upper rear hull surface.

The secondary combustion chamber:
The 3 kg/sec air supply for the secondary combustion chamber was bled off from the diffuser immediately after the compressor into a circular space or annulus which throttled the air velocity of 110 m/sec (360.8 ft/sec) down to 80 m/sec (262.4 ft/sec) and brought about a further pressure conversion. The air emerged from the annulus via a discharge nozzle connected to ducting leading to the secondary combustion chamber. This combustion chamber and the separate working turbine it supplied were mounted side by side, below the compressor unit and with their axes at right angles to the compressor unit axis. This arrangement gave a sufficiently long secondary combustion chamber for an even temperature at the turbine inlet. (From Figure 3.7 it will be noted that the axis of the compressor unit was at right angles to the Panther tank's longitudinal axis, unlike the axis of the GT 101 unit, which was parallel to the vehicle's longitudinal axis.) The design of the secondary combustion chamber was not finalised because of lack of time, but its combustion gases were ducted around 180° and then into a collection annulus at the separate working turbine inlet.

The separate working turbine:
For the working turbine (see Figure 3.9), two stages instead of three were chosen to achieve the minimum rotating mass (in order to give rapid acceleration and deceleration), to avoid a bearing on the hot discharge side and to reduce the production effort. These considerations carried weight because the loss in efficiency with a two-stage turbine was of little import, since an efficiency loss in the separate working turbine affected the overall unit efficiency very little. For the rotor blades, two discs were provided and bolted together; the disc on the intake or high-pressure side extended into a shaft which was splined for connection to the tank's transmission and was carried by two bearings – one a roller race and one a ball race.

The turbine design borrowed considerably from BMW turbo-supercharger practice, and only the rotor blades were air-cooled. The cooling air (which, incidentally, had a temperature of about 230 °C at this stage and could not therefore be ducted near any bearings) was fed by a pipe which passed through a hollow, streamlined fairing mounted inside the exhaust discharge housing. From the pipe, the cooling air entered the hollow turbine shaft, the hollow space between the two turbine discs and thence into the rotor blades. The connection between the fixed pipe and the rotating hollow shaft was made in a housing with a suitable sealing gland. A much more difficult sealing problem concerned the prevention of hot gases (at a pressure of 3.5 atmospheres absolute) leaking from

GAS TURBINES FOR LAND TRACTION

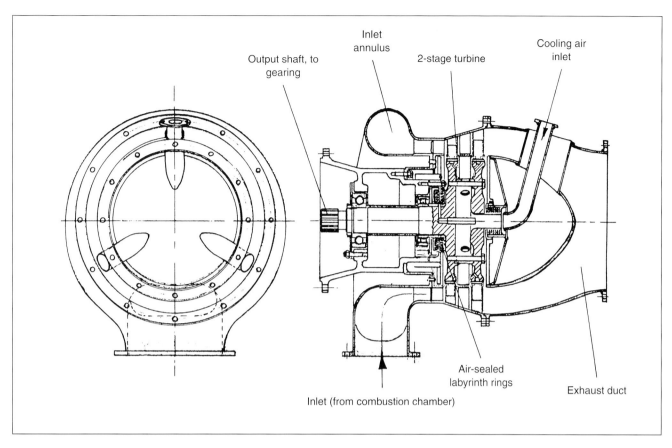

Fig 3.9 Separate working turbine for the GT 102.

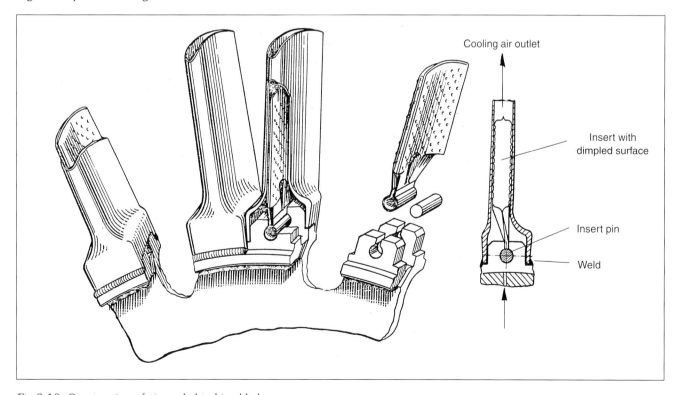

Fig 3.10 Construction of air-cooled turbine blade.

169

the turbine into the adjacent bearing housing. This was solved by using two radial labyrinth rings in series and supplying them on the inside with sealing air tapped from the cooling air in between the rotor blade discs.

Running and control:
Preliminary studies indicated that the working turbine shaft could be connected to a fairly simple ZF Type 305 transmission system. This consisted of a two-speed epicyclic gear operated by two stationary electromagnetic, multi-plate clutches. There was also a second two-speed gearbox, which could be operated only when the vehicle was stationary, which gave suitable reductions for either road or cross-country conditions.

With the GT 102 installation, the most difficult problem, which had not been fully resolved by the end of the war, was that of control. The control system had to maintain a correct balance between the fuel requirements of the compressor unit and the separate working turbine section to avoid excessive temperatures and speeds in either, with consequent waste of fuel. Centrifugal governors were to play a prime part in the various control schemes under consideration, this work being the responsibility of Dr Viktor Speiser. An electric motor was to run the compressor unit up for starting, which was done on gasoline fuel, and once idling, the fuel was switched to low grade. The low-grade fuel pump was a gear type of BMW pattern (but smaller) and gave a full-load burner pressure of about 60 atmospheres. Auxiliaries, such as fuel and oil pumps, were all driven from a gearbox at the intake end of the compressor unit and were arranged to afford easy access from the tank fighting compartment.

Data for the GT 102 installation are:

Compressor unit:	
Turbine power	2,600 hp
Compressor power absorption	2,600 hp
Maximum speed	14,000 rpm
Pressure ratio	4.5 to 1
Airflow	10 kg/sec (22.05 lb/sec)
Airflow to combustion chamber	7 kg/sec (15.43 lb/sec)
Air temperature before combustion	180 °C approx.
Combustion gas temperature	800 °C
Separate, working turbine:	
Output shaft power	1,150 hp
Max speed	20,000 rpm
Max. airflow	3 kg/sec (6.615 lb/sec)
Max. pressure before turbine	4.3 atm. absolute
Max. temperature before turbine	800 °C

Preliminary development work for both GT 102 and GT 101 installations had been completed early in 1945, and completion of the detailed drawings for these was at one time scheduled for 15 February 1945, but was never fully realised. Although the fuel consumption for GT 102 was estimated to be little better than GT 101 (i.e. about double that for the standard 700 hp Maybach gasoline piston engine) there was sufficient room to double the fuel capacity of the Panther tank to 1,400 litres (308 gallons) if necessary in order to maintain the operating range. The elimination of the cooling installation for the conventional engine provided space for the extra fuel. However, another part of the programme entailed the use of a heat exchanger under the designation GT 103.

The GT 103 unit

The GT 103 unit was simply the GT 102 unit with the addition of a regenerative heat exchanger, which was considered essential if the fuel consumption was to be reduced (see Figure 3.1D). The idea was that by feeding exhaust heat back as far as possible to heat the incoming compressed air, less fuel would be needed to maintain the temperature in the combustion chamber. Such a heat exchanger had to be carefully designed to avoid excessive pressure loss, and was to obtain heat either from behind the compressor group turbine or from behind the separate working turbine, and was to pre-heat the combustion air for the separate working turbine only, i.e. about 30 per cent of the air for the whole installation. A schematic arrangement of the proposed regenerative heat exchanger is shown in Figure 3.11, and the development was entrusted to W. Hryniszak (of Brown Boveri at Heidelberg) who worked on the GT 102 compressor and in post-war years became a leading authority on heat exchangers. Hryniszak's heat exchanger consisted of a rotating drum of ceramic material, constructed so that exhaust gases and air for combustion could not mix. A partition inside the drum allowed exhaust gases to pass from outside to inside, and combustion air (from the compressor) to pass from inside to outside, as the drum rotated. Thus, the porous ceramic material acted as a heat reservoir which absorbed heat from the exhaust during one half of a revolution and gave it up to the air during the other revolution half. An indication of the estimated performance is given in the temperature figures on the illustration. Opinion as to what improvement such a heat exchanger would give in fuel consumption varied greatly from large to insignificant, but an average estimate was that the 450 gm/hp/h (0.992 lb/hp/h) consumption of GT 102 could be reduced to about 300 gm/hp/h (0.6615 lb/hp/h), though this appears optimistic. It was also considered that the high air consumption could be reduced by the heat exchanger, later developments of uncooled turbine blades (e.g. ceramic blades), and an increase in combustion temperatures. A prime reason for reducing the air consumption was the obvious necessity to minimise openings in the tank hull.

GAS TURBINES FOR LAND TRACTION

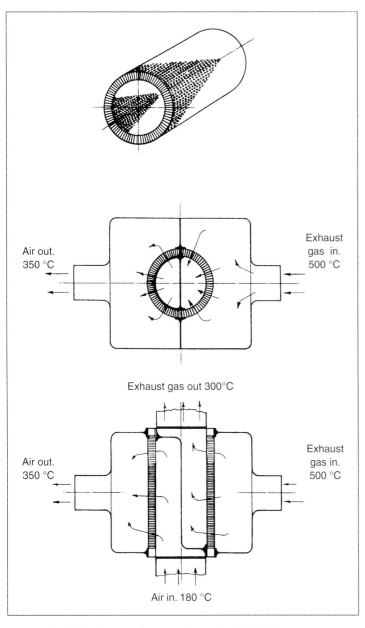

Fig 3.11 Heat exchanger scheme for GT 103.

Another scheme, elaborating the GT 103 unit, envisaged halving the specific fuel consumption by providing a second heat exchanger. Thus, the exhaust heat from the compressor group turbine pre-heated the combustion air for the compressor group, and the exhaust heat from the working turbine pre-heated the combustion air for the secondary combustion chamber. This scheme recovered waste heat to the best advantage, but at the expense of complication. The following data for the GT 103 unit with two heat exchangers were calculated prior to the design of an actual unit:

Compressor unit:
Turbine power   1,400 hp
Compressor power
absorption   1,400 hp
Maximum speed   19,000 rpm
Pressure ratio   4.5 to 1
Airflow   6 kg/sec (13.23 lb/sec)
Airflow to combustion
chamber   4 kg/sec (8.82 lb/sec)
Air temperature after
heat exchanger and
before combustion   500 °C
Combustion gas
temperature   800 °C
Separate working turbine:
Output shaft power   800 hp
Maximum speed   25,000 rpm
Max. airflow   2 kg/sec (4.41 lb/sec)
Max. pressure before
turbine   4.1 atm. absolute
Air temperature after
heat exchanger and
before combustion   500 °C
Combustion gas
temperature   800 °C
Specific fuel consumption
for whole unit   Between 150 and 230 gm/
                                    hp/h (0.33 to 0.5 lb/hp/h)
Overall thermal efficiency   28.8 per cent

The above thermal efficiency compares with about 16 per cent for a gas turbine traction unit without heat exchangers and about 35 per cent for a good internal combustion piston engine of the period.

The GT 102 Ausf. 2 unit

With the GT 102 installation, the compressor unit could only be accommodated at right angles to the Panther longitudinal axis and by taking the ends of the unit up to the sloping hull sides in space formerly occupied by fuel tanks. Thus in an aim to shorten the compressor unit, the GT 102 Ausf. 2 (Ausführung, or Version, 2) compressor unit was designed as a secondary development. The reduction in length was to be achieved by a shortening of the combustion chamber and compressor. Previously, a long combustion chamber was adopted to achieve sufficient fuel mixing and even heat distribution with low-grade fuels, but it was hoped that these criteria could still be met by a return to rotating burners. Reduction of the number of compressor stages was accompanied by an increased load on the separate stages, so that a loss in efficiency was also likely, the extent of such loss being ascertained by empiric experiment. The proposed compressor was of a seven-stage axial type with a blade-tip velocity approaching Mach 1, and was to be designed by Dr Friedrich of the AVA, Göttingen, for development by Brückner-Kanis of Dresden.

A gas turbine unit for commercial use

Because of the war, almost all the German effort expended on projects for a gas turbine traction unit were directed towards the application to military vehicles, and more specifically, armoured fighting vehicles. Although there was no official sponsorship for a commercial vehicle gas turbine unit, the designers did give some preliminary thought to such a unit and laid down some practical requirements. The basic scheme favoured was that shown in Figure 3.1E, whereby the working turbine was mounted coaxially behind the compressor's two-stage turbine and utilised a common hot gas stream. A six-stage axial compressor was planned, while a heat exchanger preceding the combustion chamber was considered essential.

Whereas the power of the units so far discussed was in the region of 1,000 hp, the power envisaged for a commercial vehicle gas turbine was only 320 hp. This, for a gas turbine, was very small, and the problem of obtaining acceptable component efficiencies (and hence overall efficiency) after considerably scaling down was not the least of the developmental problems. Another was the requirement for rapid and effective power-plant control within close limits in order to deal with varying road conditions, and in this connection an important feature planned was adjustable inlet nozzles for the working turbine. Despite the formidable difficulties facing the development of a successful commercial vehicle gas turbine unit, the following requirements were laid down as being within eventual attainment:

Output shaft power	320 hp
Primary turbine and compressor speed	21,000 rpm
Separate working turbine speed	7,500 rpm
Pressure ratio	3.7 to 1
Airflow	6.5 kg/sec (14.33 lb/sec)
Temperature at compressor outlet	172 °C
Temperature before primary turbine	750 °C
Exhaust temperature after both turbines	540 °C
Specific fuel consumption	400 gm/hp/h (0.88 lb/hp/h)
Maximum diameter with heat exchanger	1.0 m (3 ft 3 in)

Conclusion

The project to develop a gas turbine power-plant for land traction had reached the stage of detailed design work by the end of the war, but no construction work had been undertaken and much development work lay ahead. In February 1945, Dr Alfred Müller, chief architect of the power units designed for AFVs, was replaced by Dr Max Adolf Mueller, who paid lip service at least to membership of the Nazi Party. As related in Section 2, Dr M.A. Mueller had been in the forefront of the early Junkers turbojet engine development, and had subsequently moved to the Heinkel AG to continue such work, notably in connection with the promising Heinkel 109–006 (HeS 30) turbojet. This replacement of Dr Müller was in line with Hitler's policy, beginning in July 1944, that a certain proportion of all officers and scientists should join the SS Hauptamt, and those who did not comply were removed from their posts. The Waffen SS (SS units forming part of the German Armed Forces and used only at the war fronts) also had a great interest in projects for improving AFVs.

The main effort of the gas turbine AFV project was to be directed towards building and installing the unit GT 101 in a simple fashion for the purpose of furnishing experimental data and then aiming towards producing the GT 102 unit with separate working turbine for service use in the Panther tank. The Panther tank, together with the Russian T.34, was already the best then extant, and greatly influenced subsequent international tank development. On paper, therefore, a Panther successfully fitted with a gas turbine would have presented the Allies with a formidable problem, but in addition to various possible technical snags, there was the big drawback of high fuel consumption, which was probably one of the main reasons why the development was not continued immediately after the war. It is of interest to note that both Prof. W. Kamm and Prof. F. Weinig of the Stuttgart Technisches Hochschule considered the AFV gas turbine project to be far-fetched, a view that would not be shared to-day. Although neither of them was intimately involved in the work, they were associated with the design of the Brown Boveri nine-stage axial compressor for the GT 102 unit.

Since, by the end of the war, the work of the German engineers had not proceeded beyond detailed study, it could form no more than a useful guide for other engineers interested in developing the gas turbine for land traction purposes. Outside Germany, engineers had already considered such purposes and were thinking along the lines of the German schemes. In England, Rover Gas Turbines Limited were soon in the forefront of automotive gas turbine development, producing, for example, the world's first gas turbine car in 1950. Today, the gas turbine is finding increasing employment in powering various forms of land transport, such as locomotives, trucks and cars. Of particular interest to this history is the gas-turbine-powered truck independently developed by Leyland Gas Turbines Limited in England. The power unit involved resembles in its basic essentials the projected German 320 hp gas turbine unit for commercial use. Leyland's unit is of 400 hp and features a centrifugal compressor, a single can-type combustion chamber, a separate coaxial working turbine, twin glass ceramic heat exchangers and adjustable working turbine inlet nozzles for braking. To compare with the German project, the following data are available for the Leyland unit:

Output shaft power	370 hp (normal) to 400 hp (maximum)
Primary turbine and compressor idling speed	19,000 rpm
Separate working turbine maximum speed	30,000 rpm approx.
Output shaft speed	2,960 rpm
Pressure ratio	4.0 to 1
Airflow	1.70 kg/sec (3.75 lb/sec)
Combustion temperature	1,000 °C
Minimum specific fuel consumption	178 gm/hp/h (0.392 lb/hp/h)
Height	1.125 m (3 ft 8 in)
Width	0.713 m (2 ft 4 in)

It says much for the advances made in this type of power unit that, although the Leyland unit is of only 400 hp, its efficiency is such as to allow commercial production and use in a field hitherto dominated by the diesel engine. Since the war, many other companies have worked on or developed automotive gas turbines for cars, trucks, tanks, bulldozers, etc. To mention but a few, there were, or are, Austin, Budworth, Centrax and Standard in Britain, AIResearch, Boeing, Chrysler, Ford and General Motors in the USA, Turbomeca in France and Fiat in Italy. In addition, main battlefield tanks, with amazing manoeuvrability, are now usually powered by the gas turbine, and they include the Russian T-80, American M1 Abrams and Ukrainian T-84. The power unit for the M1 Abrams is the AGT-1500 gas turbine designed in the 1960s by the team under Anselm Franz of Junkers turbojet fame. This unit is a three-spool, 1,500 shp design. It is possible that today's 50- and 60-ton tanks will be replaced at some future date by machines using much lighter and stealthier plastic composite armour, and that, still using the gas turbine, these will be much faster and longer ranging.

Section 4

Marine Gas Turbine Units

Introduction — Brückner-Kanis GmbH — Blohm und Voss Schiffswerft — Maschinenfabrik Augsburg-Nürnberg AG (MAN) — Conclusion

Introduction

While the reasons for the German efforts to apply the gas turbine for land traction purposes, as dealt with in the previous section, are fairly clearly understood, the same cannot be said of German work to apply the gas turbine for marine propulsion. Since the reasoning behind this effort is now obscure, we are forced back on conjecture, and a few observations based on the general type of vessel to which the gas turbine was to be applied. These vessels were the small, fast naval craft such as torpedo and patrol boats.

Although Helmut Weinrich performed work on a marine gas turbine unit in the years just before the Second World War, no real official incentive was given to such work until about 1942. At that stage of the war, German naval strength lay chiefly in U-boats, but these were being forced away from the dense Atlantic shipping lanes by superior Allied electronic measures and air patrols. By the beginning of 1943, the Allies had the upper hand, and the U-boats, now largely forced into the Indian Ocean and the Gulf of Aden, were fighting a losing battle. As these events unfolded, so Germany's Schnellboote (fast boats), or S-boats, assumed increasing importance to her naval operations. These vessels, in various versions, took part in a great number of actions and were operational in theatres such as the Mediterranean, Baltic, Black Sea and the English Channel.

A typical Schnellboot, the type S-100 (see Figure 4.1), had a displacement of 105 cu m (approx. 103 tons) and a crew of about 23 officers and men. Armed with torpedoes and an assortment of guns (up to 40 mm calibre), it was capable of a range of 700 sea miles cruising at 30 knots, and had a maximum speed of 42 knots. The power of this vessel was derived from three diesel engines which gave a total of 7,500 shp. The planned German gas turbine units varied in power between 5,000 and 10,000 shp, but in most cases it is not known if such units were to replace some or all of the diesel engines. No doubt both cases were considered.

Where gas turbine power was to replace one or two of the diesel engines, the chief aim must have been to use the extra power to boost speed during combat and pursuit. In this case, however, an increase in the already considerable speed (42 knots for the S-100) would have been gained at the expense of fuel consumption and range. When, in addition, the increasingly critical German shortage of fuel is considered, the advantage of the gas turbine seems dubious.

Another, perhaps more important, factor must also be considered, however. The production of a single gas turbine unit promised to be far speedier and cheaper than diesel engines of equivalent or less total power. This would give not only more S-boats for the same production time and manpower requirements but would make the vessels more expendable.

Unlike the Allies, the Germans did not, for most of the war, consider their smaller armed vessels expendable. However, as the Kriegsmarine's capacity to operate in strength dwindled and large numbers of sailors required re-employment, so new methods of warfare were developed. Following the Italian example, new weapons were developed, transforming existing equipment wherever possible. One new weapon was the remotely controlled boat filled with explosive (code-named Linse) for driving against enemy ships, etc. in coastal actions. In this case, an expendable vessel needing only a short range but maximum speed to avoid the defences was required and appears to have been a suitable application for the gas turbine unit.

The Waffen-SS worked in co-operation with the Kriegsmarine to develop new weapons (such as the Linse) and to train the men for their operational use. An establishment of the Waffen-SS, the Kraftfahr Technische Versuchsanstalt der SS (Motor Technical Research Establishment of the SS) at St Aegyd, Niederdonau, near Vienna, carried out research on the application of gas turbines to fast boats and tanks. This research appears to be the only work in Germany where the different types of automotive gas turbine (land, marine, etc.) were considered together, and apart from this, there was little or no contact between the men working on marine gas turbines and those working on land traction gas turbines. In any case, the interest of the Waffen-SS came at a late stage and was confined to application studies.

Sponsorship, where forthcoming, for the companies performing the basic design and development of the marine gas turbines emanated from the appropriate department of the Oberkommando der Kriegsmarine (OKM), or High Command of the Navy (see Introduction). The companies whose work is discussed in this section are Brückner-Kanis, Blohm und Voss and MAN, these being the known companies active in the marine gas turbine field.

Fig 4.1 A typical Schnellboot of the Kriegsmarine, this is the S-128. Already amongst the fastest craft of the Second World War, they were considered for even more power from gas turbines.

Fig 4.2 The cramped engine room of the Schnellboot S-116 in 1943. This shows the normal Mercedes-Benz diesel engines which would have been replaced by a gas turbine unit. *(Photo Deutsches Museum, München)*

Brückner-Kanis GmbH

Vorkauf's Drehkessel gas turbine unit — The 5,000 shp Drehkessel gas turbine unit — The K229 gas turbine — Progress made towards the 5,000 shp unit — Weinrich's gas turbines — The 10,000 shp marine gas turbine unit — Progress made towards the 10,000 shp unit

The engineering and steam turbine company of Brückner-Kanis, Dresden, worked on two main schemes during the Second World War to apply the gas turbine to fast marine craft such as torpedo boats. These two schemes were quite different from each other in concept, and both were more elaborate than the schemes of Blohm und Voss and MAN, described in this section. First described is the so-called Vorkauf Drehkessel gas turbine unit, which combined both steam and gas turbines; some may argue that this scheme has no place in this book, but it should be noted that more than 60 per cent of the unit's power was to be derived from the gas turbine section. The second scheme described is purely of the gas turbine type and was probably the most elaborate devised by the Germans for motive power.

Vorkauf's Drehkessel gas turbine unit

The rotating boiler (Drehkessel) with capillary tubes was projected in the early 1930s and possibly before. A Dr Vorkauf wrote a paper on the subject in 1932, while P. Huettner made similar proposals at about the same time. Offering the possibility of an economic power plant having low weight and compact dimensions, the rotating boiler with its internal steam turbine required no water feed pump or regulator since centrifugal force automatically governed the water supply. In Huettner's scheme, the integral steam turbine was of the contra-rotating type and drove both the capillary boiler and the power shaft, whereas in Vorkauf's scheme the turbine drove only the boiler, and the steam was led to a separate power turbine. Applications for the Drehkessel were envisaged as electric generator drives and a gas turbine starter unit. The manufacture of tubes for a 20 hp Drehkessel starter unit (K294) for gas turbines was begun by Brückner-Kanis, but the whole scheme was dropped, possibly during 1940.

Several disadvantages of the Drehkessel contributed to its abandonment. The chief disadvantage, or rather unsolved problem, was that of maintaining dynamic balance of the rotor assembly when the boiler tubes were subject to varying stresses and temperature. Even the problem of obtaining static balance under cold conditions had not been solved. Given a workable unit, it still needed a minute or so to start it, using a separate electric motor.

Already, however, Vorkauf had proposed to incorporate a gas turbine into the system with the object of increasing the thermal efficiency and reducing the plant dimensions. An efficient rotating boiler without the problematical capillary tubes now became practical, since these could be replaced by using the revolving, cylindrical inner wall of the combustion chamber and the moving (hollow) blades of the gas turbine. Thus, the gas turbine was not only efficiently cooled but the steam produced was used to drive a steam turbine. The steam turbine drove a compressor for the gas turbine and both turbines were geared to an output shaft.

Figure 4.3 gives a flow diagram of the scheme, which will now be described in more detail.

The 5,000 shp Drehkessel gas turbine unit

For the application of the Drehkessel gas turbine as a power unit for marine (or other) uses, the Brückner-Kanis company worked in collaboration with La Mont Kessel Herpen & Co. of Berlin, which specialised in boilers and steam plant. The chief engineers from La Mont concerned in the project were Dr Vorkauf and Heinrich Peters, both knowledgeable engineers.

For marine use, the latest rotating boiler gas turbine, known as K229, was to be incorporated into a unit giving 5,000 hp at the propeller shaft. Figure 4.4 shows diagrammatically the method of connecting up the unit mechanically. The K229 rotating boiler gas turbine consisted of an annular combustion chamber, turbine rotor and a collection annulus which led to an exhaust gas duct. Compressed air to feed the combustion chamber was supplied by the steam-driven axial compressor. The hot gases from the turbine, before being exhausted to the atmosphere, could be first passed through a heat exchanger, which added heat to the compressed air supply for the purpose of reducing fuel consumption. Water was fed into the gas turbine, cooled the various hot parts either as water or steam and was then piped as superheated steam to the steam turbine, whence it passed to a condenser. Most of the gas turbine's power went into the propeller shaft (via gearing), whereas when running at full speed all of the steam turbine's power went into driving the air compressor.

The K229 gas turbine

The rough layout of the K229 gas turbine is shown on the right-hand side of Figure 4.3 Here, the output shaft is shown projecting from the exhaust end, which

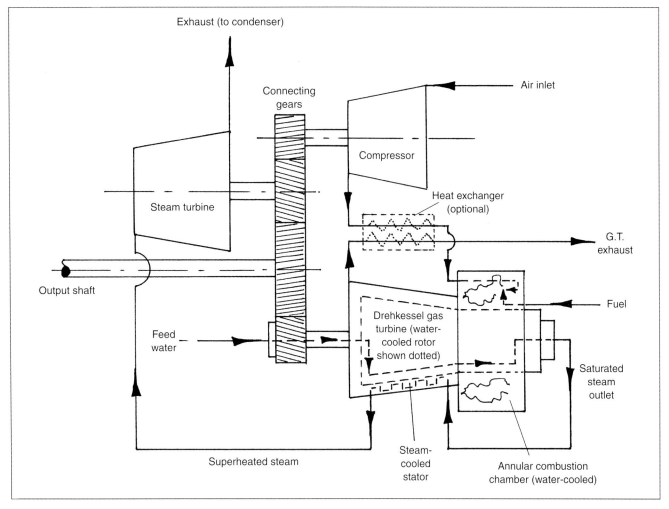

Fig 4.3 Drehkessel (rotating boiler) gas turbine unit flow diagram.

necessitated a collection annulus and duct for the gases, but another layout of K229 projects the output shaft from the intake end so that an exhaust gas nozzle could be used at the opposite end.

Combustion chamber:
Air entered by a tubular duct, passed around an air-cooled outer casing and entered the combustion chamber via short, radial, curved vanes designed to give optimum mixing. The inner wall of this annular chamber was formed by an extension of the rotor, and had a corrugated surface to relieve stresses. Low-grade fuel entered the hollow rotor shaft at one end and was sprayed into the combustion chamber from two diametrically opposed, rotating nozzles. Thus, the fuel was centrifugally thrown out at right angles to the incoming air and was effectively atomised. To ensure thorough mixing before the combustion products entered the turbine, there was a row of short baffles fitted to the rotor just before the chamber outlet. These baffles also served to shield the first turbine blades from the considerable radiation heat.

With an air–fuel ratio of 16 to 1, very high combustion temperatures were expected, giving an inlet temperature to the turbine of 1,800 °C. Thus, in addition to the inner, rotating wall being water-cooled and the outer casing being air-cooled, it was anticipated that the inner, fixed surfaces of the combustion chamber would need a refractory lining – as used in furnaces.

Turbine:
The turbine was of the 13-stage type with an inlet temperature of 1,800 °C and an exhaust temperature of 400 °C. Water entered the rotor at the cooler, exhaust end (beyond the shaft gears), passed in and out of the moving blades, through the hollow, corrugated inner wall of the combustion chamber and out through the intake end of the hollow rotor as saturated steam. The saturated steam was then piped into the hotter, intake end of a hollow jacket surrounding the turbine, and after cooling the turbine stator blades, emerged at the outer end as superheated steam. An interesting point is that

steam-cooling of the last stage of stator blades was not feasible because at this point the gas temperature was only 400 °C, compared with the superheated steam temperature of 475 °C. A possibility was that, because of the high turbine inlet temperature, water instead of steam cooling would prove necessary for the first stage of stator blades. Although the exact number of stator stages to be steam-cooled was not finalised, a steam pressure drop of about 15 per cent of the initial pressure was anticipated for this work.

Concerning steam pressure, up to 100 atmospheres was anticipated. Consequently, the original designs of sheet-metal turbine blades were considered inadequate, and various designs of solid and fabricated blades were investigated. In the case of water-cooled blades for the rotor stages, the blades had drilled holes or in-built channels for water circulation. Water entered via the smaller holes and passed via the larger, central hole back into the rotor or spaces between the rotor discs. Circulation was maintained by differences in centrifugal force on the incoming and outgoing columns of water, which had differing temperatures and densities. This system differed from later Schmidt designs (see Section 6) in that connecting passages between the holes were provided at the blade tips. Since some of the cooling holes might be as small as 2 mm in diameter, there was a serious risk of clogging from solids in the water. If this occurred, disaster would follow because of the high operating temperatures. One proposed possible solution was to use a closed circuit of water or a mixture of diphenyl and diphenyl-oxide.

Although the turbine blades were intended to be in mild steel, a Brückner-Kanis test rig (K290) had blades in Böhler SAS 8 chrome-nickel, heat-resisting steel. Rotor blades were welded onto a separate turbine wheel for each stage, each wheel being shrunk onto a bush which was keyed onto the rotor shaft. Spaces between the turbine wheels formed reservoirs for the cooling water, and holes in the wheels interconnected each reservoir and led into the rotor blades. The turbine wheels formed a rotor having a constant 0.450 m (1 ft 5^i in) diameter, but the blades increased in length towards the exhaust end. Roller bearings were provided for the rotor shaft at each end, and the output shaft was connected to the propeller shaft via reduction gearing of 5 to 1. The stator blades were connected to carrier rings which were fitted inside the outer casing to form an annular duct for the cooling steam. In order to prevent the steam-cooled stator blades becoming flooded with water before steam was raised, a spring-loaded valve was fitted between the water-cooled and steam-cooled sections.

Starting and control:
Although not finalised, the later ideas on starting the 5,000 shp unit (or other similar unit) were as follows:

(i) Water was admitted to the rotor of the K229 gas turbine.
(ii) The clutch connecting the steam turbine to the gas turbine and output shaft gears was withdrawn.
(iii) Compressed air from storage bottles was admitted to the combustion chamber, and combustion was initiated with petrol or paraffin fuel and a spark.
(iv) Following the rapid build-up of steam and pressure, the spring-loaded valve opened and passed steam to the steam turbine via the stator blades of the gas turbine.
(v) The steam turbine was then run up to speed, driving the compressor to supply air to the gas turbine, which was then run on a crude fuel, starting air and fuel (petrol or paraffin) being cut off.
(vi) The clutch was then released so that any steam turbine power surplus to the compressor requirements would be fed onto the output shaft. However, consideration was being given to a scheme whereby the steam turbine was not mechanically coupled to the output shaft since there was the possibility of difficulties occurring under sudden changes of load.

It should be noted that, where the unit was for an installation where steam was already available, storage bottles, etc., for starting air would be unnecessary, and the whole starting procedure would be simplified. Specific details for controlling the unit, once running, had not been worked out.

Data for the 5,000 shp Drehkessel gas turbine unit are as follows:

K229 gas turbine:		
Power	5,500 shp	
Speed	7,250 rpm	
Inlet temperature	1,800 °C	
Exhaust temperature	400 °C	
Rotor diameter	0.450 m (1 ft 5^i in)	
Combustion chamber diameter	0.975 m (3 ft 2 in)	
Overall length (inc. gearbox)	2.460 m (8 ft 0^m in)	
Compressor:		
Power absorption at full speed	3,250 hp	
Pressure ratio	16 to 1	
Steam turbine:		
Power	3,250 hp	
Steam temperature	475 °C	
Steam pressure	80 atm.	
Output shaft:		
Speed	1,450 rpm	
Power	5,000 hp	
Complete unit:		
Weights	kg	lb
K229 gas turbine (with gearbox)	1,863	(4,108)
Compressor	350	(772)
Steam turbine	565	(1,246)
Steam turbine piping, etc.	85	(187)

Auxiliaries and fittings	222	(489)
	3,085	(6,802)
Specific weight	0.617 kg/hp (1.36 lb/hp)	

Progress made towards the 5,000 shp unit

The prime requirement for finalising the design of the 5,000 shp Drehkessel gas turbine unit was the development of the essentially new element, the K229 gas turbine which doubled as a steam generator. Towards this end, Brückner-Kanis planned their preliminary development in the following data-producing stages:

(i) Separate combustion chamber tests to eliminate inefficient combustion before adding a turbine.

(ii) Tests with a single-stage turbine rig (K290) to investigate the efficacy of blade cooling. This turbine rig merely had a duct leading hot gases from a separate source.

(iii) Tests with a gas turbine unit (K236) to develop the water-cooled combustion chamber and blading. This unit had a two-stage turbine and an annular combustion chamber measuring 0.80 m (2 ft 7 in) diameter × 0.625 m (2 ft 0^1 in) length.

By the end of the war, unit K236 had been built and was about to be tested, so one could assume that the preceding two stages were dealt with. This meant that, while the gas turbine section (at least) of the 5,000 shp unit had, as far as possible, been designed in detail by the La Mont company, it was subject to alteration pending the final test results from Brückner-Kanis, so that manufacture of only a few parts, if any, for the prototype could begin. At least one other unit (designated T134) similar to the one just described was also projected, but to give 7,500 shp.

Before concluding this glance at Drehkessel systems, one other project should be mentioned, that designated T109. This was a compact power unit measuring 0.640 m (2 ft 1 in) long × 0.225 m (8^m in) diameter, which was to develop 450 hp to drive a torpedo. Its rotating boiler consisted of a cylinder which sucked sea water into its internal periphery. Inside this cylinder of water was injected hydrogen peroxide which was decomposed into oxygen and steam, the considerable heat producing further steam from the surrounding water. The resulting mixture of gases and steam passed through baffles and ports to the turbine, which was of the contra-rotating type, five stages driving the rotating boiler and five intermediate stages being connected to a rotor. A gearbox connected the contra-rotating elements and took the drive to the torpedo's propellers.*

---

*Virtually all the world's navies have abandoned such systems since, if hydrogen peroxide leaks onto certain metals such as brass or copper, a build-up of high-pressure gas in an enclosed space will cause an explosion. This is thought to have been the cause of the tragic sinking of the Russian submarine *Kursk* in 2000, probably the worst accident of this type.

Weinrich's gas turbines

Whereas the previously described Drehkessel gas turbine scheme had its origins in the steam industry, a second gas turbine scheme, also under development by Brückner-Kanis for marine applications, had its origins in an aircraft turbojet project, which was in turn derived from earlier work by Helmut Weinrich.

During 1939, as related in Section Two, various German aircraft engine and airframe manufacturers were either beginning or continuing their efforts to develop turbojet engines. As early as 1936, however, the independent engineer Weinrich of Chemnitz had submitted to the RLM some plans for a very advanced, even revolutionary, scheme for a turboprop engine (PTL) using a contra-rotating compressor and turbine. Unfortunately, nobody with sufficient foresight and/or influence within the RLM at that time heard of Weinrich's ideas, and he failed to obtain any support from that quarter. However, with commendable persistence and faith, he next submitted a new scheme to the OKM.

Weinrich's new scheme proposed the use of a contra-rotating compressor and turbine as before, but the gas turbine unit was redesigned as an auxiliary power plant for fast naval patrol boats. In this form the unit was rather small and is said to have been designed for a power of 100 hp at the propeller shaft. This scheme received the support of the OKM, and an experimental gas turbine unit was built which progressed far enough to develop 50 hp on the test bed. The subsequent history of this project is unknown, but it is known that the Brückner-Kanis company worked on the designs at least for an aircraft turbojet employing a contra-rotating compressor, a contra-rotating combustion chamber, and a contra-rotating turbine. This amazing turbojet fairly bristled with innovations and novelties, and as might be expected, originated with Weinrich and was probably connected with his 1936 turboprop scheme and his 1939 work on the BMW 109–002 turbojet (see Section Two). How much official backing the turbojet at Brückner-Kanis had, if any, is uncertain, but no active development of the scheme is known. However, later on (possibly at the end of 1943 or early in 1944), the Brückner-Kanis company adapted Weinrich's turbojet unit as one of four gas producers in a very large gas turbine unit, the development of which was ordered by the OKM.

The 10,000 shp marine gas turbine unit

The gas turbine unit was to be employed as an emergency power plant in high-speed naval craft during attack and/or pursuit, and was to develop up to 10,000 hp at the propeller. Design and development proceeded at Brückner-Kanis under the leadership of Dipl.-Ing. Heinrich Holzapfel, although Helmut Weinrich continued to be consulted on the work.

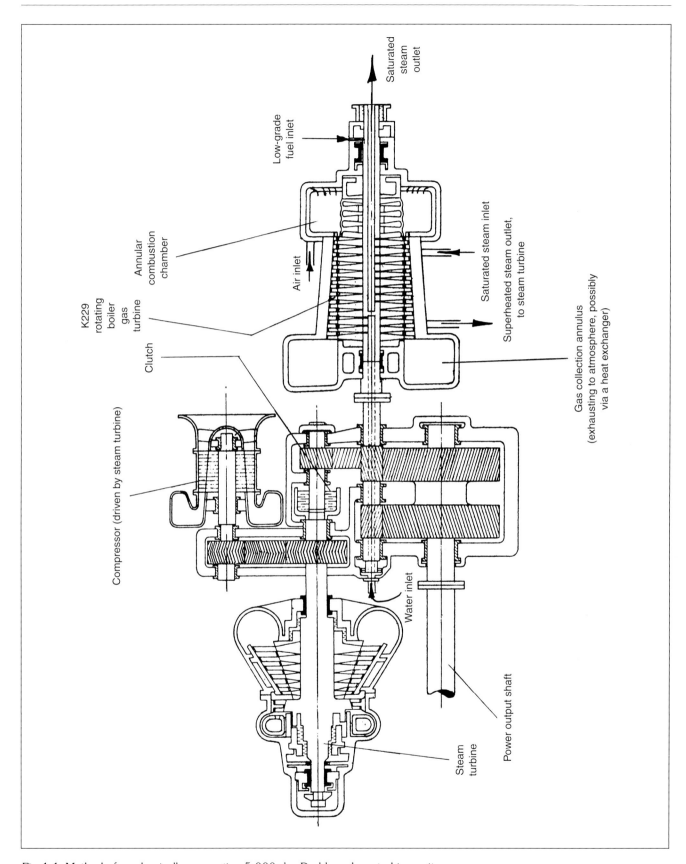

Fig 4.4 Method of mechanically connecting 5,000 shp Drehkessel gas turbine unit.

Holzapfel had some useful experience with turbines and had begun his studies at the Technisches Hochschule, Hannover, under a Prof. Roeder. (Roeder was noted for his contributions towards turbine design, so that Holzapfel, who was his assistant for some time, must have learned much during these early years.) He was then employed by BMW in the exhaust-driven turbo-supercharger department until engaged by Brückner-Kanis to assist in the development of hydrogen peroxide turbines and gas turbines for torpedo boats and the like. The hydrogen peroxide turbines (which originated with the H. Walter KG of Kiel) included types such as the Drehkessel torpedo motor already mentioned.

For the 10,000 shp power unit, the arrangement of using four separate gas turbines to supply hot gases to a separate power turbine was chosen partly to meet the limitations on installation space and partly to employ components of a size and type at least partly already known. Other requirements to be met in the design were ease of assembly and maintenance, together with some elasticity between major components such as the gas producers and the power turbine. For the general arrangement of the power unit, reference should be made to Figure 4.5. This shows the four gas producers grouped together in a circular shell and with a flexibly connected common air intake duct. Following the gas producers was the main or working turbine driving the craft's propeller via epicyclic gearing which reduced the turbine speed of 4,500 rpm down to 500 rpm. Exhaust gases from the whole unit were ducted to avoid the transmission shaft and then upwards. Each gas producer, while fixed at its rear end to the power turbine casing, was mounted on rollers at its forward end to allow for expansion and other movements, hence the need for a flexibly connected air intake duct. The components of the complete gas turbine unit will now be examined in detail, showing the many unusual and novel features of the scheme.

The gas producers:
Each of the four gas producers was a self-contained gas turbine with sufficient power to drive its own compressors only, its sole purpose being to generate hot gases to work the power turbine. Each gas producer was identical to the turbojet engine designed by Brückner-Kanis (after Weinrich's scheme) for aircraft use, except that its intake and outlet ducts were modified, and because there would no longer be an intake ram effect caused by high air speed, a pre-compressor was added before the unit's main compressor. Figure 4.6 shows a half-section of one gas producer, and its main components of pre-compressor, contra-rotating compressor, contra-rotating combustion chamber, and contra-rotating turbine are described in the following.

The pre-compressor:
The pre-compressor (or inducer) consisted of a conventional five-stage axial compressor with a pressure ratio of 1.75 to 1 at a speed of 6,600 rpm. Both stator and rotor blades had similar profiles (based on aerofoil sections) in order to simplify production. A built-up construction was used employing disc and hoop blade

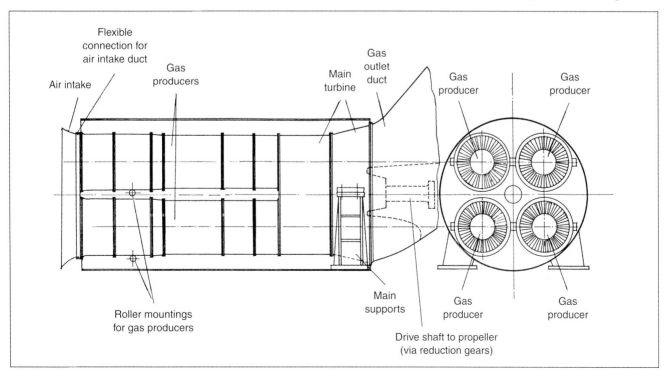

Fig 4.5 General arrangement of a 10,000 shp emergency power unit (gas turbine) for naval craft, by Brückner-Kanis.

mountings and a central shaft carried by a ball race at the front and a roller race at the rear.

Each stage of stator blades was constructed in the following manner. Blades were inserted into slots in a ring which was in turn mounted onto a disc. The loosely inserted blades were then surrounded by a ring to temporarily hold them for dynamic balancing. Following this, the spaces between the blades were filled by a mass consisting of paraffin kolophonium to give a solid disc which could be machined. After machining, the filler mass was removed and the temporary holding ring was replaced by a blade supporting ring, the completed assembly then being finally machined. Since the blade supporting ring was heated and shrunk over the blades, it had to be accurately made in order to avoid buckling of the blades. Projections on the stator blades locked into recesses on the inside of the ring. A special feature of each stator blade supporting ring was that it had a long lip which extended over and surrounded the tips of the succeeding rotor blades. This feature permitted a small rotor blade tip clearance, and in the case of a seizure the rotating blades would only damage the surrounding lip and not the whole pre-compressor casing. By this means, speedier and cheaper repair was made possible and the feature was utilised for all compressor and turbine blades throughout the unit. (The feature does not appear to have been employed in any other German gas turbine unit, with the possible exception of the Blohm und Voss 7,500 shp marine gas turbine.)

Each rotor stage for the pre-compressor used a disc and blades machined from a solid duralumin forging, the considerable amount of material cut away being used for bolts and other small items. Following the series of sawing and milling operations to shape out the rotor blades, a surrounding ring was shrunk onto the blade tips using a similar process to that described for the stator blade assemblies. Finally, an adaptor was pressed into the centre of the disc and pinned.

Labyrinth seals were provided on the outside of the rings surrounding the rotor blade tips and on the inside of the stator blade discs. The hollow steel shaft upon which the rotor blade discs were mounted was connected at its front end to an epicyclic gear unit and at its rear end to the rotating housing of the contra-rotating compressor. The epicyclic gear unit and front bearing were enclosed in a housing supported by vanes in the air intake.

Contra-rotating compressor:
Air left the pre-compressor at a temperature of about 80 °C and passed via a 0.15 m (5^m in) long annular passage to a contra-rotating, four-stage, axial compressor having a pressure ratio of 2.4 to 1. This compressor consisted of a housing with four stages running at 6,600 rpm and a drum-type inner rotor, also with four stages, but running at 8,400 rpm and in the opposite direction to the housing. The compressor housing was driven directly from the rear of the pre-compressor shaft, while the compressor inner rotor was driven by a shaft which passed through the hollow pre-compressor shaft to the epicyclic gear unit.

The four stages of rotating blades connected to the housing were guide blades only, and the first stage of these blades had thick, symmetrical profiles since they acted as the front support for the housing. These first-stage blades were machined out of the solid with their disc, and had a ring shrunk on over the blade tips. This ring was fitted inside the housing and pinned while a central flange on the first-stage disc was bolted to an assembly connecting with the pre-compressor shaft. The other three stages of guide blades were also machined from the solid, and had shrunk-on rings for attachment to the rotating housing, but terminated at the centre in discs with labyrinth seals around the compressor rotor.

Following the four guide blade stages were four stages of pressure blades. These blades were again machined from the solid with their carriers and were fitted to discs keyed to the inner rotor. Between the blade tips and the rotating housing were labyrinth seals. In between the blades of all rows except the first were fitted smaller auxiliary blades, pinned into position. These were necessary to avoid uneven airflow distribution caused by the large pitch of the main blades.

At the front of the contra-rotating compressor was provided a large roller bearing for the front supporting assembly of the rotating housing, while inside this bearing was a smaller roller bearing for the front of the inner rotor shaft. The support of these bearings, and also the rear bearing of the pre-compressor, was given by a casting which connected the outer casings of the two compressors. This casting also formed the annular passage leading to the contra-rotating compressor and had supporting vanes passing through this passage.

Contra-rotating combustion chamber:
Air for the combustion chamber left the contra-rotating compressor at a temperature of about 200 °C and a pressure of 41,300 kg/sq m (8,466 lb/sq ft), the latter also being the pressure in the combustion chamber. The annular combustion chamber consisted of the following principal parts: outer and inner air inlet guide grids with a dividing, conical, combustion chamber casing; these were connected to an extension of the compressor's rotating housing and so revolved at 6,600 rpm; combustion chamber rings, supporting fan blades and fuel injectors; these were inside the conical casing and attached to an extension of the compressor's drum-type rotor so that they revolved in an opposite direction to the conical casing and at 8,400 rpm.

The use of a contra-rotating combustion chamber was planned in the first place to ensure efficient fuel and air distribution and mixing, and hence good combustion conditions and even gas temperature and velocity distribution at the entrance to the turbine section. In any case, there was little difficulty or extra complication in providing such an unusual combustion chamber, since the principal members served to connect the rotating members of the compressor and turbine (fore and aft).

The outer and inner air inlet guide grids imparted with their vanes some swirl to the incoming air, the two air streams being swirled in opposite directions. In the zone between these two flows the swirl velocity was zero, and it was here that fuel was injected and initial combustion occurred. The outer inlet guide grid was directly connected with the rotating housing, while the inner inlet guide grid was also connected to the housing but via a stage of vanes forming the exit from the contra-rotating compressor. The conical casing, with its air holes, was attached to the outer guide grid in a manner permitting a certain amount of distortion.

Inside the conical casing were five rings (decreasing in diameter towards the chamber exit) which were carried by fan blades attached by carriers to the drum-type rotor. These profiled fan blades had the purpose of making up the pressure losses incurred by the air flow into the combustion chamber. Some of them also carried the nozzles for the fuel, which was injected by centrifugal force. The first row of fan blades had four nozzles which injected petrol for starting purposes only; the petrol sprayed into a small annular chamber which was supplied with air from the front, the resulting mixture being ignited by an electrical device (fitted through one of the vanes of the outer guide grid) with a sliding contact. The second and third rows of fan blades each had 12 nozzles which injected the main running fuel (crude oil or gas oil). Fuel reached the rotating nozzles through a tube which passed through the hollow shaft and rotor of the unit. After starting, the flow to the petrol nozzles was shut off, leaving the unit to run on the 24 main nozzles.

Air from the inner guide grid was guided through the combustion chamber rings into the separate combustion zones. The fuel, finely dispersed by centrifugal force, burned within the shortest possible distance. Air from the outer guide grid passed through holes (5 to 8 mm in diameter) in the conical casing into the outer zones of the combustion area. This air, together with that induced at the inner combustion zones by the fan blades, diluted the hot gases and reduced the average combustion temperature to about 650 °C, which was easily handled by the turbine section. The length of the combustion chamber, from air inlet to gas exit, was about 0.320 m (1 ft 0^1 in). Where the hot gases left the chamber, a divergent annulus was provided formed by contra-rotating, inner and outer, curved walls. For the combustion chamber conical casing, sheet steel was used, while the guide rings were made from Ruhr D.712 heat-resisting steel (10.0 Cr 18.0 Mn, blce Fe).

Contra-rotating turbine:
The contra-rotating turbine was designed to develop sufficient power to drive the pre-compressor, main compressor, combustion chamber and auxiliaries. There were five stages of blades attached to the rotating housing, which had a speed of 6,600 rpm, and three stages of blades attached to the drum-type rotor, which had a speed of 8,400 rpm. In addition, there was a row of fixed guide blades which preceded the final row of rotating blades, and a second row of fixed guide blades after the final rotating row. Unusually, there were no inlet nozzles or guide vanes directing the hot gases from the combustion chamber into the turbine, and the first row of rotating blades (attached to the housing) followed immediately after the divergent exit annulus of the chamber. The hot gases passed through the turbine at high axial velocities of up to 350 m/sec (1,148 ft/sec), and so the blades were set at coarse pitches. Aerofoil blade profiles were used and between 30 and 40 blades per stage. Blades increased in length in the direction of the gas flow.

For each of the turbine stages, the blades were machined with their carriers or wheels as one from the solid, in a similar manner to the compressor stages. During machining of the inner and outer circumferences, the spaces between the blades were again filled with a compound, and when this was removed carrier rings for the rotor or hoops for inside the rotor housing were shrunk on. The various stages were pinned to the housing and rotor to give a construction which allowed for expansion and contraction. Unlike most German gas turbines, the turbine blades of this unit were not air-cooled but made solid from heat-resisting steel – either Böhler FBD (17.6 Cr, 15.2 Ni, 2.2 Mo, 1.06 Ta or Nb, 1.8 Cu, blce Fe) or Ruhr D.712 (10.0 Cr, 18.0 Mn, blce Fe). In view of Germany's acute shortage of metals for heat-resisting steels, one wonders if an air-cooled turbine using economical hollow blades would not have been used in later production units. Temperatures aimed at in the turbine blades were 550 °C near the rotating housing, 750–800 °C at the blade centres and 500 °C near the rotor. It is known that Weinrich (presumably working independently at Chemnitz) undertook preliminary investigations on the use of cooled blades for the turbine, but the prime aim seems to have been to allow a higher working temperature rather than to permit a more economical use of blade material. However, for unknown reasons, Weinrich's contract for this particular work was cancelled towards the end of 1943.

The rotating housing of the turbine was connected at the front to the rotating housing of the combustion chamber, while at the rear it was connected via the blades of the seventh row to a disc and shaft having a roller bearing. The rotor for the turbine was connected to the rotor for the combustion chamber at the front, while at the rear it led into a disc and shaft with a roller bearing. A cast housing carried the two roller bearings and was mounted inside a welded structure with a sheet-metal, streamlined dome. The whole bearing section was carried by three hollow, streamlined struts inside the exhaust nozzle. These struts also contained air and oil pipes and were designed with aerofoil sections to eliminate any

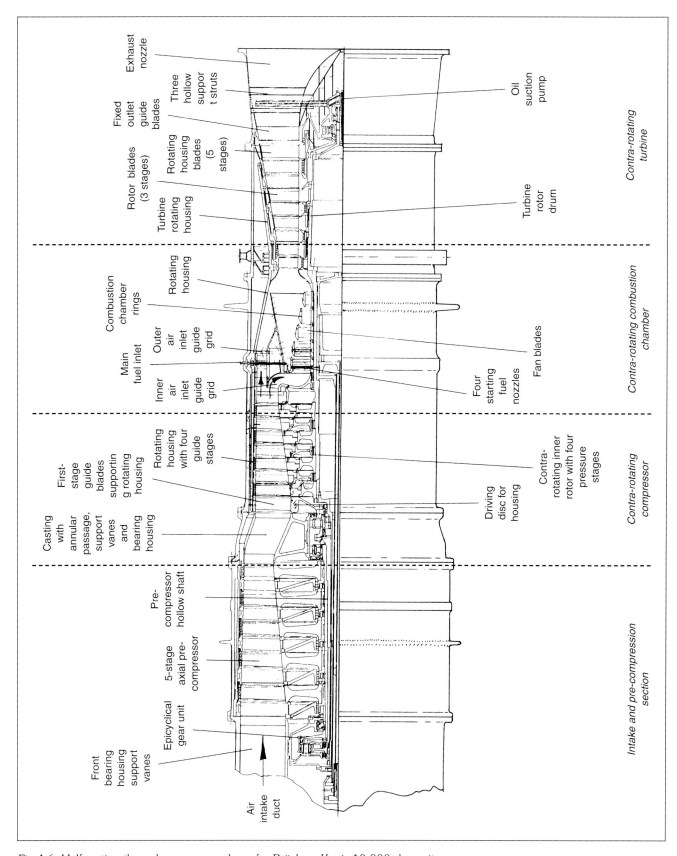

Fig 4.6 Half-section through one gas producer for Brückner-Kanis 10,000 shp unit.

swirl in the hot gases leaving the unit. The exhaust housing (which was flange-mounted to the working turbine section), together with the internal bearing housing, was flexibly connected, and the front of the whole unit was mounted on rollers.

Auxiliaries:
Starting of the unit was accomplished with a 15 hp DC electric motor which, once the gas turbine was running, was then used as a generator. The fuel pumps were driven from the epicyclic gear unit (which connected the rotor and contra-rotating housings), and electric relays were fitted to the fuel lines. A gear-type suction oil pump was fitted inside the rear bearing housing and driven from the turbine rotor. Oil entered at the front bearing and gear housing, travelled through a hollow shaft to the other bearings, and was sucked out from the front bearing housing. Cooling of the bearings was achieved with the oil and also partly with an air supply which was used to seal off various labyrinth seals around the rotor.

The working turbine:
Four of the above-described gas producers were fitted to a gas-collecting housing from which the hot gases passed into the working turbine. There was nothing elaborate or novel about this turbine since its working conditions were very moderate; it worked in a temperature of only 450 °C and had a maximum speed of 4,500 rpm. Under these conditions, the turbine developed 12,000 hp, which, after transmission losses, left a minimum of 10,000 hp available at the boat's propeller.

The working turbine was of the two-stage type and presumably had solid steel blades. The blade carriers formed discs to which stub shafts were bolted, and thrust from the turbine was absorbed by a multi-stage Radiax bearing. Epicyclic gearing having a reduction ratio of 9:1 was fitted between the working turbine rear shaft and the propeller shaft, and a duct carried the exhaust gases into the atmosphere.

Estimated data for the 10,000 shp marine gas turbine unit are:

For one gas producer (four required):	
Inner rotor speed	8,400 rpm
Contra-rotating, outer drum speed	6,600 rpm
Pressure ratio	4.13 to 1
Airflow	30 kg/sec (66.15 lb/sec)
Mean combustion temperature	650 °C
Compressor efficiency	83 per cent
Diameter	0.70 m (2 ft 3 in)
Length (excluding intake duct)	2.010 m (6 ft 7 in)
Diameter of cowling for four gas producers (and whole unit)	1.80 m (5 ft 10^m in)
Working turbine:	
Speed	4,500 rpm
Efficiency	87 per cent
Power	12,000 hp
Casing diameter	1.675 m (5 ft 6 in)
Propeller shaft speed	500 rpm
Propeller shaft power	10,000 hp
Overall length of complete unit	3.50 m (11 ft 5^i in)

Progress made towards the 10,000 shp unit

By the end of the war, the 10,000 shp marine gas turbine unit planned at Brückner-Kanis was designed in detail as far as possible, and some testing of components had begun. How much manufacturing of parts had been carried out is unknown, but the project appears to have had official backing up to the war's end in May 1945.

Obviously, the first task was to develop the gas producer design, and for this the first step was to perfect the contra-rotating combustion chamber to ensure the correct temperature distribution over the turbine blades. Combustion chamber tests necessitated modification of the original design. Chiefly, this meant an alteration in the number and positions of the air supply holes in the combustion chamber conical casing to obtain the requisite temperature distribution. Despite the very short flame fronts, all fuel was burned inside the chamber and the general results were excellent. In one test, the airflow was increased by one-third over the designed requirements without any adverse effects on combustion. Redesign of the combustion chamber in the light of tests also brought about the repositioning of the turbine a little further aft. Details of any other progress made are unknown, but time ran out before a complete unit could be built.

Blohm und Voss Schiffswerft

The 7,500 shp marine gas turbine unit (Auftrag 353) — Progress made on Auftrag 353

The Blohm und Voss company, with various works near Hamburg, was chiefly concerned with shipbuilding and maritime engineering, including steam turbine work. In addition, an aircraft designing and building section was formed in July 1933, but although the design section was very prolific and active, wartime orders were comparatively small. Most aircraft production centred on the BV 138 flying boat and, increasingly, sub-contract work for other aircraft companies.

By March 1942 at the latest, Blohm und Voss were working on the marine gas turbine unit described below, and its design was substantially complete by the end of March 1943. Backed by the OKM and known by Blohm und Voss as Auftrag (order or commission) 353, the unit was designed in Büro Tu.107 at the company's Hamburg-Steinwärder works. The only name known to be connected with the project is Hermann Schepler, and it seems likely that he was the chief engineer concerned.

The 7,500 shp marine gas turbine unit (Auftrag 353)

The unit simply consisted of a gas turbine with its shaft connected via a gearbox to the boat's centre propeller shaft. Two other outboard shafts of the vessel were driven by diesel engines, and although the vessel type is unknown, we might assume that it was of the S.100 type. In this case, the normal 2,500 hp centre diesel engine was replaced by the gas turbine unit of about three times the power. While the Blohm und Voss gas turbine was doubtless to be used only for boosting speed during pursuit and combat, it is interesting to note that it was designed for continuous operation. Figure 4.7 shows the layout of the unit, which is described as follows:

Intake and compressor:
The intake section consisted of a cast annular housing with a bell-shaped mouth at the top. Also at the top were attachment points, while an integrally cast bearing housing was formed at the centre. A flange at the rear of the intake casting was bolted to the front of the compressor housing, which was cast with deep, external circumferential ribs and was split longitudinally for assembly. At the top rear end of the compressor housing there were further attachment points.

For the compressor, the AVA at Göttingen made alternative studies of 11-, 12-, and 14-stage axial compressors based on the same rotational speed but with varying diameters. From these studies, a 14-stage axial compressor was chosen using impulse-type stator blades whereby most of the pressure rise occurred across the rotor blades, leaving the stator blades to act as guide vanes. This compressor gave a compression ratio of about 5.0 to 1 with an overall efficiency of 85 per cent. Geometrically similar curved aerofoil blade sections were used for all rotor blades, while the stator blades had fairly thick, round-nosed sections of different series at certain stages. Progressing from the intake end, the blades decreased in size, and at certain stages the number per stage was increased. The rotor was of a constant 0.530 m (1 ft 8^m in) diameter, leaving the blades to progressively decrease in height.

Following Blohm und Voss low-pressure steam turbine practice, the compressor rotor consisted of a hollow drum with bolted end discs and stub shafts. To accommodate the substantial axial thrust from the compressor, the front shaft fitted into a ball thrust bearing housed inside the intake casting. The rear shaft was carried by a roller bearing. Inside the drum rotor was a steel shaft which was supported at intermediate points by radial steel stays to the drum wall. This internal shaft carried thermocouple wires from the turbine to obtain temperature readings, electrical sliprings making the external connection inside the intake housing.

All compressor blades were made from steel, with substantial feet for inserting into grooves with separate packing pieces. For the rotor blades, the feet were of a rhombus-shaped section. Just before the eighth stage of rotor blades, the compressor casing was provided with a lip and annular chamber from which was ducted about 7 per cent of the total airflow for use in sealing and in cooling the turbine. Since the unit was for marine use, the castings of the intake and compressor casing (and certain other items, such as blade packing pieces) were made of a special corrosion-resistant aluminium alloy (4.5/5.0 Mg, 0.6/1.5 Si, 0.1 Zn, 0.1/0.5 Mn, 0.05 Cu, 0.5 Fe, 0/0.3 Sb, 0/0.3 Cr, 0/0.5 T, blce Al).

The rear of the compressor casing was bolted to the flange of a central casting which provided a housing for the compressor rear bearing and also diffuser ducts leading air to the combustion chambers. Air left the compressor and entered the ducts at a temperature of 190/200 °C, the air mass flow being about 3.3 times the theoretical combustion needs.

Combustion chambers:
The combustion chambers were designed by the DVL, Berlin, and were of the separate, cannular type. Further details are lacking, other than the fact that the peak combustion temperature was 900 °C, but we could assume from the general arrangement drawing that six

combustion chambers were to be used, although there was room for seven. The chambers were bolted at their inlet ends to the diffuser ducts of the central casting, and at their outlet ends to a cast inlet nozzle ring.

Turbine section:
The cast inlet nozzle ring carried, at its top, rear attachment points for the unit, while inside a circular flange was attached by close-clearance, radial keys to a tubular housing for the turbine shaft bearings. This method of key attachment reduced heat conduction to a minimum and also permitted expansion. The tubular housing was flexibly connected to the rear centre of the central casting, and was fitted with two roller bearing sets for the turbine shaft, there being no other bearings (or shaft) on the other side of the turbine wheel. A connection between the turbine and compressor shafts was made just behind the compressor rotor's rear housing.

Turbine inlet nozzles, or guide vanes, were dovetailed and welded into inner and outer supporting rings which were then fitted and bolted into the inlet nozzle ring. Perhaps surprisingly, the turbine itself was of the so-called two-row Curtis type (having about 10 per cent reaction), which had two stages of rotor blades on the same wheel with a single stage of stator blades between the rotor stages. The mean turbine diameter was 0.990 m (3 ft 3 in). For the inlet nozzles and turbine blades, a heat-resisting steel by DEW was used and known as ATS (18.0 Cr, 9.5 Ni, 1.5 W, 0.60 Mn, 0.50 Si, 0.14 C, blce Fe). The turbine rotor blades had hollow feet which fitted over spigots on the turbine wheel and were pinned, while the stator blades were dovetailed and notched into inner and outer supporting rings, the outer ring being bolted into an extension of the inlet nozzle ring forming the turbine housing. Between the stator blade inner support ring and the turbine wheel rim there was labyrinth sealing. An ingenious but simple method of sealing at the tips of the turbine rotor blades was employed whereby thin plates were welded inside the turbine housing to give a very small clearance annulus and yet retain flexibility in the event of blade-tip rubbing.

The turbine wheel was made from an alloy steel by Bochumer Verein and known as BVT 90 (1.43 Cr, 1.10 Mo, 0.55 Mn, 0.93 Si, 0.55 W, 0.10 C, blce Fe), and was bolted to a flange on the hollow turbine shaft. On each side of the turbine wheel were vanes which supported plates to form cooling-air ducts for the rotor blades. The cooling-air system for the turbine rotor blades and exhaust guide vanes is described further on, but since no evidence can be found concerning cooling for the inlet nozzles and turbine stator blades, it is assumed that these items were in solid, heat-resisting ATS steel.

Exhaust section:
Just past the turbine section were air-cooled guide vanes which directed the exhaust gases into an angular duct. This duct was of relatively light, welded construction and had an attachment point at its top. The duct directed the exhaust gases upwards and incorporated a channel for drawing in atmospheric cooling air.

Cooling and sealing air system:
Air bled off from the compressor was piped to a connection point at the rear of the central casting (just past the point where the turbine and compressor shafts were joined). This connection point led into an annulus from which the cooling air was distributed fore and aft into the bearing housings and then into the atmosphere. However, the bulk of the air was led from the annulus into holes in the hollow turbine shaft. It then passed along inside the shaft and emerged into the channels formed on the forward face of the turbine wheel. From these channels the air entered the first row of rotor blades via root slots and emerged through the blade tips into the hot gas stream. The second row of rotor blades was similarly cooled, but in this case the air was drawn from the atmosphere through a separate channel in the exhaust duct and was encouraged to enter the channels on the rear face of the turbine wheel by short impeller vanes attached to the wheel. Also from the atmospheric air channel was drawn cooling air for the hollow exhaust guide vanes; this cooling air emerged from the trailing edges of the guide vanes, and by lowering the temperature of the exhaust gases made a lighter exhaust duct construction possible.

On the forward, or inlet, side of the turbine wheel there were three sets of radial packing to seal out hot gases from the internal bearing housings. The innermost of these packings was around the turbine wheel attachment flange, and this packing was pressure-sealed by air tapped from the compressor. This air leaked out through the other seals into the hot gases (passing into the turbine) and also into the turbine bearing housing.

Output shaft:
The turbine developed 20,000 hp at full speed, of which 12,500 hp was used to drive the compressor and auxiliaries. The remaining 7,500 hp was delivered to the vessel's propeller shaft via a gearbox, of which no details are known. Sensibly enough, power was taken from the shaft of the gas turbine unit at the cool intake end.

Auxiliaries:
The types and method of connecting items such as starter, generator and oil pumps are unknown. For lubrication, oil was piped to the bearings and then drawn off from the bottom of the front and centre bearing housings by scavenge pumps for filtering and recirculation.

Estimated data for the Blohm und Voss marine gas turbine unit (Auftrag 353) are:

Turbine power	20,000 hp
Compressor power absorption	12,500 hp

Output shaft power	7,500 hp
Speed	5,400 rpm
Weight complete	11,000 kg (24,255 lb)
Pressure ratio	4.98 to 1
Specific fuel consumption	450–500 gm/shp/h (0.992–1.1025 lb/shp/h)
Airflow	44.95 kg/sec (99.11 lb/sec)
Specific weight	1.46 kg per hp (3.23 lb per hp)
Peak cycle temperature	900 °C
Compressor efficiency	85 per cent
Turbine efficiency (incl. cooling losses)	82 per cent
Overall unit efficiency	15 per cent
Height (over combustion chambers)	1.630 m (5 ft 4 in)
Overall length (from output shaft flange to end of exhaust duct)	5.060 m (16 ft 7 in)

Progress made on Auftrag 353

By the end of the war, the design of the unit was substantially completed and a model combustion chamber had been tested. The rough-machined turbine wheel was made but not delivered to the Steinwärder works. Blohm und Voss themselves only reached the stage of making the compressor castings and some of the turbine blades.

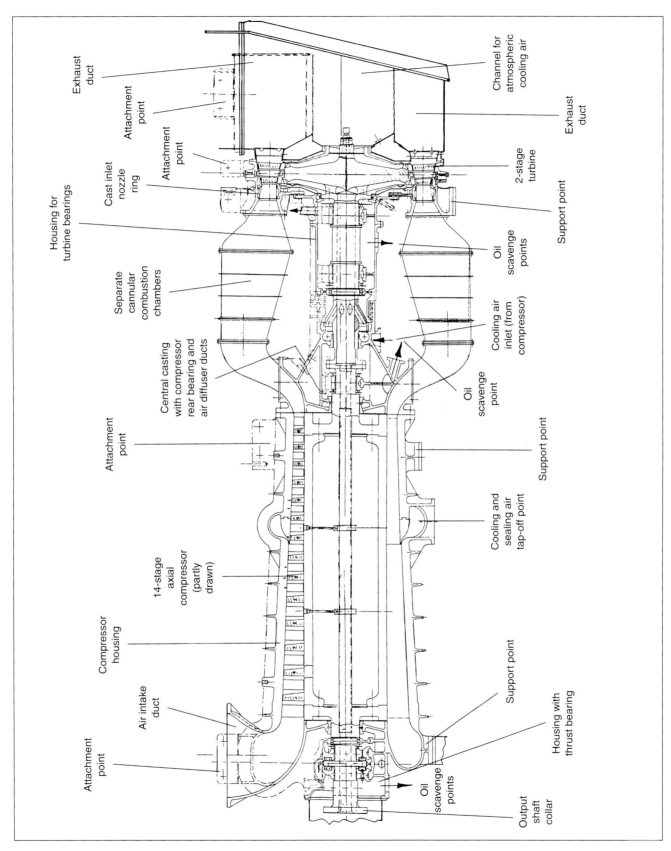

Fig 4.7 Blohm und Voss 7,500 shp marine gas turbine unit (Auftrag 353).

Maschinenfabrik Augsburg-Nürnberg AG (MAN)

The 7,500 shp marine gas turbine unit — Conclusion

The large organisation of the Maschinenfabrik Augsburg-Nürnberg AG (MAN), as its title implies, was centred chiefly at Augsburg and Nürnberg. A great deal of MAN work was connected with the Kriegsmarine, production covering diesel engines and large guns, development including two-stroke warship engines up to 16,500 hp and research covering such items as the supercharging of U-boat engines. Other work in the heavy engineering field included tank production. In the Augsburg area alone, MAN had about 10,000 employed on production, while others worked in their vast research centre in the same area. This research centre was sponsored chiefly by the OKM, and dealt mostly with reciprocating engines.

The 7,500 shp marine gas turbine unit

When, in 1942, MAN offered a project for a marine gas turbine unit, it was not accepted. Attention then turned to the design of a gas turbine plant to run on producer gas for driving a 12,000 kW generator set (see Section 5), but the marine gas turbine was returned to following the company's dispersal from Augsburg. The dispersed design office was located in the offices of a cement factory at Harburg (about 50 km north of Augsburg). Here, under the leadership of a Dr Scheutte, a 7,500 shp marine gas turbine unit was designed during 1943, but in March 1944 it was once again turned down by the OKM. During an air raid on the Augsburg area in February 1944, the drawings for the power unit were destroyed so that, although the OKM still held their copy of the material, the only details of the scheme left to us are from Scheutte's memory.

The unit consisted of a compressor set supplying air to a gas turbine which drove the propeller shaft through reduction gearing. For the compressor set, a gas turbine consisting of an axial compressor, annular combustion chamber and axial turbine was used, the turbine having sufficient power to drive the compressor only. A proportion of air was tapped off from the compressor and ducted to a second, separate annular chamber which supplied hot gases to the power turbine. The turbine blades used Krupp's Tinidur heat-resisting steel (29.2 Ni, 14.9 Cr, 2.1 Ti, 0.8 Si, 0.7 Mn, 0.13 C, blce Fe), which, if no cooling was used, meant that the mean temperature at the turbine could not exceed about 600 °C.

Figure 4.8 gives a schematic layout of the unit, for which the only known data are:

Output shaft power	7,500 hp
Power turbine speed	6,000 rpm
Power turbine maximum inlet gas temperature	800 °C
Overall unit efficiency	25 per cent
Output shaft speed	1,000 rpm

Conclusion

As we have seen, none of the marine gas turbine schemes worked on by the Germans reached fruition by the end of the war, when development ceased. This failure, where schemes were backed by the OKM, was apparently because of a lack of time, although prototypes at least of the Brückner-Kanis and Blohm und Voss units could have been built had more resources been switched to the work. Also responsible, in the case of the Brückner-Kanis units, may have been an official directive to 'change horses in mid-stream', whereby the Drehkessel gas turbine (a cross between the steam and gas turbine) was put on a low-priority, experimental basis in order to concentrate on the pure gas turbine system.

Notwithstanding the considerable developmental difficulties inherent in the Drehkessel gas turbine, its pursuance was considered worthwhile because it promised a thermal efficiency at least as good as gas turbines and better than marine steam turbines of the day. An important claim was that of high power-plant efficiency (probably about 30 per cent) without the use of acutely scarce heat-resisting steels in the gas turbine. By using water-cooled parts, very high combustion temperatures were possible. However, as post-war development rapidly improved the gas turbine without resorting to cooled parts, so the importance of the Drehkessel gas turbine unit diminished, and no further work on it is known apart from theoretical study in immediate post-war years. Subsequent use of combined steam and gas turbine (COSAG) machinery has been made though; for example, the Royal Navy commissioned a COSAG frigate from AEI in 1959.

The Brückner-Kanis pure gas turbine unit, whereby four adapted turbojets were to act as gas producers for a single power turbine, must now be seen as an expedient for building a large power unit within a comparatively short time. Although the whole unit is historically

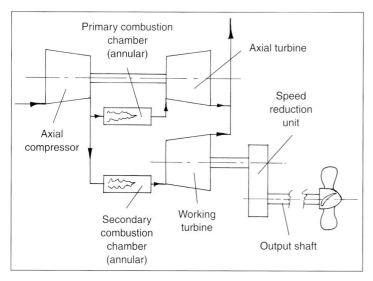

Fig 4.8 Schematic layout of MAN 7,500 shp marine gas turbine.

interesting, the chief interest must lie in the gas-producing element, i.e. the contra-rotating gas turbine. Such a gas turbine offered high power and very high component efficiency for a minimum installation volume, which was probably why a combination of four such units was seen as practical for the marine unit. Used as a turbojet, the original purpose, the Brückner-Kanis contra-rotating gas turbine would in all probability have been superior to all other German turbojets of similar size, assuming, of course, successful development. Given this probability, it is hard to understand why the RLM failed to back it for aircraft use, especially since it was to be built (in modified form) for marine use. Somewhere amongst the varied types of modern turbojets will be found various of the elements planned for the Brückner-Kanis turbojet, although one would hesitate to say which particular items owed their origin to the German work. Certainly, Helmut Weinrich, who was behind this particular work, was a very forward-thinking engineer.

Of the marine units discussed in this Section, that proposed by Blohm und Voss was based on the most simple scheme of a direct mechanical connection between the gas turbine rotor and the propeller shaft. However, although displaying some ingenious design features, such as those concerned with sealing and turbine cooling, their unit was very heavy for its power output, was low stressed and had a low overall efficiency. Furthermore, even for a gas turbine unit, the specific fuel consumption was very high, the figures quoted being the same as those for the straightforward GT 101 gas turbine unit (for armoured fighting vehicles and described in the previous section), which was only about one sixth as powerful. A quick calculation shows that the Blohm und Voss unit at full power would gobble up one ton of fuel in about eighteen minutes! Despite these factors militating against the unit, it seems logical that the OKM should have continued to back its development as the least complicated first step in breaking into the new field of marine gas turbines. With the Blohm und Voss unit, built without too much difficulty, experience on the water could be gained before proceeding towards a more sophisticated and efficient gas turbine unit.

The scheme upon which the MAN 7,500 shp marine gas turbine unit was based was a progression of that adopted by Blohm und Voss, and was similar, in principle at least, to the scheme for the GT 102 unit described in the previous section for armoured fighting vehicles. Although not sponsored by the OKM, there was possibly a plan to develop the MAN unit once experience was gained with the more simple Blohm und Voss unit. Too few details are available fully to assess the MAN unit, but the good overall efficiency expected is interesting, even if optimistic. Although both Brückner-Kanis and MAN units had separate power turbines, the feature was chosen for different reasons. In the Brückner-Kanis units, a separate power turbine formed an essential part of a system (the steam turbine in a gas/steam system) or of an adaptation (a 'unifying' turbine in a multi-gas generator system). In the MAN unit, a separate power turbine was freely chosen in preference to a simple layout such as that used by Blohm und Voss. The reasons for the MAN choice are unknown but must have included the following. Use of a mechanically separate power turbine results in a simple gear transmission to the propeller shaft, a greater speed range without deviating too far from the optimum speed of the compressor unit and high torque at low speeds or when stalled by debris around the propeller.

Sure enough, one of the first gas turbines to be applied to marine propulsion used a mechanically separate power turbine. In 1950, two 200 bhp T.8 gas turbines developed by the Rover company in England were installed in the motor yacht *Torquil*, and others were purchased by the Royal Navy for evaluation. The T.8 unit (which, incidentally, owed nothing to previous German work) had its power turbine arranged coaxially, behind the turbine driving the compressor. Other post-war events of note, and the companies responsible for the power-plants concerned, were in 1947 when the first gas turbine vessel put to sea (Metropolitan-Vickers), in 1952 when the first Atlantic crossing by a ship powered solely by the gas turbine was made (British Thomson-Houston) and in 1959 when the US Navy tested a gas-turbine-powered hydrofoil (Lycoming).

Despite their quite different natures, the projected German marine gas turbine units all had one thing in common – high fuel consumption. As in the case of gas turbines for armoured fighting vehicles, the need to reduce the specific fuel consumption was strongly felt, although not to the same extent, since, in general, the marine units were only to be used for short-time speed-boosting purposes. Once again, the answer lay in increasing efficiency by the use of a heat exchanger and/or by increasing combustion temperatures.

Section 5

Gas Turbines for Industry

Holzwarth Gasturbinen GmbH — Brown Boverie & Cie, AG (BBC), Mannheim — Maschinenfabrik Augsburg-Nürnberg AG (MAN) — Allgemeine Elektrizitäts-Gesellschaft (AEG) — Conclusion

In this section, a look is taken at a cross-section of German companies which worked on stationary gas turbine units to provide power for industry. The power produced took the form of electricity in the Holzwarth, MAN and AEG schemes discussed, but from Brown Boveri came more diverse uses of the gas turbine in the generation of air and steam. Often, turbine schemes were employing heat which was going to waste anyway.

At first glance, the development of a stationary gas turbine power unit posed far fewer problems than the development of lighter, more compact mobile gas turbines, but a vast number of schemes up to about 1930 came to nothing, largely through a lack of suitable heat-resisting materials. Then, once acceptable gas turbines became possible, there were still the problems of building them either sufficiently large or rugged and ascertaining whether or not they had a worthwhile place in industry.

Most companies interested in steam turbines, large piston engines and so forth, at some time or other made studies of the gas turbine, sometimes combining it with the more familiar power-plants. Naturally, not all German companies or organisations that had anything to do with the industrial gas turbine can be discussed here. Many had barely begun their work before 1945. It should be realised, however, that power-plants discussed elsewhere in this book were sometimes projected as the gas generator/s for an industrial turbine. An example of this is the multi-bank version of the Lutz swing-piston gas generator (see Section 6) which was to supply a 40,000 hp turbine. Again, versions of the Drehkessel gas turbine (see Section 4) were planned for stationary applications, while multiple arrangements of turbojets as gas producers for a large power turbine were also possible (e.g. see under Brückner-Kanis, Section 4). Regarding the Drehkessel (rotating boiler) or water-cooled combustion chamber, an interesting comparison can be made between this and Holzwarth's gas turbine. In the former, steam produced by cooling the combustion chamber and turbine was used to power a steam turbine driving the compressor and giving a large amount of the power output. Holzwarth also used steam produced by cooling his gas turbine to drive the compressor, but in this case the production of steam was almost incidental and not one of the prime aims. In the MAN 5,000 kW unit described later, we find steam produced by cooling the gas turbine used in a fundamentally different way in a process to produce gas as the fuel for the combustion chamber.

Holzwarth Gasturbinen GmbH

Holzwarth's constant-volume gas turbines

Dr-Ing. Hans Holzwarth, one-time chief protagonist in Germany of the constant-volume or explosion gas turbine, had the first such turbine built to his designs in 1908. His aim was to develop and market the turbine for industrial use, although, in later years, at least one patent (575,054) was filed in February 1930 proposing the use of the turbine to drive vehicles. Although Holzwarth's turbines were built by other companies, the Holzwarth Gasturbinen GmbH was eventually formed at Mülheim (Ruhr) to carry out design, development and exploitation, and there was, before the Second World War, a New York office also. However, because of inherent difficulties with the turbine units, efforts to introduce it into commercial use met with only limited success. Two principal engineers of the company were G. Schaper and F. Hofmann.

Holzwarth's constant-volume gas turbines

The incentive for beginning work on the constant-volume or explosion gas turbine arose because of the difficulties involved in developing a compressor efficient enough for the constant-pressure or continuous-combustion gas turbine. In the latter type, the compressor has to deliver a large amount of air to mix with the combustion gases in order to reduce the gas temperature to one acceptable by the turbine; at the same time, this compressor must be efficient enough not to drain too much of the available turbine power. Holzwarth sought to avoid these difficulties (which were then thought insurmountable) by using the constant-volume turbine, since the compressor for such a unit had to supply a relatively much smaller amount of air, and its efficiency could be low without disastrous results.

From 1905, Holzwarth worked on his ideas, and between 1906 and 1908 his first gas turbine was built by the firm Körting of Hannover. This experimental unit furnished much useful data to aid further designing. Subsequent designs employed hydraulically operated valves to admit an explosive mixture of fuel and air into the combustion chamber, and spring-loaded (later, hydraulically operated) valves to admit hot gases from the subsequent explosions to the turbine inlet nozzles. The combustion chamber and turbine was water-cooled, and the steam derived from this process was used to drive a steam turbine which drove a Rateau compressor supplying air under pressure to the combustion chamber. In addition, the steam turbine also drove an exhauster stack which sucked residual explosion gases from the combustion chamber prior to the entry of a fresh fuel/air charge.

Because water was used to take care of the cooling problem, the compressor only had to deliver an amount of air slightly above combustion requirements, and because of the large volume of the combustion chamber, this air needed only slight compression. Following an explosion, the pressure in the combustion chamber rose to about 4.5 times its original value, and it was thought that such a pressure would result in good efficiency. However, this advantage was offset by the intermittent torque and reduced turbine efficiency caused by the explosive or pulsing nature of the cycle. In addition, there were other disadvantages, such as plant complication and cost brought about by the automatic valves, elaborate castings, equipment to handle the water and steam system, and so forth. Altogether, Holzwarth and his team needed much fortitude and tenacity of purpose in order to tackle the formidable problems of development.

Between 1909 and 1913, Holzwarth's second explosion gas turbine was constructed, this time by Brown Boveri & Cie (BBC) in their Mannheim workshops. This unit, possibly planned for commercial use, had a designed rating of 1,000 hp, but only gave a net output of about 200 hp and was not very satisfactory. Contributing to this were technical weaknesses and the nature of the materials available at the time. One of the most important limiting factors was that the materials available for the turbine restricted its temperature to a mere 450 °C or so, and efficiency consequently suffered.

From 1914 to 1927, Holzwarth continued to develop his turbine and worked in co-operation with the Maschinenfabrik Thyssen Company at Mülheim (later merged with DEMAG, Duisburg). During the last couple of years of this phase, combustion tests were made with various fuels such as coke-oven gas, oil and both bituminous and brown coals, the latter type of coal forming a high proportion of Germany's solid fuels. Pulverised coal burned very smoothly and without failure, but the effects of coal ash on the turbine blades were severe and resulted in deep pitting within an unacceptably short time. (Nevertheless, Holzwarth patents were filed as late as 1931 concerned with coal dust feeding equipment (566,740) and explosion chamber valve control when operating with pulverised coal (578,254), to give two examples.) Experience and better materials enabled general improvements to be made to the Holzwarth turbine, and an order was obtained from the Prussian State Railways for a turbine to drive a 350 kW generator. This unit was built by the Thyssen company and put into service in 1923. It ran successfully for a number of years and was for some time claimed to be the only gas turbine in regular commercial use.

1928 saw Holzwarth working once again in co-operation with the Brown Boveri company, which built an explosion turbine for driving a 2,000 kW three-phase generator. This unit was installed at the very large steel works of August Thyssen Hütte at Hamborn-Brückhausen (north of Duisburg) in 1933, and used a two-chamber, two-stroke cycle.

In the meantime, a large number of patents were filed covering Holzwarth turbine unit design and development. To give some examples, there were various schemes for turbine blade cooling using oil, steam, etc., and equipment for blade cooling was attributed to F. Hofmann and filed under patent 597,556 in 1933; a patent application (H 143,820) for a heat exchanger working with combustion exhaust gases was filed in 1935 but not granted. Figure 5.1 gives a schematic arrangement of a Holzwarth gas turbine based on patent 619,216 filed in August 1932 and granted in September 1935. The actual patent drawing indicates features planned or developed by then. These features included hydraulically operated valves throughout, hollow, cooled walls for the combustion chamber and turbine housing and cooled turbine blades. The turbine had a shaft and bearings on the power output side only, while its blades followed steam turbine practice with two rows of rotor blades and an intermediate row of stator blades. From the same specification, a turbine speed of 8,500 rpm is indicated, together with an explosion or combustion chamber having an overall length (including conical portions) of six internal diameters and a central parallel section length of 2.3 diameters. An alternative scheme specified the use of two combustion chambers, which presumably would do much to even out the explosions by arranging for alternate firing.

A second gas turbine unit, built by Brown Boveri in 1938, was installed at the August Thyssen Hütte steel works. Designated type 4706, the gas turbine drove a 5,000 kW generator, and had its steam turbine supplied from a BBC exhaust heat boiler. Fuel for combustion was blast-furnace gas compressed to about 6 atm., and the hydraulically operated valves worked at between 60 and 100 cycles per minute. This unit was regarded as experimental and not as part of the regular steel works equipment. Consequently, it was only run from time to time during day shifts, the last run being made in 1943, when it suffered heavy bomb damage. After repairs, and just before further trials were planned, a bombing raid on 22 January 1945 once again damaged the unit. As late as February 1944, designs were drawn up for a large 20,000 kW unit (type 4710), but the end of the war in May 1945 saw an end to the Holzwarth constant-volume gas turbine, and no further interest in it is known.

GAS TURBINES FOR INDUSTRY

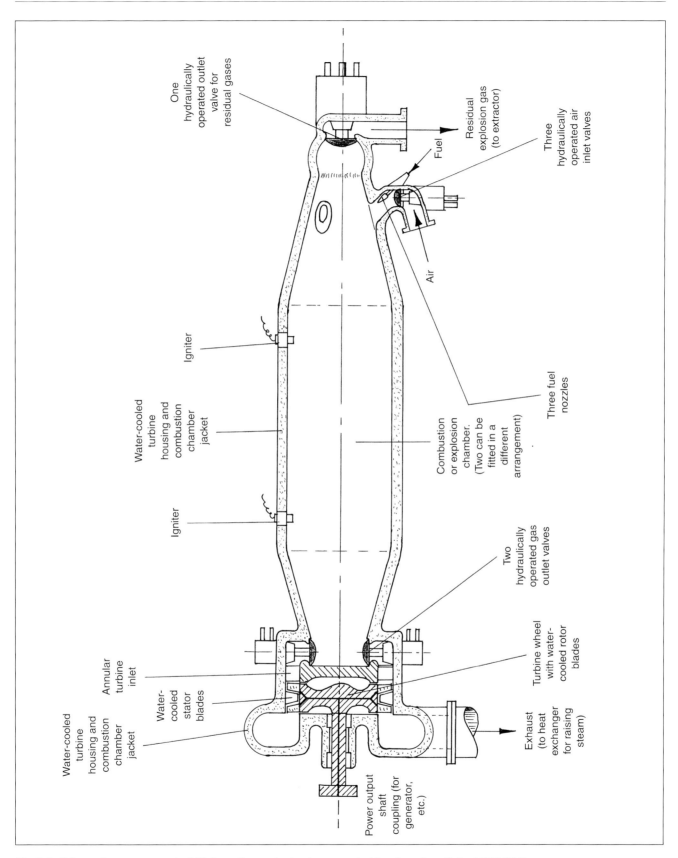

Fig 5.1 Schematic arrangement of Holzwarth constant-volume gas turbine (based on Patent 619,216). *(Author)*

Brown Boveri & Cie, AG (BBC), Mannheim

The Velox steam generator — The blast furnace gas turbine blower — The regenerator gas turbine — Combining steam generation with air heating and blowing — The gas turbine blower with supercharged air heater — Quantitative comparison of BBC systems

The experience of the famous Brown Boveri company with gas turbines goes back to the early 1900s, and if one considers both the German and Swiss branches of this company, an impressive list of first achievements appears. The German BBC was based at Mannheim-Käfertal and the Swiss BBC at Baden, but the extent of co-operation and information exchange on gas turbines between the two companies during wartime is uncertain. Here, because of the terms of reference of this book, we are concerned with the efforts of the German BBC to introduce the gas turbine into industry, but a few comments on both German and Swiss early work should be of interest.

Probably the first acquaintance of BBC with the gas turbine occurred when the Swiss built a Rateau-designed compressor in about 1905 for the gas turbine built by Armengaud and Lemale in France. Between 1909 and 1938, BBC in Germany gained some experience with their work on Holzwarth's explosion gas turbines, while in 1922 they began making exhaust turbo-superchargers for diesel engines. The type of supercharger concerned was based on the ideas of Dr Alfred J. Büchi, who invented the exhaust gas turbo-supercharger for piston engines in 1905 in Switzerland, and subsequently held a great many patents for such equipment. (Dr Hoss in the USA independently thought of the same idea as Büchi at about the same time.) The BBC superchargers were made to operate in exhaust gases having a temperature in the region of 550 °C, the turbine blade material being selected to withstand 575 °C continuously and up to 600 °C for short periods.

From 1928 up to about 1940, a great variety of patents in the gas turbine field were filed in Germany under the name of BBC, Baden. Engineers connected with those patents and some of the items which these men covered were:

Hans Pfenninger	Gas turbine air generation
Walter G. Noack	Heat exchangers
Erich Schmitt	Gas mixing
Claude Seippel	Control and construction
Adolf Meyer	Processes and equipment

In 1936, BBC in Switzerland built the first process gas turbine unit for what became known as the Houdry process. In this, hot air from a petroleum cracking plant was used to power a turbine which drove a compressor supplying air for catalyst regeneration. The same turbine also drove an electric generator. This first unit was installed at Marcus Hook refinery at Philadelphia, USA, and the Houdry process was subsequently extensively developed there. Three years later (1939), BBC in Switzerland produced another two gas turbine firsts. At Neuchatel, the first successful open-cycle, constant-pressure gas turbine generator set was installed in a bomb-proof housing for emergency war-time use; it used a compressor and turbine of the axial type and ran on a light distillate fuel. Also in 1939, the company produced the first gas turbine for a locomotive, already mentioned in the Land Traction Section.

In the meantime, BBC in Germany were not idle in the gas turbine field. Although in 1941 the company set up a department to work on turbojet problems, its main effort was made much earlier to utilise the gas turbine in industry (e.g. in foundries). This work, which is examined next, appears to have had its foundation or starting point in the patents, already mentioned, held in the name of BBC, Baden. At Mannheim, BBC's chief engineer concerned with industrial gas turbines was Max Schattschneider. During the Second World War, his department (and others) was dispersed to the Robert Bunsen school at Neuenheim, across the river from Heidelberg.

The Velox steam generator

The construction by BBC of Holzwarth's 2,000 kW gas turbine, which was installed by 1933, gave the stimulus to develop the Velox (fast) steam generator. In those days, steam generation was of paramount importance to heavy industries, such as the iron and steel industry, since the steam turbine could efficiently drive electric generators and blowers of sufficient size to supply the quantities of electricity and air needed. The aim of the Velox steam generator was to obtain rapid and efficient steam generation within a comparatively small space by employing combustion under pressure. By using the hot exhaust gases from the combustion within the steam generator to drive a turbine, a supercharger could be driven for the Velox combustion chamber. Patents for the overall scheme and its main elements were filed by April 1932 and had been granted by January 1938.

Converting these ideas into useful hardware entailed intensive development work, especially as only rugged and reliable equipment can be tolerated in the type of heavy industry in mind. BBC's previous experience at Mannheim with diesel-engine superchargers gave them some useful data on turbines, but data on the centrifugal compressors of these superchargers were of little use since axial compressors were considered necessary for the Velox unit. Furthermore, where the steam generator

GAS TURBINES FOR INDUSTRY

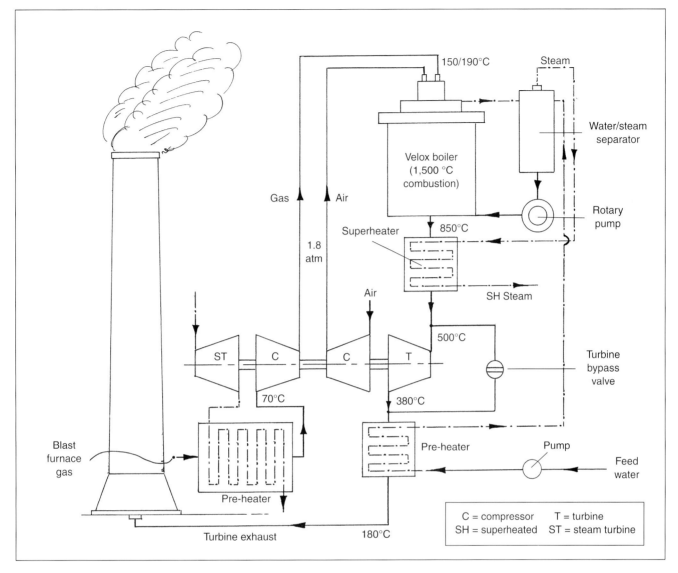

Fig 5.2 BBC Velox steam-generating system.

was to use blast furnace gas as fuel, a compressor was needed for this gas in addition to the air compressor. Since both compressors were driven by one gas turbine, it was necessary to design the output of both to give the requisite air/fuel ratio at all speeds. The development of these axial compressors, with their exacting specifications, was aided by newly gained knowledge of aerodynamics. Regarding power requirements, the combustion chamber of the steam generator had to provide hot exhaust gases sufficient (after steam raising) to run the gas turbine at a power range of between about 1,350 and 9,350 hp.

Figure 5.2 shows the operating scheme of the Velox steam-generating system. After passing through a gas pre-heater unit, condensed water was fed into a steam/water separator. Using a turbulence principle, the separator passed steam out via a superheater, and the remaining water proceeded on to the Velox steam generator. In this, a combustion temperature of up to 1,500 °C operated on the vaporiser elements to produce saturated steam which was pumped back to the separator for steam extraction and recirculation. As steam reached the required condition, it passed from the separator into and out of the superheater.

Some heat from the water/steam circuit was used to pre-heat washed blast furnace gas up to about 70 °C. Gas and air then passed through their respective compressors and reached the Velox combustion chamber with a pressure of 1.8 atm. and a temperature of 150/190 °C. This delivery pressure and temperature was a prime factor in the speed and temperature of combustion and therefore the speed of steam raising.

After the work of steam raising, the hot exhaust gases left the Velox generator and were piped to the steam superheater, which they entered with a speed of 200 m/sec (656 ft/sec) and a temperature of 850 °C. From here, the gases passed to the gas turbine, which was entered with a pressure of 1.5 atm. and a temperature of 500 °C. Leaving the gas turbine, which drove both compressors, the gases were exhausted to the atmosphere via the feed-water pre-heater. The final speed and temperature of the gases was about 60 m/sec (197 ft/sec) and 180 °C.

Automatic regulation of the system was as follows. If steam pressure dropped owing to an increased demand, the steam turbine (or other type of auxiliary motor used for starting up the gas turbine) was started and speeded up both compressors and gas turbine, all connected together. Thus, more fuel and air flowed to the combustion chamber and the steam pressure rose. Should a reduction in steam pressure be required, a valve opened to bypass the hot gases past the gas turbine and direct to the feed-water pre-heater. This virtually cut off the fuel/air supply to the steam generator until the requisite lower steam pressure was obtained.

Advantages of the Velox steam generator over the conventional water tube boiler (with, say, a steam turbo-blower) of the day can be summarised as follows:

(i) Heat, previously largely wasted after steam raising, acted as the combustion system for a constant-pressure gas turbo-supercharger system supplying the steam generator. The result was high efficiency combined with smaller space and material requirements.
(ii) Efficiency was 88–90 per cent with blast furnace gas and 91–93 per cent with oil fuel. Such efficiencies were maintained between a quarter and full loading.
(iii) Less material meant new and modernised plant more easily obtained. For one ton of steam, the blast furnace gas-fired Velox generator needed about two tons of material, compared with six to eight tons for conventional water-tube boilers.
(iv) Quick starting up. From cold to full running in ten minutes, or even five minutes in special cases.
(v) Quick repairs. After switching off, the Velox combustion chamber was ready in one hour for working on.
(vi) Reduced danger from explosions, because of smaller gas and water spaces.

By the end of 1940, BBC's success with their Velox steam generator was already attested to by the fact that they had a total of 75 units on order, which, altogether, could generate some 2,000,000 kg (4,410,000 lb) of steam per hour. By that time, German iron and steel works had received, or were about to receive, six of the blast furnace gas-fired Velox units.

The blast furnace gas turbine blower

During 1940, BBC began construction of a pure gas turbine unit for application to a blast furnace. Inasmuch as the Velox system contained a steam element, the blast furnace pure gas turbine represents a further step in BBC's gas turbine development. Previously, air had been supplied to the blast furnace by means of a gas piston blower or a steam turbo-blower, the latter possibly being supplied with steam from a Velox generator. Until more recent years, the blast furnace was the backbone of the iron and steel industry, and therefore of industry in general. Once charged with iron ore, coke and limestone, and brought to working temperature, a continuous blast of air through the furnace is all that is needed to produce the liquid pig iron, which can in turn be transformed into various steels. A large quantity of gas, together with a good deal of slag, is also produced in the process. The process cannot be speeded up beyond a certain limit by raising the temperature or the quantity of air blown, since the furnace will resist by uneven working and/or an increase in the specific coke consumption.

The aim of BBC, therefore, was to make the process more efficient and reduce the necessary plant space, the latter always being at a premium in iron works, especially old works. By using gases from the blast furnace to drive a gas turbo-blower, it was foreseen that boiler and condensation plant could be dispensed with, less material and space would be needed and the air generating plant could be sited directly on to the blast furnace, thereby eliminating flow losses in ducts.

Figure 5.3 shows the scheme of the gas turbine blower system for blast furnaces. Essentially, the elements used in the Velox system could be adapted, except that a combustion chamber specially for the gas turbine was needed. Two axial compressors, driven by one turbine, delivered air and blast furnace gas to the combustion chamber. The gas/air ratio was such that an excess of air kept the combustion gases on their way to the turbine down to about 500/550 °C. After leaving the turbine and before exhausting to the atmosphere, the hot gases passed through a heater which pre-heated the air between compressor and combustion chamber. From the same compressor, air was tapped off at an intermediate pressure point and passed to the blast furnace via a heater which derived its heat from steam. Normally, about 100,000 cu m (3,531,000 cu ft) of air per hour at a pressure of 1.2 atm. could be delivered to the furnace, but the pressure could be raised up to 1.85 atm. by increasing the rpm.

Two gas turbine blower units, each of about 5,350 hp, had been built by BBC at Mannheim by 1945. They were destined for the Hermann Goering steel works at Wattenstadt, near Braunschweig, but only one was partly installed before the end of the war.

The regenerator gas turbine

The next piece of iron works equipment that BBC sought to improve with application of the gas turbine was the blast furnace regenerator, or heat accumulator. This equipment, first applied by Wilhelm and Frederick

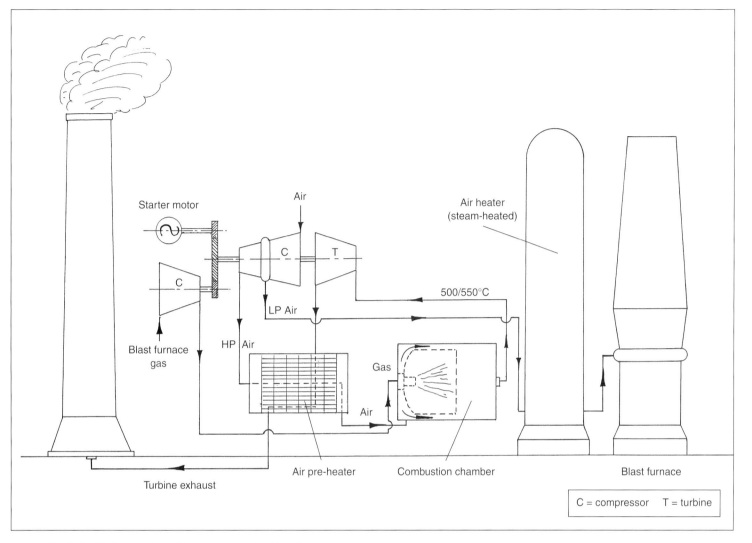

Fig 5.3 BBC gas turbine blower system for blast furnace.

Siemens in 1857, consisted of a refractory brickwork chamber which stored furnace heat and transferred it to the air (and, later, gas also) being fed to the furnace. By using two regenerators, one being heated up by furnace gases while one gave up its heat to the air supply, a continuous process was obtained and the consequent conservation of heat saved much fuel. As far as possible, the regenerator had to act as an equaliser between blast furnace gas generation and consumption. Unfortunately, enormous quantities of brickwork and metal were required to build regenerators. When one considers that there were normally three (including one standby) per furnace, and that each measured some 6 to 8 m (19.7 to 26.2 ft) in diameter by 25 to 30 m (82.1 to 98.4 ft) in height, a reduction in size was obviously most desirable.

To this end, BBC's first scheme proposed to use a gas turbine driving axial compressors to supercharge the regenerator with blast furnace gas and air for combustion. By thus heating under higher pressure, instead of only at near-atmospheric pressure, the brickwork channels in the regenerator could be made much narrower, so that the whole unit was reduced in size to about 3 to 4 m diameter (9.8 to 13.1 ft) by 6 to 8 m (19.7 to 26.2 ft) height. After combustion of the gas and air in the regenerator, the resulting hot gases passed to the turbine driving the compressors. Leaving the turbine, these gases were used to heat the gas and air on their way to the regenerator. A separate steam turbo-blower (consisting of a steam turbine driving a four-stage centrifugal blower) was employed to push air through the regenerator and deliver it to the blest furnace with a temperature of over 700 °C. Switching between two regenerators was made every ten minutes. Figure 5.4 gives the main elements of the scheme.

While this scheme did much to improve refractory brick-lined regenerators, such regenerators were not ideal, since their heat-accumulating capacity was not great and the heat conductivity of refractory brick was

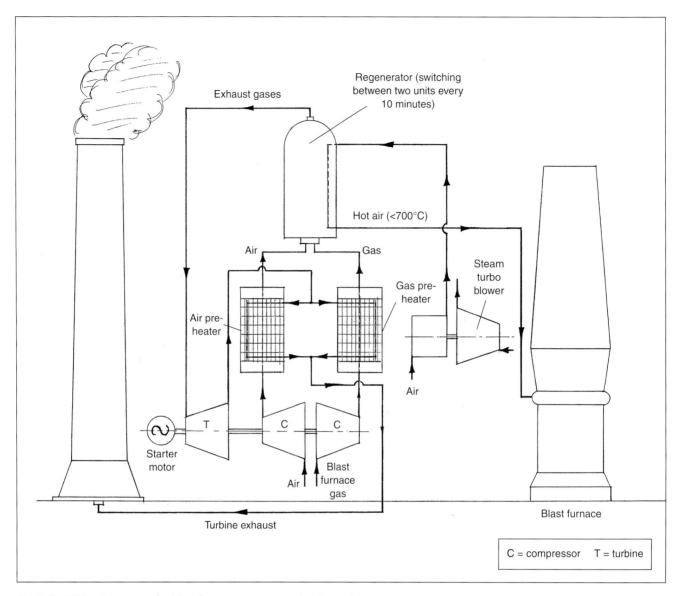

Fig 5.4 BBC turbo system for blast furnace regenerator (air heater).

slight, these shortcomings becoming more noticeable as temperatures were increased. In Figure 5.5, a second scheme is shown in which refractory brick regenerators are replaced by all-metal heaters but with the limitation that the air blown to the blast furnace has a temperature below 700 °C. A gas turbine drove two compressors supplying blast furnace gas and air to a combustion chamber. The resulting combustion gases were cooled to about 900 °C by an excess of air as they left the combustion chamber, and flowed through the high-temperature section of a heat exchanger before entering the gas turbine. The exhaust gases from the turbine then flowed out through the low-temperature section of the heat exchanger before exhausting to atmosphere. Air for the blast furnace was supplied by a separate steam turbo-blower and fed through the low- and high-temperature sections of the heat exchanger before reaching the furnace inlet.

With these arrangements, the air blowing was independent of the air heating, and whatever volume of air was delivered by the steam turbo-blower could be brought up to the maximum temperature of 700 °C.

Combining steam generation with air heating and blowing

So far, steam generation and air heating/blowing schemes have been treated separately, but BBC recognised the economic advantages of combining both processes, and despite the expected difficulties in control, planned a scheme shown in essence in Figure 5.6. A

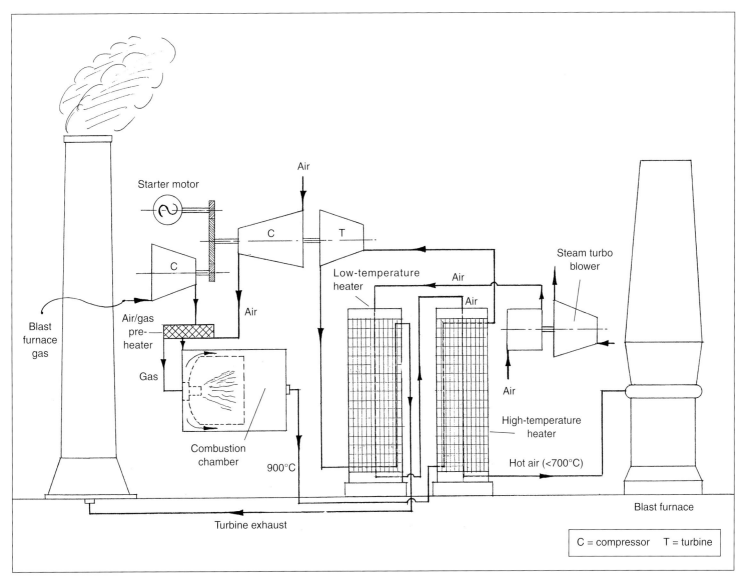

Fig 5.5   BBC gas turbine system for blast furnace air heating.

Velox steam generator was supercharged with blast furnace gas and air from compressors driven by a gas turbine. From the Velox steam generator, the hot gases flowed through the steam superheater and then the high-temperature section of an air heater before entering the gas turbine. From this, the gases flowed through the low-temperature section of the air heater before exhausting to atmosphere. In the meantime, a steam turbo-blower, driven by some of the steam from the Velox generator, supplied air to the blast furnace via the low- and high-temperature sections of the heater. This air left the blower with a compression heat of about 100 °C, and attained temperatures of 275 °C and 700 °C respectively as it passed through the two sections of the air heater.

The air temperature limitation of 700 °C was set by materials then available, and in the scheme just described, temperature limitations governed the volume of steam generated and thus the amount of air blown. A way round this drawback was to use a second Velox generator coupled to the first.

The gas turbine blower with supercharged air heater

The final goal of BBC in applying the gas turbine to the blast furnace was to supply heated air using a gas turbine blower system running on blast furnace gas alone. The scheme (shown in Figure 5.7) differed from others in that no steam turbo-blower was required, and also differed

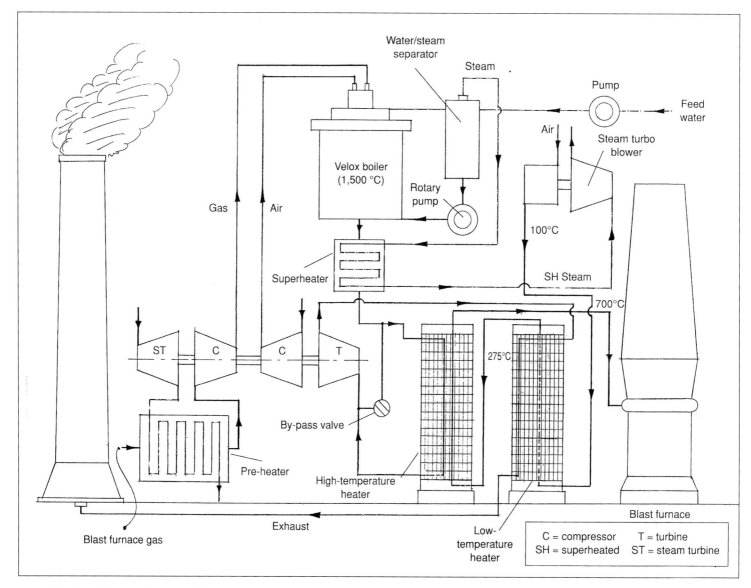

Fig 5.6 BBC combination of Velox steam generation and blast furnace air heating.

from that shown in Figure 5.3 in that no steam was required for air heating, and thus the system formed a unity with the blast furnace. As before, one compressor supplied air to the blast furnace and to the combustion chamber for the gas turbine, while a second compressor supplied blast furnace gas to the combustion chamber. However, the hot gases from this combustion chamber were allowed the increased temperature of 900–1,000 °C so that they could pass through a heater for the blast furnace air before proceeding to the gas turbine. The special air heating equipment for this system was developed in co-operation with Rekuperator GmbH of Düsseldorf.

Quantitative comparison of BBC systems

The following figures give one basis for comparing various systems of providing air for blast furnaces. Systems (A) and (B) were those current at the time of BBC's developments, while (C) and (D) were BBC's schemes for introducing steam gas and then gas systems.

In each blower house there were five blowers (4,500 kW each) – four for supplying air to the blast furnaces and one in reserve. The gas piston blower plant included cooling boiler, piping, valves, foundation bolts and special switching plant. The steam turbo-blower plant included feed-water preparations, pre-heater, evaporator, feed-water container, piping, etc. For steam turbo-blower houses either three Velox generators or three water tube boilers were included, allowing one for reserve. For the different systems, therefore, the following figures were given for space and material requirements of the blower house:

	Plant	Floor space (sq m)	Cubic space (cu m)	Mechnical plant, building, foundations, crane, etc. (tonnes)
(A)	Gas piston blower	3,700	70,000	8,450
(B)	Water tube boiler with steam turbo-blower	2,200	34,600	2,350
(C)	Velox steam generator with steam turbo-blower	1,500	24,000	1,850
(D)	Gas turbine blower	1,900	26,500	1,600

When both air blowing and heating for the blast furnace were considered, the various systems were compared in the light of the blast furnace gas volume they consumed (there being other uses to which the remaining gas could be put):

	Plant	Percentage of total generated blast furnace gas required
(A)	Gas piston blower	31
(B)	Water tube boiler with steam turbo-blower and regenerator	35
(C)	Velox steam generator with steam turbo-blower and regenerator	32
(D)	Gas turbine blower with supercharged air heater	26^n

From these and other estimates, plus capital and running cost analyses, it emerged that the gas turbine blower was superior to other systems, and was followed, in order of merit, by the steam turbo-blower with Velox generator, the steam turbo-blower with conventional boiler and, finally, the gas piston blower.

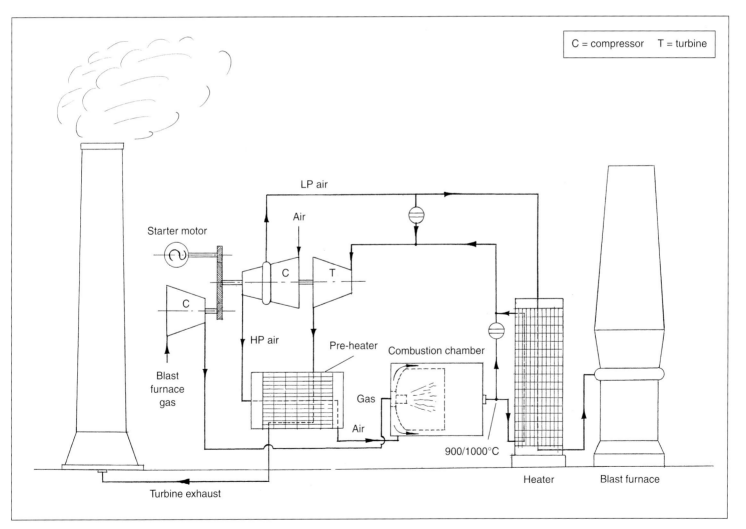

Fig 5.7 BBC blast furnace gas turbine blower with supercharged air heater.

Maschinenfabrik Augsburg-Nürnberg AG (MAN)

The 12,000 kW compound gas turbine installation L760 — The 5,000 kW gas turbine generator

Studies for the application of the gas turbine to industrial use were begun by the MAN organisation in about 1939. Other work on gas turbines, as related in Section 4, covered the development of exhaust turbo-superchargers and a project for a 7,500 shp marine power unit. The director of this, and all other development work in the MAN organisation, was Prof. Dr-Ing. Emil Sorensen, who was also vice-president of the company and a member of the board of the DVL.

By the time the marine gas turbine unit was finally turned down by the OKM in March 1944, design work had begun under Dr Alfred Scheutte at Harburg (50 km north of Augsburg) on an extensive gas turbine installation to drive a 12,000 kW generator. Through the efforts of Sorensen, a contract was obtained from the Electro Werke, Berlin, to develop the installation which was to run on gas produced from brown coal and was to have a guaranteed efficiency of 32 per cent, exclusive of the gas-producing process. Designation of the plant, at least in its project form, appears to have been L760.

The 12,000 kW compound gas turbine installation L760

A schematic layout of the MAN gas turbine installation is shown in Figure 5.8. Four axial-flow compressors (in series and with intercooling between each) driven by three turbines were used, with a fourth turbine driving the electrical generator. After passing through the four compressors in turn, air emerged with a pressure of 20 atm., and was delivered via an exhaust heat exchanger to two combustion chambers connected in parallel. Hot gases from these were supplied to two turbines driving the last two, high-pressure, compressors.

After passing through the two turbines, the hot gases were then delivered to a third combustion chamber, mixed with more fuel and supplied to a third turbine driving the first two, low- and intermediate-pressure, compressors. Exhausting from the third turbine, the hot gases entered a fourth combustion chamber, were again mixed with more fuel and then delivered to the final turbine driving the 12,000 kW generator. Thus, while fuel and air were both passed into the first two combustion chambers, only fuel was passed into the last two, there being sufficient air left in the hot gases to support burning at each stage. The aim in all cases was to make up the heat lost through each turbine so that a gas temperature of 700 °C was maintained before the next turbine.

Upon exhausting from the final turbine, the hot gases, now having a temperature of about 465 °C, passed through a waste-heat boiler and steam/air heat exchanger (for the gas producer) and an air heat exchanger before emerging into the atmosphere. The air heat exchanger, fitted between the final compressor and the first two combustion chambers, fed some of the exhaust heat back into the system and thereby saved fuel. However, there was nothing unusual in its design, it being of tube bundle type, with an efficiency of about 80 per cent.

The first two, low-pressure, compressors were each of the six-stage axial type, while the third, intermediate-pressure, compressor had eight stages and the final, high-pressure, compressor had 15 stages. These compressors were designed with low-solidity blades, and for a single-stage efficiency of 90 per cent. Each turbine had six axial stages and was designed for an efficiency of 88 per cent. The turbine blades, cooled with air tapped from the compressors, were made from Krupp's chrome-nickel Tinidur heat-resisting steel and had a maximum tip speed of about 200 m/sec (656 ft/sec). No details are known about the combustion chambers or the considerable amount of ducting needed for the installation, but it is known that MAN carried out very careful investigations and measurements on airflow and pressure losses in pipes and ducting, and so were aware of the problem involved.

Fuel system:
Fuel for the installation was gas, produced by the high-pressure gasification of brown coal in equipment designed by the Lurgi Gesellschaft für Warmetechnik of Frankfurt (Main). Gas from the Lurgi equipment was expected to have good heating value and to leave the producer and cleaner with a temperature of 145 °C. However, one problem in its use was the residual ash produced, which would damage the turbine blades. Various schemes were considered for the removal of this ash, such as centrifugal separation, electrical precipitation and washing with water, but until better methods were developed, the washing method was to be used.

The Lurgi equipment consisted of a gasifier, gas cleaner and water separator, waste-heat boiler and steam/air heat exchanger. Using this equipment, brown coal (dried to a moisture content of 30 per cent) was gasified by reaction with high-pressure air and steam in the gasifier. Water was circulated through the waste-heat boiler to produce steam which was mixed with the air supply, the steam/air flow then passing through a heat exchanger before entering the gasifier. After washing, the resulting gas was ducted to the four combustion chambers, apart from a small quantity

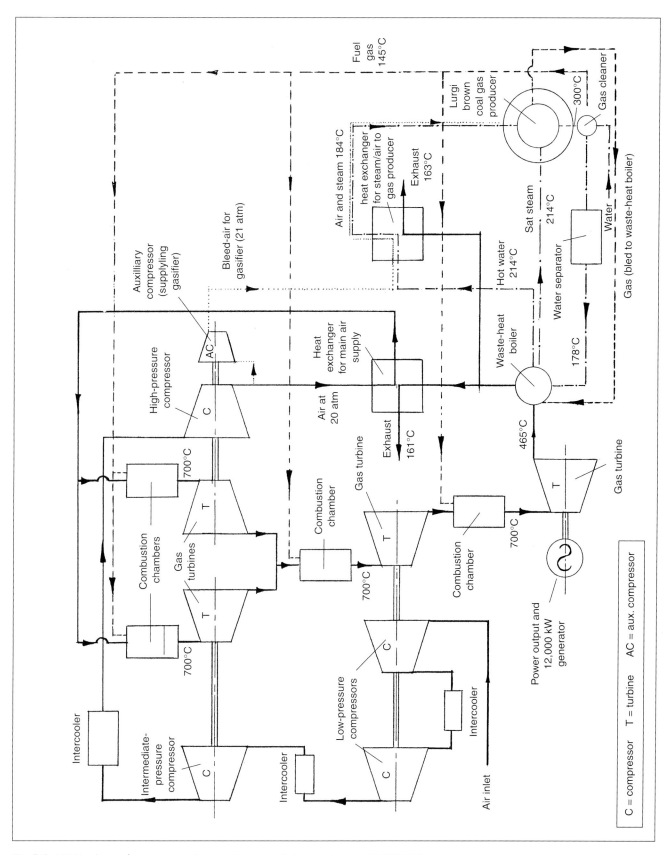

Fig 5.8 MAN scheme for compound gas turbine installation to drive a 12,000 kW generator.

(2.5 per cent) which was bled off for burning in the waste-heat boiler, this being necessary to raise the heat-exchange temperature above that possible with the turbine exhaust alone.

The air supply for the gasifier was derived from a small auxiliary compressor which was driven on the same shaft as, and tapped air from, the final, main compressor. This air, tapped off at a pressure of 20 atm., reached the gasifier with a pressure of 21 atm. after passing through the auxiliary compressor and heat exchanger. Considered as necessary for the gasification process, the high working pressure was also seen as conferring advantages by eliminating the need to pass the gas through compressors or blowers prior to entering the combustion chambers; this meant reduced equipment costs and the elimination of fine cleaning of the gas. It was estimated that the Lurgi gas producer would deliver 1.65 times the weight of coal consumed in gas. Since this gas had about 49 per cent of the calorific value of the coal, a fuel conversion efficiency of 81 per cent was expected. By way of a small bonus, the gas producer yielded a small amount of tar and benzine as by-products.

Data for the MAN-Lurgi 12,000 kW installation are:

Power output of final turbine	16,090 hp
Speed of final turbine	3,000 rpm
Speed of low-pressure compressors	4,600 rpm
Speed of intermediate- and high-pressure compressors	9,000 rpm
Pressure ratio after four compressors	20 to 1
Fuel (gas) consumption	5.70 kg/sec (12.57 lb/sec)
Airflow (at inlet)	47.30 kg/sec (104.3 lb/sec)
Air bled for gasification	2.80 kg/sec (6.17 lb/sec)
Air bled for cooling	5.40 kg/sec (11.9 lb/sec)
Peak cycle temperature	700 °C
Overall installation efficiency (ex coal drying)	20.5 per cent (assuming fuel consumed with 35 per cent overall efficiency)

By the end of the war, this installation, which was scheduled for completion within two years after delivery of the materials, had not been built because of non-delivery of material. Reports vary as to the actual stage the work had reached by 1945, but it appears that tests on such components as single compressor stages had been carried out, compressor and turbine design was about 50 per cent complete, and a start had been made on manufacturing. Full details of the Lurgi gasifier were not available and undoubtedly delayed finalisation of the complete plant design.

The 5,000 kW gas turbine generator

Another MAN gas turbine project planned to use the special T-4 turbine designed under Dr E. Schmidt of the LFA (see Section 6). This turbine was to drive a 5,000 kW electrical generator and two compressors connected in series, with intercooling. For fuel, gas was produced by Lurgi equipment (as previously described), using brown coal, high-pressure air and steam. The air was tapped from the main compressor supply but the steam was obtained from the water used to cool the turbine blades. Exhaust gases from the turbine passed through a heat exchanger to pre-heat the main airflow on its way to the combustion chamber.

The complete installation was to form one of a series in a power station, and it was hoped to obtain an overall efficiency of 34 per cent, using a compressor efficiency of 80 per cent. Although the design work was finished, no construction of the installation had been started by the end of the war. A simplified schematic arrangement of the installation is shown in Figure 5.9.

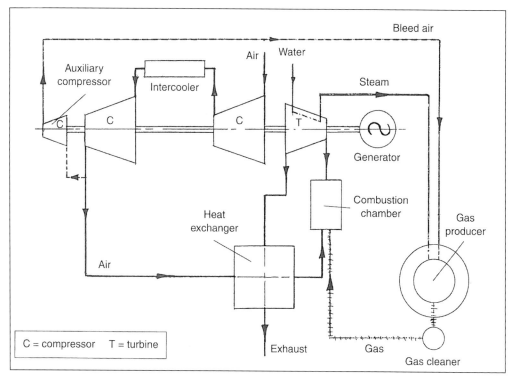

Fig 5.9 MAN scheme for a 5,000 kW gas turbine using a Schmidt water-cooled turbine.

Allgemeine Elektrizitäts-Gesellschaft (AEG)

As one would expect, the main interest of AEG (or General Electric Company) in the gas turbine lay in its use for driving electric generators, but by the end of the war in 1945, only studies of different schemes had been made. One of the earliest of a great many AEG gas turbine patents was that of Dr Ernest A. Kraft (438,782), which was filed in July 1924 and concerned a construction for equalising the axial pressure in steam or gas turbines. (Later, Kraft played a part in the design of the turbine for the Junkers 109–004 A turbojet.) Subsequent patents and studies covered such items as gas flow control, filtering, regulation, thermodynamics and turbine cooling, and the engineers concerned included Otto Rosenloecher, Arthur Rosch, Heinrich Treitel and S.R. Puffer. In 1937, one of the first AEG schemes for an actual plant was put forward and was described as a compressor plant driven by a condensed steam engine and a heated air turbine connected in behind the compressor. This scheme, by Willy Kuehne, was granted a patent (664,018) in 1938. In the following year, a patent application (L99,185) was made for a gas turbine electric generator designed by S.R. Puffer, and in 1940 another application (L99,853) was made for a gas turbine with cooled blades designed by Richard Stroehlen.

Of the various possible gas turbine cycles, AEG studied open cycles using gas or oil, compound open cycles with cooled compressor (the MAN installation used such a cycle) and closed cycles. Particular attention was paid to the closed cycle because it appeared to have certain advantages when using solid fuel, and the use of such fuel, especially in power stations, was economically essential. The scheme of a closed-cycle gas turbine is shown in Figure 5.10. Its working medium is air (or other gas) which flows in a closed circuit through the compressor and turbine. Before entering the turbine, the air is heated indirectly by an external combustion system and also by a heat exchanger using the heat from the turbine outlet. Before re-entering the compressor, the air is cooled down. A small compressor is used to charge or pressurise the system, and a valve to discharge it when required. Theoretically, the closed cycle could operate on anything combustible, and with solid fuels the problem of ash eroding the turbine blades was avoided, since no combustion products flowed through the circuit. In addition, good part-load efficiency was possible since both the working pressure and fuel input could be altered to suit. While the compressor, turbine and ducting can be small because of the high working pressure, the heater must be very large owing to

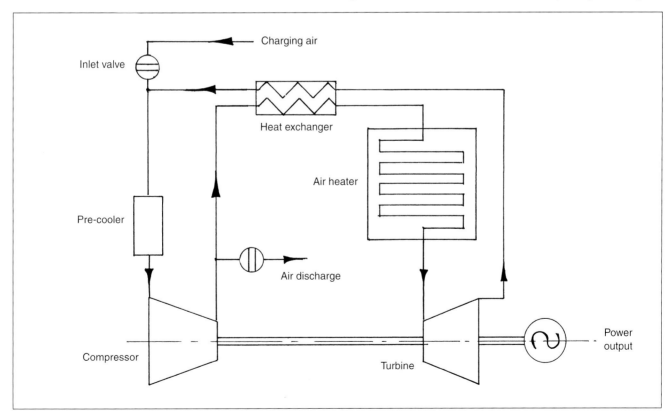

Fig 5.10 Closed-cycle gas turbine scheme.

combustion taking place in comparatively low pressures. Also, control of the system presented difficult problems. To some extent, AEG felt that the efficiency of the cycle was over-estimated because of its expected susceptibility to pressure loss.

Pioneering of the closed-cycle gas turbine was, in fact, left to the Swiss company of Escher-Wyss, which made the first test runs in 1940. By 1945, AEG had reached no definite conclusions but were directing their attention to several lines of future research. One item, already mentioned, which held considerable promise in the opinion of AEG was the Lurgi brown coal gasifier.

Conclusion

Judging from the early introduction of Holzwarth and BBC gas turbines, at least some sections of German industry did not suffer from over-conservatism and were willing to experiment with the new power form. To some extent, this was aided by the fact that steam turbines were already familiar in industry. It was the misfortune of Holzwarth in choosing the constant-volume gas turbine cycle that the inherent difficulties in the utilisation of such a cycle caused its abandonment by 1945.

In the case of BBC, their diverse application of constant-pressure gas turbines, especially for air and steam generation in the iron and steel industry, met with full success, and developments of their schemes are still very much with us today. The increasing predominance of the gas turbine in air blowing and heating systems for the blast furnace appeared vindicated by quantitative comparisons, and with the promise of other types of systems planned, new possibilities for centralisation were opened up.

In the field of electrical power generation using the gas turbine, we have the examples of the Holzwarth 5,000 kW unit and a few 1,500 kW units built by BBC. However, industries at the time needed generators between 10,000 and 50,000 kW in size, and a straightforward gas turbine with sufficient power (13,400 to 67,000 hp) to drive such generators was considered out of the question. Generators of up to 10,000 kW size driven by steam turbines could not then be surpassed, and for such turbines the BBC Velox steam generator was considered ideal where fuel such as blast furnace gas was available. Nevertheless, increasingly larger gas turbine units were being planned, not only by BBC, but by MAN, AEG and others.

These larger installations, which used compound open and closed cycles were, however, far from straightforward. In the case of the MAN 12,000 kW installation, for example, there were some doubts as to whether the high capital cost of building such a complicated installation would not offset any advantages. As a potential source of energy from low-grade, solid fuel, the MAN-Lurgi schemes were nevertheless of great interest, and various compound gas turbine installations, specially tailored to their environment and available fuel, are in use today. During the war, work on the interesting closed-cycle gas turbine was in its infancy in Germany, although Escher-Wyss of Zürich were making practical progress on such a system.

In post-war years, Escher-Wyss GmbH of Ravensburg, Germany, and Gutehoffnungshütte (GHH) of Sterkrade built the world's first commercial closed-cycle gas turbine unit, which drove a 2,000 kW generator, and from 1956 ran entirely successfully on bituminous coal. Other German companies that became active in the industrial gas turbine field included AEG, BBC, Klöckner-Humboldt-Deutz, MAN Turbo (formerly BMW), MAN, Siemens Schuckertwerks and VEB Starstrom-Anlagebau. In several cases, however, licensed production of British and Swiss designs is undertaken.

Section 6

Gas Turbine Research and Development

Compressors — Combustion and fuel — Turbines: E. Schmidt's water-cooled turbines, T.2 and T.3; Ceramic turbine blades; F.A.F. Schmidt's hollow turbine blades; DVL turbine blade temperature measurement; LFA turbine testing methods — Jet unit installation — The regenerative heat exchanger — The Lutz swing-piston gas generator — Conclusion

So far, the work of individual companies to develop specific gas turbine engines for application in the air, on the land and on the sea has been examined. Where appropriate, help given by various German research institutions was mentioned, but now a closer look will be taken at the work of these institutions insofar as it related to the gas turbine. Very often, individual companies were treated to over-rigid control by the military technical departments, such as the Technisches Amt and Heereswaffenamt, which laid down features for development far exceeding a general specification. On the other hand, the research institutions were very much left to their own devices, while, at the same time, they had little contact with industry or companies needing their help. Sometimes, contact with a company was arranged officially only after the company had been wrestling with a problem for some time. Other contacts were made on a personal level following, for example, the industrial employment of men trained at the research establishments. Some news travelled between institutions and industry by the usual means of reports, year books and meetings, the latter being held, for example, under the aegis of the Deutschen Akademie der Luftfahrtforschung (DAL), or German Academy for Aeronautical Research.

The first of these jet propulsion meetings (actually, a symposium) was organised by the RLM for January 1941. The top men from Junkers, BMW, Heinkel and various institutions all got to know each other very well, but interchange of information was not good.

Frustrated by this lack of interchange, the RLM tried to improve matters in all fields, not just jet propulsion, by direct order. An Arbeitsgemeinschaft für Triebwerksplanung was created, and at its first meeting on 11 April 1942, Eisenlohr declared: 'I have recently had several opportunities to observe that one firm had no notion of what other firms had contributed in similar fields, although the RLM has explicitly emphasised – and not just once – that this information should be exchanged. Now, compulsory exchange has been introduced.' (The words were actually written by Helmut Schelp.) Meetings were at least every six months, but Eisenlohr's directive had little effect, and it was not until late in the war when Speer's Armaments Ministry ordered more co-operation that this happened. One noteworthy result of this was that BMW at last produced a usable turbojet control system by having access to, and adopting, the designs of Junkers.

Nevertheless, one gains a general impression that research workers and scientists were invariably free to follow programmes of their own inclination. The lack of a fully active co-ordinating body, combined with somewhat remote contact with industry, resulted in the research institutions sometimes duplicating each other's efforts, which were in any case often directed along academic lines bearing little relation to the practical gas turbine problems of the day. In time, of course, much of the fundamental research would have borne fruit, but this was lost to the Germans (or at least their war effort) with the end of the war in 1945. Towards the end of the war, some effort was made to co-ordinate research activities and to turn them onto the most urgent problems.

A few notes on the more important research institutions may now be of interest before looking at their gas turbine work. One of the oldest institutions was the Deutsche Versuchsanstalt für Luftfahrt (DVL), or German Experimental Establishment for Aeronautics, which was founded in Berlin in 1912. It was steadily expanded over the years into all branches of aeronautical research, and in the engine field played a major part in the development of piston engine superchargers in the 1930s. During the Second World War, the DVL had its main dispersals at Garmisch-Partenkirchen and at Altes Uppenborn just outside Moosburg. Gas turbine research, mostly concerned with combustion and turbines, was directed by Prof. F.A.F. Schmidt.

Almost as old as the DVL was the Aerodynamische Versuchsanstalt (AVA), or Aerodynamics Experimental Establishment, which was built up as part of Göttingen University. Its first wind tunnel was in operation by about 1915, and, to mention one facet, aerofoil profiles evolved there became universally accepted. Expansion led, in 1927, to the formation of the Institut für Strommungsmaschine (Department for Turbomachinery). The leadership of this department was given to Dr Encke during the early 1930s, and from about

1935, work on compressor problems was tackled which became of fundamental importance to German gas turbine progress. The first systematic investigation into the aerodynamic problems connected with jet engine installation was undertaken by a department under Dr D. Küchemann, while another department pioneered important heat exchanger studies under Dr Ritz.

Following the Nazi rise to power in 1933 and the steady expansion of German industrial and military power, various new research centres were established, while smaller, existing centres (usually forming part of the technical universities, or Technische Hochschulen) were gradually given more tasks. At the same time, all the research and technical establishments had to fight a constant battle against losing their best men to industry and the military services.

The largest of the new establishments was the Luftfahrtforschungsanstalt Hermann Goering (LFA), or Hermann Goering Aeronautical Experimental Establishment, but despite its lavish equipment it was not officially brought into contact with gas turbine companies until about December 1943. The engine department of the LFA at Völkenrode was begun in 1936 and was under the direction of Prof. E. Schmidt. Dealing mainly with reciprocating engines, this department also tackled gas turbine work and eventually had about forty highly qualified engineers, backed by about forty-five less-qualified engineers and designers. It was organised into four sections, dealing with thermodynamics (Prof. Eckert), heat conduction (Dr Hilpert), combustion (Dr Sellschoppe) and chemistry (Dr Edse). As with all the larger institutions, it had its fair share of wind tunnels, test rigs, laboratories, workshops, offices and other facilities.

Another new research institute was the Technischen Akademie der Luftwaffe (TAL), or Technical Academy of the Air Force, which was involved at an early date in gas turbine combustion and fuel research which particularly helped the BMW and Junkers companies. Originally based at Berlin-Gatow, the TAL had a number of dispersed establishments during the war, including the following. At Schussenried bei Ulm, fuel research for all types of engine was performed under Prof. Holfeder, while at Eckertal-Hartz, Dipl.-Ing. Nägel headed combustion chamber research. Under Prof. Schardin, at and near Biberach bei Ulm, fuel research was performed. These and other stations of the TAL were under the general direction of Prof. Walter Hermann.

Mention need not be made of the many other research centres which played small parts, or attached little importance to gas turbine research, but one other institute is worthy of mention. This was the Forschunginstitut für Kraftfahrwesen und Fahrzeugmotoren at Stuttgart University (FKFS), or Research Institute for Motor Transport and Vehicle Motors. We have already seen how this institute, under Prof. W.I.E. Kamm, assisted the Ernst Heinkel AG with their ducted-fan engine development, worked on supercharger development and so on. In addition, from 1939, work aimed at deducing basic compressor theories and formulae was performed by F. Weinig and B. Eckert. Of course, the research institutes mentioned, together with others, worked on all kinds of problems connected with science and engineering but which do not concern us here. Other institutes, such as the famous DFS and KWI, are mentioned in the sections dealing with pulsejet and ramjet engines. In the following, various aspects of gas turbine research and development will now be looked at.

Compressors

So far as gas turbines were concerned, German compressor development was directed almost exclusively towards the axial compressor. A couple of exceptions, such as the diagonal-flow compressor for the Heinkel-Hirth 109–011 turbojet, and the centrifugal compressor for the BMW P.3303 turbojet, have already been noted in Section 2, but strangely, the double-sided centrifugal compressor favoured by the British was totally ignored by the Germans. Centrifugal compressors figured largely in the research programme pursued by the RLM in the field of aircraft superchargers, but this side was rather specialised. Straying further afield, we find that companies such as AEG, BBC, DEMAG, Escher Wyss and GHH manufactured both internally and externally cooled centrifugal compressors for mines and industrial applications, but no important advances were made in this field, which had practically no connection with gas turbines and concerned the research institutes hardly at all.

A major stimulus for axial compressor development came in 1939 following the official backing of the turbojet, for which the general consensus of opinion favoured the axial compressor, largely on the grounds of its minimum frontal area. Much of this development work was later carried into the marine and land gas turbine fields.

For much of the period covered by this history, German axial compressors were of the impulse type, in which most of the pressure rise occurred in the moving rotor blades. Professors Betz and Prandtl, together with Dr Encke (who was also an engineer) were generally acknowledged to be the fathers of this type of compressor, and their work particularly dictated the development of BMW and Junkers turbojets. Betz and Prandtl had close ties with Keller in Switzerland, from where Stodola's books on turbine dynamics and so forth became the bible for German universities on such matters. Following general axial compressor research, the first official project worked on by the AVA's Institut für Strommungsmaschine concerned an axial supercharger compressor (or blower) for a Junkers piston engine. It turned out to have a performance superior to the corresponding centrifugal blower, but was not pursued on the grounds of its greater weight and cost. Attention then turned to the contra-rotating axial blower for piston-engine superchargers, but this was soon dropped because of unwarranted extra complication. For the work on supercharger blowers, a simple test rig with a low-power electric drive sufficed. A larger test rig for compressor work was, however, commissioned in about 1935 and derived its power from a 90 hp two-stage turbine which was supplied from a compressed-air reservoir. This power was sufficient since only small-scale models of proposed designs were tested, but because of the limited capacity of the air reservoir, tests were of short duration. It was therefore imperative that the data from each test were recorded by high-speed photography for examination later. About eighteen pressure tubes backed by a translucent scale were located around the test model, and a camera recorded the fluctuating pressures and other data.

This test rig supplied not only data for the original Junkers 109–004 and BMW 109–003 turbojet compressors, but a mass of other data also, and because only model compressors were used, only a moderate financial outlay was involved. Most of the model compressor rotors had an outside diameter of about 150 mm (6 in) and a disc diameter of only 75 mm (3 in), and usually had two or four stages only. Nevertheless, stage efficiencies of over 88 per cent were recorded at blade tip speeds of 200 m/sec (656 ft/sec), and the soundness of the AVA work was later confirmed when the full-size compressors and engines were built.

Because of the small scale of the AVA model compressors, much care was needed in their accurate manufacture and checking. In order to eliminate the errors and variables possible with an assembled rotor, it was decided to face the great difficulties of machining model compressor rotors from solid magnesium alloy forgings. The responsibility for designing and developing the requisite special machines fell to the AVA's chief manufacturing engineer, Karl Grothey, who was both intelligent and capable. Several machines were developed and used in the AVA's Göttingen workshops, but were later transferred to the AVA's dispersal at Reyerhausen, where the surface buildings of a disused salt mine were used as workshops. There, development was continued. One machine cut out and profiled twenty compressor blades at a time on rotor discs, while another machined blades on the inside of rings.

The products of these machines still required much hand work to finish them to the required accuracy, and a special method of measurement and checking was used during their manufacture. In essence, a low-powered microscope was focused on a blade at set stations, and the readings were plotted on a large scale to produce a magnified drawing of the blade profile, which could then be compared with another large drawing of the profile as designed. Using this checking and correcting method, it was claimed that the model compressor could be made to an accuracy of 0.025 mm (0.001 in) on all dimensions.

With the above equipment and methods, the AVA obtained data of an accuracy not previously attained with such small models. However, during 1944, a second, improved test rig was commissioned with the object of obtaining longer test runs and with a constant speed throughout. The drive for the test model consisted of a turbine and two electric motors on the same shaft, the second motor permitting the testing of contra-rotating compressors. After passing through the model compressor, the air was expanded through the turbine (to reduce the power needed from the electric motor/s) and, after cooling, returned to the compressor inlet. Because of electrical difficulties, little work was done on this second rig before the war ended.

Another new test rig running by the end of 1944 was for the testing of full-size compressors. This rig used a 5,360 hp DC electric motor with a step-up gearbox at each end of its shaft. One gearbox drove a high-speed shaft for the compressor under test, while the other was intended to drive an air compression plant for other work. In the event, the only full-size compressor tested at Göttingen was the diagonal-flow compressor for the Heinkel-Hirth 109–011 turbojet. The design of this compressor, of course, had nothing to do with Encke and his staff, and in fact their dealings with most companies were very distant. Only in the case of Junkers was Encke allowed to witness a full-scale test of a compressor based on his work; this test was on an early 109–004 turbojet compressor carried out at Dessau in March 1942. Later, there was actually co-operation between the AVA and Junkers on compressor design.

As far as application of results was concerned, the AVA's compressor research under Encke surpassed all other German research institutes, and all credit is due here. It is certainly no disparagement to reflect that Encke was more of an experimental worker than a theoretician, which possibly explains why little deviation was made from the impulse (as opposed to the reaction) type of axial compressor. Attractions of the impulse type of compressor were that stator blades could be made of sheet metal since they only had to guide the airflow, and fine clearances between the rotor and stator blades were not necessary. Such factors led to simple construction. On the other hand, high efficiencies were not obtainable, while the high-end thrust from the rotor meant keeping the compression ratio fairly low (about 3 to 1) and the use of large, complicated bearings. Finally, such a compressor was heavy.

The answer to these problems lay in the compressor with 50 per cent or more reaction whereby the work of air compression was shared between the rotor and stator blades. In Germany, the first designer of such compressors was the aerodynamicist Rudolph Friedrich, who, before 1938, designed a 50 per cent reaction axial compressor for a turbojet at Junkers, this engine later evolving into the promising Heinkel 109–006 turbojet. He later designed compressors along similar lines for the BMW 109–003 D turbojet (1944) and the GT 102 Ausf.

2 land traction gas turbine (early 1945); these compressors were to be developed by the Brückner-Kanis company, and by 1945 at least, Friedrich appears to have been working at the AVA, Göttingen. Even so, the impulse type of axial compressor was still being developed at the AVA and elsewhere for new engines, and so reaction compressors were not universally adopted. No doubt this was partly through a lack of official support which originally stemmed from a peculiar mistrust of reaction compressors on the part of Helmut Schelp of the Technisches Amt. Later, this mistrust was belied by the promising progress of the Heinkel 109–006 turbojet and the success of the work under Hermann Reuter of the Brown Boveri company in developing new compressors for the BMW 109–003C turbojet.

At the FKFS, F. Weinig and B. Eckert considered the most important task in compressor development to be the achievement of higher-stage pressure rises without, of course, impairing efficiency. Therefore, from 1939, various designs of compressor were experimented with and much fundamental data collected. One type of axial compressor had split rotor blades with adjustable flaps at their trailing edges, while other forms of axial compressor included a contra-rotating type and a cut-out type in which one or two stages could be cut out if not required. The multi-stage radial compressor was also studied, since, while its number of stages was limited by the diameter, it was considered useful where the compressor length had to be minimised. A further reduction in length was achieved with the contra-rotating radial compressor, and Figure 6.1 gives the schematic layout of two types of radial compressor. (Incidentally, radial turbines were used with success many years previously in steam practice.) Other experiments at the FKFS provided information on the effects of compressor intake and hub, adjustable rotor and stator blades, various blade profiles and so forth.

Chief protagonist of the contra-rotating axial compressor was the independent engineer Helmut Weinrich, who designed such a compressor for a turboprop engine before 1936. Engines which used his designs included the BMW 109–002 turbojet, and further work of Weinrich's has been discussed under Brückner-Kanis in Section 4. The contra-rotating axial compressor for Daimler-Benz's 109–007 turbojet may have had its origins at the DVL with Karl Leist, but in general this type of compressor appears to have been studied by companies rather than research institutes.

Very little work on compressors can be recorded for the DVL, LFA and other German institutes. For a while, compressors and other parts for a proposed DVL gas turbine were tested at the DVL Altes Uppenborn

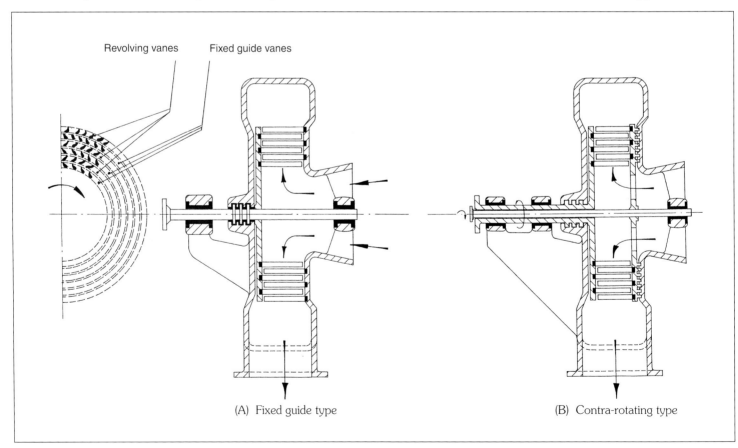

Fig 6.1 Schematic layout of two types of FKFS radial compressors

dispersal, but this work was discontinued in favour of work on turbines. (Unfortunately, no details are available for the DVL gas turbine, or even its planned purpose.) Since there was a disused hydro-electric station at Altes Uppenborn, a 4,000 hp test rig was easily supplied for compressor tests. Another DVL test rig, at the MAN works at Augsburg, tested the diagonal compressor for an early Heinkel-Hirth 109–011 turbojet, but no other compressor work there is known. At the LFA, Völkenrode, BMW compressor stator blades were tested using interferometer techniques, but this work was mostly concerned with turbine blades and is described in the turbine section.

Combustion and fuel

In common with most people, the Germans at first under-estimated the problems of gas turbine combustion. For much of the time, companies working on gas turbines had to solve both combustion and fuel system difficulties on their own. Often the combustion research performed by the research institutes gave little practical help on specific engines and was not related to actual projects. The instances where the TAL assisted BMW and Junkers with their initial turbojet combustion difficulties are exceptions. At TAL, combustion chamber research with both liquid and gaseous fuels was performed, while Schlieren high-speed photography was used to record information on fuel, flame propagation and detonation.

The LFA and DVL were drawn more actively into combustion research once turbojet engines were more extensively flown and the problem of flame-out at high altitudes became troublesome. The Technisches Amt acquainted certain institutes with the problem, which was threatening the Luftwaffe's jet aircraft with tactical disadvantage. Ignition limits were explored over a range of sub-atmospheric pressures and fuel/air ratios at the LFA, where propane was used for fuel and spectroscopy for analysis. The test rig used was for a single cannular combustion chamber (of the Junkers 109–004 turbojet type), which was mounted vertically to discharge upwards; it was supported on knife edges connected to weighing gear in an adjoining room where thrust was measured. In order to avoid interference with the thrust measurements, the air supply pipe was pivoted and connected by flexible bellows to the combustion chamber. Compressors provided an air supply of up to 4.0 kg/sec (8.82 lb/sec) at a pressure of 4 atm. Few conclusions seem to have been drawn from this LFA work other than that the ignition limits could be extended in low atmospheric pressures by enlarging a given combustion chamber, while a hot surface was found better than spark energy until about 0.02 atm. was reached. Another piece of LFA apparatus consisted of a quartz-windowed bomb in which flame fronts and detonations could be examined by Schlieren high-speed photography.

When the flame-out problem was presented to the DVL, tests concentrated on the BMW 109–003 turbojet, from which a segment of the annular combustion chamber with one full fuel nozzle was used. The first tests determined the ignition and burning limits under various air entry velocities and with different fuel/air ratios, these tests being normally at around atmospheric pressure where flame-out occurred. For fuel, hydrogen, methane and acetylene gas was used, the former two giving results similar to the normal turbojet fuel, while the latter gave much-improved results. For relighting the 109–003 after a flame-out, an improvement was given by fitting a streamlined section in front of one of the primary chamber fuel nozzles. Using only one sparking plug, situated by this nozzle, all three nozzles lit up successfully, but the suspicion was that the streamlined device worked by heat conduction rather than by turbulence. Once again, combustion research seems to have yielded little in the way of solutions to an immediate problem. However, in one instance at least, the DVL gave practical help to industry on gas turbine combustion when it designed the combustion chambers for the Blohm und Voss marine gas turbine unit (Auftrag 353). BMW would have made faster progress on combustion if they had been allowed access to Junkers work, but this was not forthcoming until the Luftkriegs-akademie (LKA) at Gatow arranged full access for them.

More enterprising research was conducted on fuel flow. It was found at the LFA that carbonisation in combustion chambers could often be traced to an unbalanced fuel spray distribution, which gave varying mixture strengths in the combustion zone. In order to study this problem, a spray distribution apparatus, or patternator, was constructed. This consisted of a 0.70 m (2 ft 3 in) perspex cube case which held the fuel nozzle under test in the centre of its roof to spray downwards. Six arms in the bottom of the case held rows of vertical tubes to collect some of the spray. The extent to which the tubes were filled gave information on the radial distribution of the spray, and the arms could be revolved to a new position in further tests to give a more complete picture.

To determine spray particle size, another piece of equipment allowed a spray to pass through a camera-like shutter to briefly impinge upon a soot-covered plate. Examination of the 'exposed' plate then revealed the average particle size and also the spray cone angle, and tests could be performed under pressures of several atmospheres if desired. The principle of this equipment was well known in other countries, but other fuel nozzle apparatus at the LFA included an enlarged experimental nozzle in which pressure could be measured at different points and a meter which could measure the momentum of sprays.

At the DVL also, a Dr Jung performed research on fuel atomisation, particle size and so forth. Using a special form of spark apparatus, many magnified photographs of sprays were taken and data were obtained on spray speeds, paths and evaporation rates. BMW swirl atomisers were tested amongst others.

Most of the companies working on turbojets were enlisting the aid of specialist companies (e.g. L'Orange and Bosch) to develop the duplex type of fuel nozzle as the obvious answer to the flame-out problem. It therefore seems curious that no reference can be found to the research institutes specifically developing such nozzles, but their general results were doubtless of great value to the specialist companies.

Turbines

To a great extent, the efficiency of a gas turbine engine depends on the efficiency of the turbine, which must extract the maximum amount of energy from the combustion gases to drive the compressor, and also, in certain arrangements, deliver external power. Any inefficiency in the turbine must be made up by burning extra fuel. In general, German turbines were of the impulse type based originally on steam practice, and were fairly low in efficiency. Also, turbine design was influenced by the necessity to use inferior materials, which put the Germans far ahead in techniques of turbine cooling. When companies began developing gas turbines, they either turned to steam turbine engineers or drew on previous experience with piston engine turbo-superchargers for initial designing data. Once again, little assistance came directly from the research institutes, although a number of men took their former experience with them when they joined industry.

Even so, a great deal of interesting and worthwhile turbine research was undertaken by the institutes (chiefly the DVL and LFA) on their own initiative. The only major turbine research programmes set up in the institutes by a governing body concerned turbo-superchargers, hollow metal blade production and ceramic turbine blades, the latter involving the MAN company in addition to the LFA.

E. Schmidt's water-cooled turbines, T.2 and T.3

Beginning in 1938, Prof. Ernst Schmidt of the LFA, Völkenrode, began studies which led to his water-cooled turbine. The original studies looked into the heat transfer properties of liquids at their critical points of pressure and temperature. At its critical point, the thermal conductivity of a liquid theoretically approaches infinity, and, in any case, is greatly increased. To demonstrate this, a steel bar was hollowed out over part of its length and filled one-third full of ammonia. The bar was then heated, and at 20 °C its apparent heat conductivity was equivalent to the bar being of solid copper, while at the critical temperature and pressure, the conductivity was 20 times that of copper.

Having established the principle of the phenomena, Schmidt turned to the design of an experimental turbine which would operate in the very high gas temperature of 1,200 °C by utilising water-cooled blades. The first such turbine was of the single-stage type, and was designated T.2 (T.1 being a ceramic turbine described later). Dr-Ing. K. Bammert and Ob.-Ing. P. Lechky assisted Schmidt while the steam turbine works at Brno in Czechoslovakia were commissioned to carry out the actual manufacturing under Schmidt's direction.

The T.2 turbine had a rotor blade tip diameter of 0.320 m (12.6 in), which at the operating speed of 12,000 rpm gave a tip speed of 200 m/sec (656 ft/sec). The rotor blades had sharply cambered profiles and were machined integral with the forged, low alloy steel rotor. Five holes were then drilled into each solid blade, almost to the tip, but, unlike Vorkauf's water-cooled blades (see Section 4), there was no interconnection between these holes inside the blade. It was originally proposed to drill the holes in the blades from inside the hollow rotor, but this idea was abandoned in view of the extremely accurate jigging necessary. Instead, the holes were drilled from the outside and then plug-welded near the blade tip.

Water was caused to circulate in and out of the rotor blade holes by a considerable buoyancy effect under high centrifugal force, the density of the hotter water at the circumferences of the holes being less than that at the hole centres. The thermal conductivity of water falls away sharply on either side of its critical temperature of 374 °C, but at this temperature, a high degree of heat transfer was obtained. Water was led in through a hollow shaft at one end of the rotor, was heated inside the blades and only turned to steam as it emerged back into the rotor, where the pressure was about 7.75 kg/sq cm (110.2 lb/sq in), compared with about 35.0 kg/sq cm (497.8 lb/sq in) at the blade tips. The steam left the rotor through a hollow shaft at the other end and was sufficient to drive a steam turbine of about one-tenth the power of the gas turbine. Automatic control of the water flow in the rotor and its blades depended on the depth of the water in the rotor and a reduction in the conductivity and flow characteristics brought about by a deviation from the water's critical point. In a gas temperature of 1,200 °C, the main body of the blades ran at about 400 °C, and the trailing edges, where no holes were drilled, at 500 °C.

Most difficult to cool were the stator blades (made from sheet metal) and the turbine housing, because of the absence of centrifugal force to give a natural flow. Cooling water for these items was therefore circulated by a pump. A single cannular or cylindrical combustion chamber was provided, which had three fuel nozzles and a simple flame tube directing primary and secondary air flows. This unremarkable chamber, which is shown schematically in Figure 6.2, was probably mounted separately from the turbine housing.

Turbine T.2 was completed around 1943, and performed between 10 and 15 hours' total test running very satisfactorily. Following this success, a 2,000 hp four-stage turbine designated T.3 was designed and built.

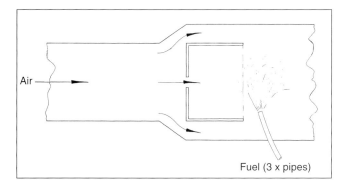

Fig 6.2 Schematic arrangement of cannular combustion chamber for E. Schmidt's T.2 turbine.

Many of its features were similar to T.2, including rotor blades machined from a solid rotor forging, but the rotor drum had strengthening ribs to permit the higher rotational speed of 20,000 rpm, and the single cannular combustion chamber had one instead of three fuel nozzles. The T.3 turbine was never run on hot gases, since the stator blades failed during a preliminary test because of weld decay, and replacements from the Brno turbine works failed to materialise. A final design of this type of turbine, designated T.4, was made for a 5,000 kW industrial gas turbine planned by MAN but never built (see Section 5).

Ceramic turbine blades
In the summer of 1944, the Technisches Amt ordered the MAN company and the LFA to begin immediate investigations into the possibility of replacing high-temperature steels in gas turbines with ceramic materials. This official move was somewhat belated, since not only had the necessary metals such as nickel and chromium been acutely short for some time, but MAN had been making ceramic investigations since 1939 and the LFA since about 1940.

At the MAN company, work under Prof. Dr-Ing. Emil Sorensen was directed towards producing ceramic gas turbine blades which could operate in a gas temperature of 800 °C (ultimately 1,000 °C) and with a blade tip speed of 150 m/sec (492 ft/sec). The aim was to satisfy the requirements for a stationary, industrial gas turbine, although, later, the company was largely responsible for the formation of a committee to deal with problems of the application of ceramics to gas turbines in general.

Under Sorensen, possible materials were placed in either sintered, quartz or porcelain groups. Each material was then evaluated on its properties regarding coefficient of expansion, thermal conductivity, Young's modulus and strength-to-weight ratio at high temperatures. Sintered materials, such as carborundum, were found to have good strength at high temperatures and to be resistant to thermal shock, but their high density made manufacture of blades difficult. Since moulding was not possible, it was necessary to grind the sintered blades to the finished size, but about 400 hours was needed to grind only 0.25 mm (0.010 in) from the root of a sample blade! Hardly surprisingly, sintered carborundum blades were abandoned.

Although quartz was incapable of taking thermal shock, MAN thought it might be used in industrial applications where it would be possible to have slow heating up and cooling down to avoid fracturing the turbine. High temperatures were also to be avoided since quartz undergoes a crystalline change at 900/950 °C. A 200 mm (8 in) diameter turbine wheel moulded in one piece from fused quartz was designed, and although it was thought this would cost something like 100,000 RM, there was a pleasant surprise when the Heraeus Glassmelze company of Hanau/Main manufactured two or three samples for only 1,200 RM each. Unfortunately, the quality of the wheels was not consistent, and in cold spinning tests one disintegrated at a tip speed of only 18 m/sec (59 ft/sec), while another withstood 45 m/sec (155 ft/sec).

Attention then turned to porcelain products, which appeared to be one of the most promising ceramics. The Heschau company of Hermsdorf supplied various test pieces, blades and a complete stator ring with 18 blades. This ring was made in one piece from an electrical porcelain, and was 230 mm (9 in) in diameter, with a rim width of 50 mm (2 in) and a blade length of 30 mm (1 in). No tests were performed on it as the MAN ceramics department at Augsburg was almost wholly destroyed in an air raid in February 1944, and most apparatus and records were lost. Probably the best ceramic material found by MAN by that time had a base of quartz, aluminium silicate and manganese oxide (or manganese silicate). Even this material had its drawbacks, such as a serious drop in tensile strength above 800 °C, and there was an inconsistency of strength between superficially identical sample blades. A turbine wheel was designed to test these blades, and a section through this wheel, showing the method of blade attachment, is given in Figure 6.3. From this, the ample

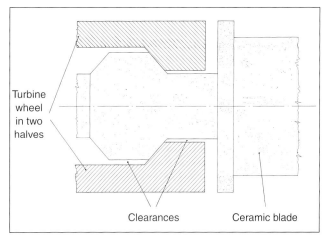

Fig 6.3 Method of ceramic blade attachment in a MAN turbine wheel.

clearances for expansion will be noted, but the mechanical properties at high temperature and stress must be doubted in view of the LFA's investigations described later. Many other samples in various materials were obtained by MAN, but by the end of the war, the sum of their ceramic work was mechanical testing only, plus a great deal of design work. Nevertheless, the company must be credited with pioneering ceramic turbine blade investigations.

At the LFA, such investigations were in the hands of a Dr Dirksen and an assistant of Dr E. Schmidt's, named Sohngen. In 1940, tests with ceramic-coated steam turbine blades proved unsuccessful, and the following three years were spent in a fruitless search for satisfactory ceramic materials for jets and nozzles. When the priority was raised in 1944, the LFA turned to ceramic manufacturers in an effort to find a new gas turbine blade material. The first materials thus obtained were ceramics already in use for household and electrical purposes, and were not found suitable for gas turbines.

Companies such as Stemag (Berlin), Hescho (Hermsdorf), Degussa (Frankfurt), Koppers (Düsseldorf), Porzellan-Manufaktur (Berlin), and Osram (Berlin) were then specifically requested to conduct research on ceramics capable of withstanding high stresses and temperatures in gas turbines. Much time was lost in waiting for materials, and by 1944 most manufacturers were handicapped by the effects of bombing. Several ceramics appeared promising, but most were limited to a temperature of 700 °C, and it would hardly be illuminating to go into a long list of these materials now. One of the most promising materials, Osram's Dug, was thought usable up to 1,000 °C, but this only came to the notice of the LFA in the last few months of the war. Methods of shaping blades, depending on the type of material, included the pressing or vibrating of powders, extruding through nozzles and moulding. Both solid and hollow blades were made.

During 1944, the LFA began work on an experimental four-stage axial turbine with the blades of both stator and rotor in solid ceramic material. The turbine was designated T.1, and it was planned to use a ceramic by Stemag known as Sipa H, containing silicon oxide and manganese oxide. In order to prevent high-tensile loads on the ceramic blades, the rotor blades were mounted inside a revolving drum so that the forces on the blades were compression and bending. Figure 6.4 gives a half section schematically showing the layout of turbine T.1. The rotor blades, moulded in one with their bases, were located inside the drum by lugs on steel rings, and the whole ceramic assembly was held in place by friction

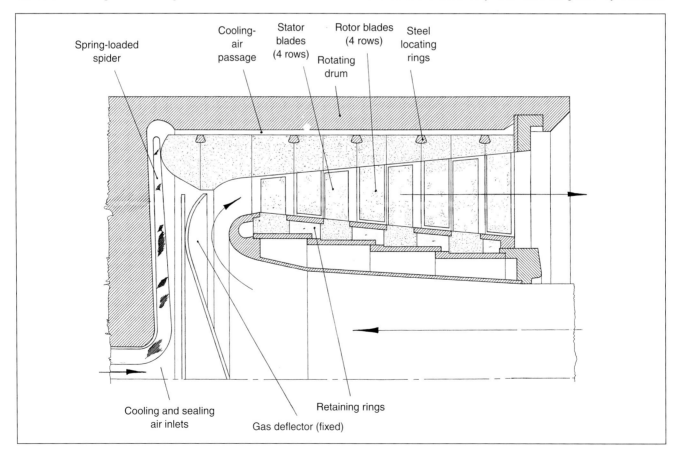

Fig 6.4 LFA experimental turbine T.1 with ceramic blades (1/2 section).

produced by the pressure of a spider with spring-loaded arms. The locating lugs and spider were designed to permit expansion of the ceramic at high temperatures. Steps were formed in the ceramic bases of the stator blades to provide locations for steel rings acting as separators and supports; again, the assembly was pulled together in a manner allowing expansion but avoiding mechanical strain on the ceramic.

Hot gases entered turbine T.1 though a central duct, were turned outwards through 180 °C by a deflector shield and entered the turbine blades. The gases exhausted in an opposite direction to their inlet so that there was an annular ring of exhaust gases surrounding the inlet duct. At the opposite end to the gas inlet and exhaust there was an inlet from which cooling air was led over the spider and around the outside of the rotor ceramic. Some of this cooling air also leaked between the edge of the fixed gas deflector and the rotating drum to act as a seal. There was supposed to be means of air-cooling the stator blade bases, but it is not clear how this was done.

Because Stemag were operating under extreme wartime difficulties, delivery of the ceramic components was not accomplished until near the end of 1944. In the interim, construction of the rest of the turbine went ahead, and the LFA conducted tests on various materials and blade profiles in a hot gas stream, but all materials then available failed at about 800 °C. Finally, because of Schmidt's success with his water-cooled T.2 turbine, it was decided to abandon the T.1 turbine before it was ready for running and to concentrate all further effort on developing water-cooled rotor blades and ceramic stator blades. With this arrangement, believed to combine the best features of both schemes, it was hoped to produce a gas turbine operating at 20,000 rpm in a 1,200 °C gas stream.

Various means of making a ceramic stator blade assembly were now explored. A solid ceramic stator ring was ruled out because of the high coefficient of expansion and the inevitability of cracking under temperature changes, and so attention focused on means of individually attaching the blades. At some point, the ceramic foot of the blade had to be attached to metal, but without any point pressure such as produced by bolts and pins. Blades with ceramic bases, such as those originally planned for the T.1 turbine, were abandoned, and attempts were made to build up a metal foot around a blade only. In the first attempt, a foot of pressed metallic powder was formed around the blade but did not succeed because of the different rates of expansion of the powder and ceramic. Next, a ceramic blade was inserted into a heat-resisting steel foot, and a copper-silver solder was poured into a clearance space around the blade, but this method also failed as the solder melted at high temperatures. Building a foot up by hot-metal spraying the end of the blade proved successful but had to be rejected on the grounds of it being too lengthy a process for manufacturing. Finally, the arrangement

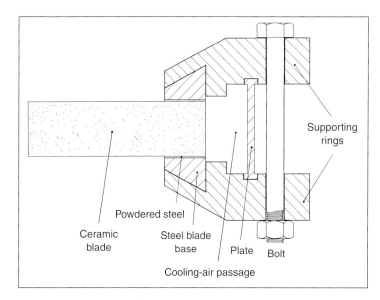

Fig 6.5 LFA method of ceramic stator blade attachment.

shown in Figure 6.5 was arrived at whereby the ceramic blade was held by powdered steel in a profile-shaped slot made in a heat-resisting steel foot. The steel powder was pressed in around the blade inside the slot and then baked. A complete ring of stator blades could then be clamped between two supporting rings, which also formed an annular channel for cooling air. Given another year's work, it seems likely that, by 1946, the LFA would have produced ceramic stator blades good for a 1,200 °C gas temperature.

The contribution of the DVL to ceramic turbines was small, and only in the last few months of the war was such work tackled. Dr F.A.F. Schmidt experimented with thin, sprayed coatings of ceramic on steel blades as a possibility for turbojet engines.

F.A.F. Schmidt's hollow turbine blades

While Dr F.A.F. Schmidt's work on ceramic-coated blades was brief, his department at the DVL (Garmisch-Partenkirchen dispersal) performed a good deal of other, earlier, turbine work. On the practical side, effort was directed towards improving the construction of hollow, air-cooled turbine blades, this work later having strong official backing from the RLM. Various unusual forms were evolved, and most of these used a main structural member of simple shape which was formed integrally with the turbine disc and thus had good strength characteristics. Around this member, or core, was formed the blade profile from thin heat-resisting sheet metal which was spot-welded in place. This left internal cooling passages between the core and the profile, while, because of poor heat conduction across the welds, the core remained comparatively cool. The central cores supporting the blades could, in some designs, be attached to the turbine disc and not integral with it. Another type of Schmidt hollow blade gave excellent cooling results, but was more expensive to manufacture. It was made from solid metal with longitudinal fins (or slots) machined out of the concave side. Over these fins was spot-welded the actual concave face in sheet metal, which could be extended to form a thin trailing edge. The resulting channels inside the blade directed the cooling air. As mentioned in the turbojet section, a meeting of manufacturers was held at the William Prym works in May 1943 to discuss problems of hollow blade production, and the DVL was the only representative of the research institutes at this meeting.

DVL turbine blade temperature measurement
Also at the DVL, Garmisch-Partenkirchen, an ingenious method for measuring the temperature of revolving rotors and turbines was devised. A Dr Gnam and Dr Küh were largely responsible for this work. Thermocouples were used to measure temperature, these being of iron/constantan for temperatures up to 600 °C and nickel/chromium for higher temperatures. (A thermocouple basically consists of two wires of dissimilar metals joined together in a circuit. Heating of the joint produces a minute current proportional to the temperature.) A thermocouple was mounted in a turbine blade or other desired part of the rotor and its wires led out through grooves to a hollow extension shaft. Protruding from the end of the hollow shaft was a second thermocouple connected with the first so that the currents were in opposition. The second, or protruding, thermocouple was surrounded by a thermostatically controlled furnace. When the turbine was running, the furnace temperature was adjusted until no current flowed from either thermocouple through the circuit. This meant that the known furnace temperature was equal to the temperature of the blade or other part with the thermocouple.

At first glance, the apparatus seems over-elaborate, but it permitted rotor speeds four times greater than was possible for accurate readings with simple electrical slip rings. In place of the sliding contacts of slip rings, current was indicated in the thermocouple circuit by an armature mounted on the rotor extension shaft. A stationary coil surrounded the armature and was connected to an oscillograph. When current flowed in the thermocouple circuit, an AC current was generated in the stationary coil and was indicated on the oscillograph. The apparatus was very sensitive and it was necessary to avoid the use of iron in the armature and to screen the whole generator from the earth's magnetic field.

LFA turbine testing methods, including the interferometer
Two main rigs were available at the LFA, Völkenrode, for testing turbines and turbine blading. For turbines, the rig was of the low-speed, cold-air type in which a 1 kW centrifugal fan sucked air from the turbine through a plastic duct. An electric brake measured power from the turbine while its speed was measured stroboscopically. Silk tufts, viewed through the plastic duct, gave a visual effect of the flow.

The other rig was for testing turbine cascades or single blades in series. Models of the blades were generally about 1.5 times full size, and made of pressed wood, the profile section being about 50 mm chord × 125 mm long (2 in × 5 in). This rig was again for low speeds (limited to Mach 0.3), and measurements of pressure loss and air deflection were carried out. Most interesting at the time, however, was the equipment used to study the airflow, and which allowed many photographs to be taken. The equipment was the interferometer made by the famous Carl Zeiss optical company. Figure 6.6 shows the scheme of operation of the interferometer. Light from a monochromatic source was split into two parts by a partially silvered mirror (i). Light passing through (i) struck the fully silvered mirror (iii), and then passed through the blade cascade under test which was mounted between optical flats (v). The remaining part of the light, which was reflected by (i), travelled precisely the same distance, but via mirror (iv), and was subjected to the same degree of optical resistance – with the exception of the air in the cascade, in which variations of air density and therefore light resistance occurred.

The two light streams were finally brought together at the partly silvered mirror (ii), and the interference resulted from the slight light phase shifts produced by air density variations in the air stream of the cascade. Regions of constant density were represented by lines of interference which could be photographed on a plate at (vi). By taking static pressure measurements it was possible to relate the interference fringes with actual pressure values, and with care the velocity at any point of the cascade could be deduced.

There were two different ways in which this equipment could be used. In the first method, all mirrors were set absolutely parallel to each other, and a uniform intensity of light surrounded the blade cascade while there was no airflow. Once the airflow began, interference fringes appeared in accordance with density changes around the blades, each fringe being a line of constant density. This method was preferred for speeds approaching and exceeding Mach 1.0, although the LFA rig was not capable of such speeds.

For subsonic flows, a more sensitive arrangement was used in which either mirror (i) or (ii) was set very slightly out of parallel with the other mirrors. This displacement caused a series of parallel fringes to appear when there was no airflow, but when the airflow began, the fringes were bent in accordance with the density changes around the blades of the cascade. It was also possible to alter the direction of the interference fringes relative to the cascade, which sometimes increased the sensitivity.

Using the interferometer, the LFA conducted seven series of separate tests on turbine blades for outside concerns, and measurements were made over a wide range of gas inlet angles and pitch/chord ratios. Four of

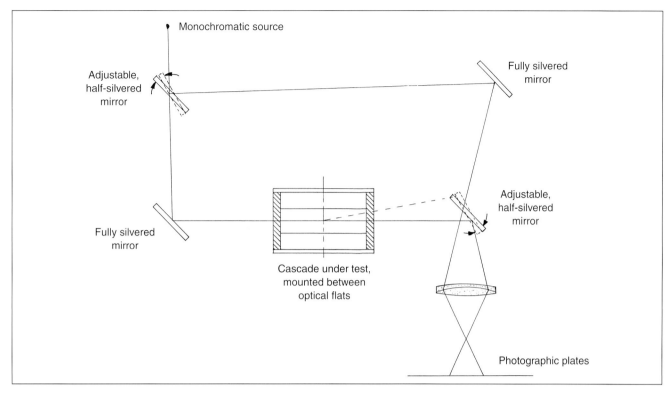

Fig 6.6 Principle of Zeiss interferometer, as used by the LFA for blade cascade tests.

the tests were for BMW-Spandau (between November 1942 and July 1943), while the others were for MAN (September/October 1943), Heinkel-Hirth (March 1944) and Junkers (May 1944). One other test series, for BMW, was conducted on compressor blading in November 1944. Only in the case of BMW does any modification of design appear to have been made in accordance with the LFA tests. On the basis of LFA findings, the pitch/chord ratio of the BMW 109-003 turbojet turbine blades was altered, but apart from this instance, the LFA seemed required merely to verify results already obtained with designs scheduled for production and not modification. The interferometer rig was used for other work, including tests on Schmidt water-cooled turbine blades. For these, steam was passed into five holes in each blade, observations were made with the interferometer and temperatures were measured at different points. Conventional blade cascade rigs were also used by the LFA.

Jet unit installation

The appearance of jet units, by which is meant chiefly turbojets, brought about problems of suitable installation in aircraft. Similar problems later presented themselves for other jet units, such as the pulsejet and ramjet, where air is drawn in at the front end and an exhaust jet of increased velocity emerges from the rear end. At first, the attempt was made to install jet units in a similar manner to airscrew units, but not only did this lead to unsatisfactory results, the special installation advantages offered by jet units were missed.

Germany's, and the world's, first aircraft to fly solely on the power of a turbojet engine was the Heinkel He 178, which made its first true flight on 27 August 1939. The engine and aerodynamic lines of this experimental aircraft were far from perfect, but a problem not fully anticipated arose from the fact that the single engine was mounted in the fuselage with a long intake duct and a longer exhaust duct. Losses in these ducts were great, because of friction with the considerable wall area, and with basic aerodynamic data lacking, other installation solutions had to be resorted to. In seeking such solutions, the German airframe industry was generally left to its own devices, and although it possessed some good wind tunnels and other research equipment, some bizarre projects were put forward.

The companies developing the actual jet engines more than had their hands full to obtain reliable operation without considering the aerodynamics of installation. On the aerodynamic side, the main aim was to obtain the minimum frontal area, and because of the war there was little time left to go into the installation question very thoroughly. Consequently, German jet aircraft appeared with their engines mounted in the simplest manner, such as beneath the wing (e.g. He 280, Me 262, Ar 234) or outside the fuselage (e.g. He 162, Hs 132, Fi 103). This partly resolved the problem into one of designing a nacelle with very short ducts and designing a good aerodynamic attachment to the aircraft. One of the few cases where the research institutes played an active part in this designing concerned the BMW 109–003 turbojet nacelle, for which the LFA provided data for the inlet duct and the AVA designed the outer form of the nacelle.

It was probably not until 1944 that a concerted effort was made to identify all the likely methods of installing jet units together with the work needed to resolve the problems in each case. A research programme was drawn up by Prof. D. Küchemann, O. Conrad, and J. Weber of the AVA, Göttingen, and a wind tunnel test programme was begun. The planned test procedure was as follows. Models with different engine arrangements were prepared as far as possible using a conventional wind tunnel to obtain fillet shapes, etc., giving a smooth airflow. This stage was possibly aided with bubbles in a water tunnel. Further tests eliminated models with high frictional drag, and the final selection was made using a high-speed wind tunnel. The engines for the aircraft models were represented by models having a simplified internal shape, but which were reminiscent of the actual engine nacelle regarding internal and external flow conditions. In order to allow for the energy imparted to the internal flow in the actual engine, the model engine was supplied from inside with a jet of compressed air. For water-tunnel tests, a mixture of water and air was blown into the model engine. Figure 6.7 shows a section through a one-fifth-scale model representing the BMW 109–003 turbojet.

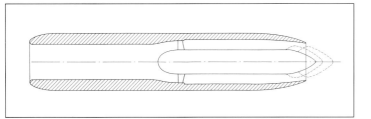

Fig 6.7 Section through AVA 1/5th scale model representing internal and external flow shape of BMW 109–003 turbojet.

Most tests by firms and research institutes did not follow the AVA procedure, and many results were treated with reservations because of the short time spent on tests, the small models used and the lack of large high-speed wind tunnels. Even for jet aircraft actually built, there was only a short time available to improve the fillet between the engine nacelle and the wing or fuselage. Since the AVA test programme had not been under way for long before the war ended, the use of a considerable amount of space is scarcely warranted even to briefly record the many installation layouts planned for investigation, together with their anticipated problems. Instead, we can conclude by mentioning two interesting aircraft projects thrown up by an Emergency Fighter competition at the end of 1944. In this, the Luftwaffe called for an improved fighter with a level speed of about 1,000 km/h (621 mph) at 7,000 m (22,960 ft), and a service ceiling of 14,000 m (45,920 ft). A single Heinkel-Hirth 109–011 A turbojet was to provide the power, and in all the designs submitted this engine was housed inside the fuselage. However, out of the eight designs submitted, only two appear to have taken the problems of air intake losses into account. The first was the Junkers EF 128, a tailless aircraft with swept-back wings to which were attached twin fins and rudders; the air intakes for the engine were positioned at the fuselage sides under the wing at about the wing mid-chord position, and were arranged to divert the boundary layer airflow to a vent outlet aft of the cockpit fairing. The second was the Messerschmitt P.1110/II, an aircraft with swept-back wings and a butterfly, or vee, tail unit; its air intake was of the annular type which surrounded the fuselage just forward of the wing trailing edge fillets. A suction fan, driven from the turbojet engine, was to draw in the boundary layer air from the forward part of the fuselage through slots in the air intake. As far as is known, these air intake features originated from the research of the companies concerned, and not from the research institutions. Certainly, both the Messerschmitt and Junkers companies maintained large aerodynamic research departments.

The regenerative heat exchanger

In previous sections, various mentions were made of the desire to obtain heat exchangers which, by feeding back to the combustion chamber some of the heat normally wasted in the gas turbine exhaust, could reduce fuel consumption. Recuperative heat exchangers (in which the hot and cold streams are divided by thin metal surfaces such as tubes) have already been dealt with in the section on industrial gas turbines, and we are concerned here with the regenerative heat exchanger. In the regenerative type, a suitable material, such as porous refractory, is exposed alternately to the hot and cold gas streams, picks up heat from one and then gives it up to the other. Such a regenerative heat exchanger was designed by W. Hryniszak of Brown Boveri for the GT 103 land traction gas turbine (see Section 3), but more intensive and possibly more fundamental studies were made by a Dr Ritz of the AVA, Göttingen. His aims were to produce a regenerative heat exchanger which not only improved gas turbine efficiency but had a weight and bulk low enough to permit use with aircraft turbojet engines.

Ritz, who had eight assistants at Göttingen, was originally interested in problems of de-icing before he was led to heat exchanger studies. These studies showed that a heat exchanger with an effectiveness (or thermal ratio) of 90–95 per cent was possible with a weight and volume only a fraction of that of exchangers then available. It was already accepted that, with a heat exchanger of only 80 per cent effectiveness, a gas turbine of the day could equal the efficiency of a reciprocating engine. The type of regenerative heat exchanger experimented with consisted basically of a slowly rotating refractory disc (or tube) with very small holes passing alternately through the hot gas and cold air streams. Immediately the holes passed from the hot to the cold stream, air began to pass through them and became fully heated to the refractory temperature within a short length of the holes. As the holes continued to be rotated through the cold air, a greater length of the holes was needed to bring the air to the refractory temperature, since some of the heat was taken by the preceding air. Finally, as the holes neared the end of the air segment, all the hole length was needed to heat the air to the prescribed temperature. This process took half a revolution, and at the same time, the other half of the refractory disc or tube was gathering fresh heat as it passed through the hot gas section.

Two arrangements of the heat exchanger were experimented with, the schemes of these being shown in Figure 6.8. In the first, an axial flow was used through discs with holes, while in the second, a radial or cross flow was used through a tubular element of woven mesh. (For Reynolds numbers of 50 to 200, the cross-flow type was found superior.) Regarding material for the rotating element, ceramics were chosen because they could store a large amount of heat for their weight and could be easily cleaned, if necessary, with chemicals or powerful heating. The correct choice of ceramic would give reasonable resistance to thermal shock and sufficiently high heat conduction into the wall thicknesses between the holes without permitting too high a conduction between the hot and cold sections directly through the rotating element. Another point requiring special attention was the sealing around the rotating element to prevent detrimental leakage between the hot and cold sections. Comprehensive tests showed that a feather type of labyrinth seal gave leakages of less than 1 per cent for compression ratios of up to 7 to 1.

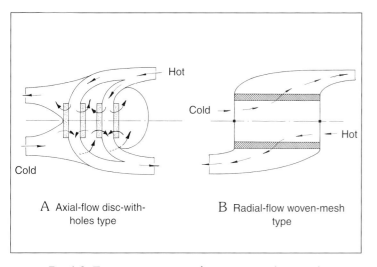

Fig 6.8 Two arrangements of regenerative heat exchanger tested by Ritz of the AVA.

After its superiority was decided, attention was focused on the design of the cross-flow type of heat exchanger. The rotating cylinder or tube for this was constructed from multi-layers of woven mesh, giving local turbulent motions to the flow and thereby increasing the heat transfer with little effect on pressure loss. The mesh formed a convenient system of rods, these being in glass for medium temperatures and quartz for higher temperatures up to 900 °C. Sectional bars along the axis of the cylinder were used to reinforce the roll of mesh. The wall thickness of the cylinder was about 60 mm (2 in), and its mesh weave was such as to give holes equivalent to a diameter of 1 mm (0.04 in), with walls between the holes of about 0.15 mm (0.06 in) thickness. It was thus extremely porous to radial flow. A rotational speed of about 30 rpm ensured that as a section of the rotating element reached the gas turbine exhaust temperature it passed into the air flowing to the combustion chamber and began giving up its stored heat.

Once it was felt that the most difficult problems of sealing between the hot and cold sections and the construction of a sufficiently rugged cylinder had been overcome, various projects to utilise Ritz's heat exchanger were begun. Proposed applications ranged over not only aircraft gas turbines, but others such as Sulzer closed-cycle gas turbines for fast boats and torpedoes. Applications to gas turbines delivering shaft power were more successful than those to jet propulsion gas turbines.

Comparative studies were made of three types of aircraft using the heat exchanger. The aircraft were the short-range Me 262 jet fighter, a projected three-hour-duration jet bomber and the long-range Me 264 project with turboprop engines and a duration of well over ten hours. The results of these studies indicated that no heat exchange should be used with the Me 262, while moderate heat exchange gave a slight improvement with the three-hour jet bomber. On the other hand, heat exchange used with the Me 264 fitted with two turboprop engines (the prototypes of this bomber/recce aircraft had four piston engines each) doubled the range owing to the phenomenal improvement in fuel consumption.

Acting on this study, the Technisches Amt gave, in 1943, both BBC (Heidelberg) and AEG (Berlin) the job of designing a turboprop engine (PTL) with Ritz regenerative heat exchanger for the Me 264. Hermann Reuter was responsible for the PTL design at BBC, and this was submitted to the Technisches Amt. The BBC unit used two compressors, each driven by its own turbine, and two power turbines to drive the contra-rotating airscrews. There was also reheat at one stage. The compressors and their turbines were underslung to the heat exchangers, which were coaxial with the airscrew turbines and were rotated by small electric motors. The layout of the BBC turboprop with Ritz heat exchangers is shown in Figure 6.9, and the following gives the

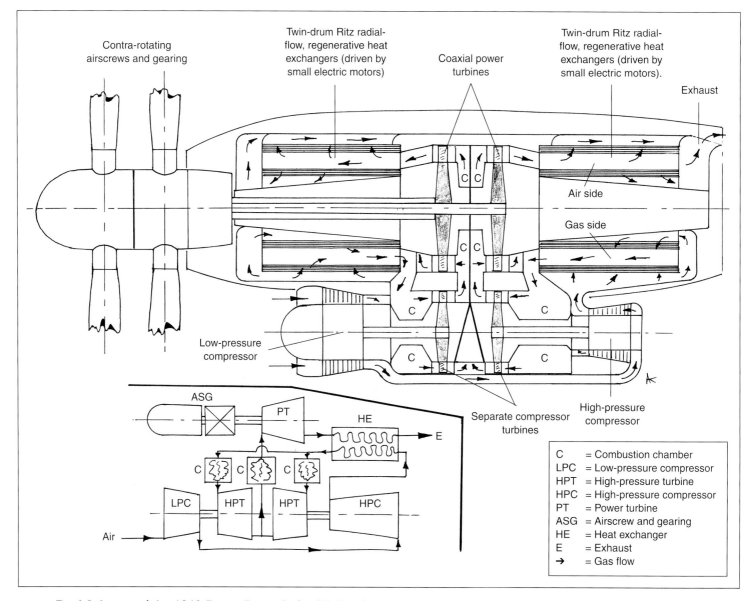

Fig 6.9 Layout of the 1943 Brown Boveri & Cie (BBC) turboprop (PTL) project using twin Ritz heat exchangers. Below is a simplified schematic arrangement. Projected to power the Messerschmitt Me 264 bomber with two of these units, a very long range of 11,000 miles was expected, making a bombing raid on New York, for example, possible. *(Author)*

specification, but estimates of performance figures varied:

Power rating at 10,000 m (32,800 ft) and 600 km/h (373 mph):	5,000 equivalent shaft hp
Maximum weight	6,000 kg (13,230 lb) including 750 kg (1,654 lb) for heat exchangers
Pressure ratio	7.0 to 1
Specific fuel consumption (assumed at 10,000 m and 600 km/h)	140.7 gm/thrust hp/h (0.3 lb/thrust hp/h)
Heat exchanger effectiveness	90/95 per cent
Overall unit efficiency	41/42 per cent
Heat exchanger section diameter	1.40 m (4 ft 7 in)
Depth with underslung compressors and turbines	1.94 m (6 ft 4 in)
Overall unit length	5.0 m (16 ft 4ⁱⁿ in)

The above estimated fuel consumption figure of 140.7 gm/thrust hp/h was compared with 207.9 gm and 181.6 gm for the alternative power-plants of four-stroke piston engine and diesel engine respectively, fitted to the Me 264 and operating under the same conditions.

GAS TURBINE RESEARCH AND DEVELOPMENT

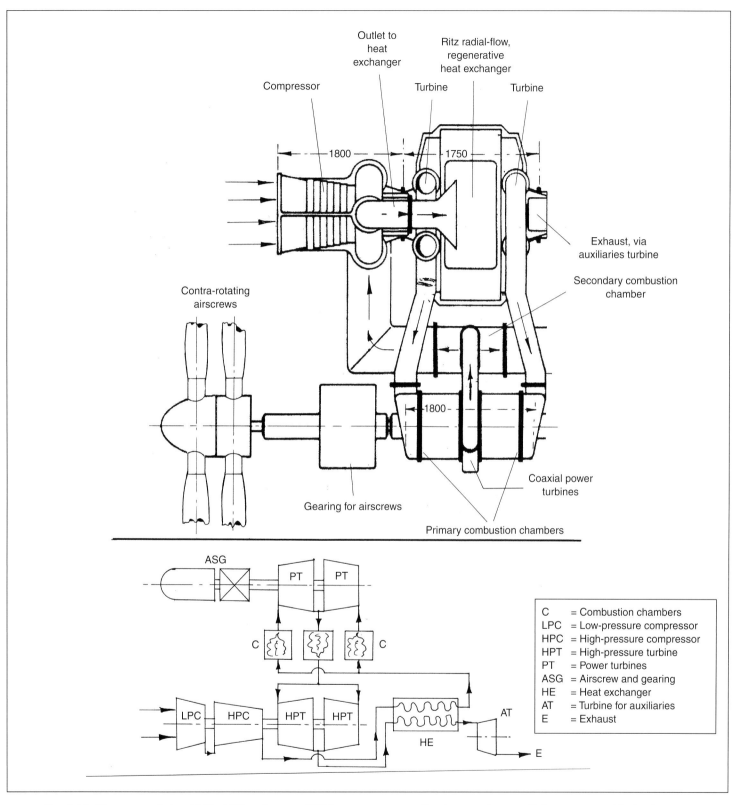

Fig 6.10 Plan view of the 1943 AEG turboprop (PTL) project using a Ritz heat exchanger. Below is a simplified schematic arrangement. A competitor of the BBC PTL, it was another projected power unit for the Messerschmitt Me 264 very-long-range bomber. As with all heat exchanger schemes, the aim was to increase efficiency and specific fuel consumption – prior to the advent of more developed turbojets. *(Author)*

Fig 6.11 A project existed to fit the ultra-long-range Messerschmitt Me 264 bomber and reconnaissance aircraft with two Ritz heat exchanger propeller turbines of around 5,000 shp each. These would have replaced the previous four radial piston engines of 1,700 hp each, and because of the much better fuel economy, the 9,315-mile range would have been doubled. Either BBC or AEG engines were to be fitted, AEG engines being shown on the drawing. *(Author)*

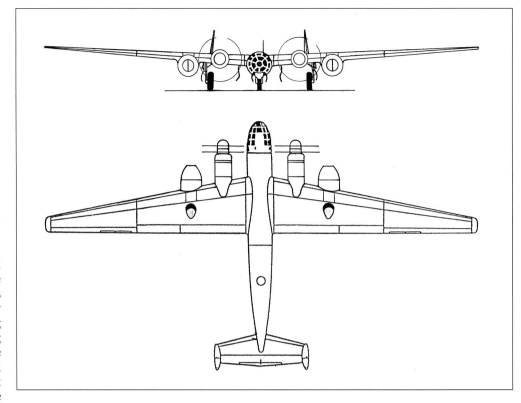

The competing AEG design was along similar lines to the BBC one but was envisaged as being arranged horizontally with the wing instead of having some of the unit underslung beneath the wing. There were other differences, but the main one was that the Ritz heat exchanger was positioned in the circuit between the compressor turbine outlets and the exhaust (whereas the BBC design had the Ritz heat exchanger position in the circuit between the airscrew power turbine outlet and the exhaust). No data are available for the AEG design, apart from the dimensions on the drawing (see Fig. 6.10).

In theory, Ritz's heat exchanger promised a phenomenal improvement in gas turbine efficiency, but by the end of the war only laboratory models of aircraft heat exchangers had been tested (with promising results) and various projects drawn up.

The Lutz swing-piston gas generator

It will be recalled how attempts were made to use piston engines and other gas generators in connection with gas turbines; in particular, the Junkers company were interested in utilising the free-piston gas generator with aircraft gas turbine units. On 2 October 1942, Franz Neugebauer, who at one time worked for Junkers, gave a lecture at the eighth scientific meeting of the DAL on the efficiency of high-pressure combustion chambers for turbines and jet units. By high-pressure combustion chamber was meant one in which not only combustion under high pressure occurred, but also the gases were *compressed* before delivery as hot pressurised gas to perform work. For the turbojet or other gas turbine this meant coupling a compressor with a combustion chamber and some engine portion where expansion could take place in order to drive the compressor by means of a turbine.

Following Neugebauer's lecture, Prof. Dr-Ing. Habil. Otto Lutz of the LFA, Munich, set about developing a high-pressure gas generator for use with a turbine. He first studied the previous work of Pescara, Junkers, Neugebauer, Kamm and others on the free-piston engine which was then the chief representative of the high-pressure combustion chamber. These and other studies led Lutz to conclude that such a combustion chamber should have the following basic requirements:

(i) Large passage areas in order to increase frequency of strokes. Valves to be replaced by piston-controlled ports.

(ii) No decrease in volumetric efficiency at full load, and therefore no series coupling of the engine with the compressor.

(iii) Avoid throttling losses to maintain the highest attainable efficiency.

(iv) Apart from design simplicity, plant should comprise a number of cylinders of 3 to 4 litre swept volume.

At first (*c.*1938), types of helical-form pistons were considered. Then, Lutz concentrated on the so-called swing-piston, the working cycle of which is shown in Figure 6.12. An annular chamber, or toroid, enclosed two piston groups, each of which comprised three pistons attached to two hubs. As the six pistons travelled around the annular chamber, planetary gearing caused them to swing to and fro in relation to each other. Thus, six spaces were formed, of which, for example, 1, 3 and 5 increased their volumes simultaneously, while the volumes 2, 4 and 6 decreased by the same amounts.

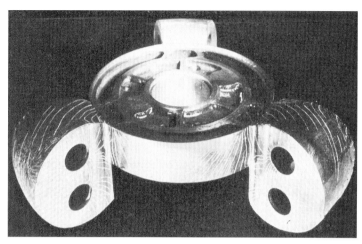

Fig 6.13 This example of the Lutz swing-piston engine was built in 1944 by the Büssing-NAG-Entwicklungsgesellschaft in Brunswick. With the forward half of the toroidal casing removed, the antifriction bearings and the two groups of three pistons can be seen. (O. Lutz)

Fig 6.14 An experimental Lutz piston assembly which has undergone test at the LFA using the 'stress coat' method. (O. Lutz)

Fig 6.15 Two views of the Lutz swing-piston gas generator in an engine test cell during 1944. (O. Lutz)

Of various projects worked out for the Lutz unit, we are chiefly interested here in its use as a gas generator for a turbine. For aircraft use, a power unit was envisaged for high-speed flight as shown in Figure 6.16. The aim with this swing-piston-TL unit was to improve on the fuel consumption of contemporary turbojets (TL) by replacing the normal constant-pressure combustion chamber/s with high-pressure combustion chambers. The design condition was an airspeed of 900 km/h (559 mph) at 10,000 m (32,800 ft) altitude.

From the section in Figure 6.16, we see the high-pressure combustion chambers or gas-generating section consisted of five blocks of swing-piston units. The five-stage axial compressor at the intake supplied part of its airflow to the high-pressure combustion chambers, which greatly increased its energy level before feeding it to the three-stage axial turbine as hot gases. The turbine used some of the energy in the gases to drive the axial compressor, while the remainder was converted into thrust by the exhaust nozzle. Compressor air surplus to combustion requirements simply passed the combustion section and entered the exhaust section.

The characteristics of this unit were expected to be as follows. The highest thrust would be in the region of the best fuel consumption conditions, but both thrust and specific fuel consumption would fall off rapidly after reaching their optimum. However, fuel consumption was expected to be only half that of a comparable turbojet unit (with constant pressure combustion) and to be especially good under part load and static thrust conditions. Of course, these and any other gains were obtained at the expense of a more complex engine and the need for a cooler or radiator for the swing-piston blocks. The only data available for the unit are:

Efficiencies	Compressor 80 per cent
	Turbine 80 per cent
	Combustion chambers 88–90 per cent
Pressure after compressor	1.5 atm.
Gas temperature at turbine inlet	650 °C
Overall length	2.80 m (9 ft 2 in)
Overall diameter	0.88 m (2 ft 10^1 in)

Another project was for a swing-piston-PTL or turbo-prop engine for long-range flight, the design condition being 540 km/h (335 mph) at 10,000 m (32,800 ft) altitude. In this scheme, the combustion chambers consisted of eight blocks of swing-piston units mounted in the aircraft wing and accessible during flight. A turbine drove the airscrew and also the axial compressor supplying the combustion chambers, and the possibility existed of mounting the latter separately from the turbine and compressor section/s. Data for this long-range unit are:

Power at 10,000 m (32,800 ft) and 540 km/h (339 mph)	4,930 hp
Pressure after compressor	1.07 atm.
Airflow	12.0 kg/sec (26.45 lb/sec)
Pressure after combustion chambers	4.80 atm.
Gas temperature at turbine inlet	650 °C
Compressor diameter	0.680 m (2 ft 2^i in)
Swing-piston block diameter	0.560 m (1 ft 10 in)
Turbine diameter	0.875 m (2 ft 10 in)
Airscrew diameter	6.0 m (19 ft 8 in)

In another field, that of industrial power generation, a 40,000 hp Lutz swing-piston scheme was proposed. In this, four banks of gas generators, each comprising eight swing-piston blocks, supplied hot gases at only 550 °C to a central turbine which drove an air blower (for the combustion chambers) and provided the shaft power. Similar schemes were proposed for ships, and the advantages claimed were compactness, convenient layout, small fuel system pipes and good fuel consumption.

Conclusion

With the end of the war, the value of much German gas turbine research could not be immediately evaluated. In many cases, research workers from the victorious powers had the co-operation of German personnel to continue or evaluate research projects at institutions such as the DVL and AVA. Project finalisation and evaluation of data and records sometimes took up to two years of co-operative effort in Germany.

It is now very difficult to give specific instances of where German research was useful to the Allies, but at least one investigator suggested the German research centres were more notable for the profusion and excellence of their equipment than their results. In the fields of compressors and combustion, the Germans appear to have had little to offer that was original or not surpassed by the British. More interest was created by the work on turbines, and ceramic, water-and-air-cooled turbine schemes came under careful scrutiny. Today, intensive development of air and other cooling systems, plus the use of new and exotic materials, has led to the enormously high temperatures of around 1,400 degrees C in the turbine. To reach even higher temperatures, ceramics are still being investigated in ongoing

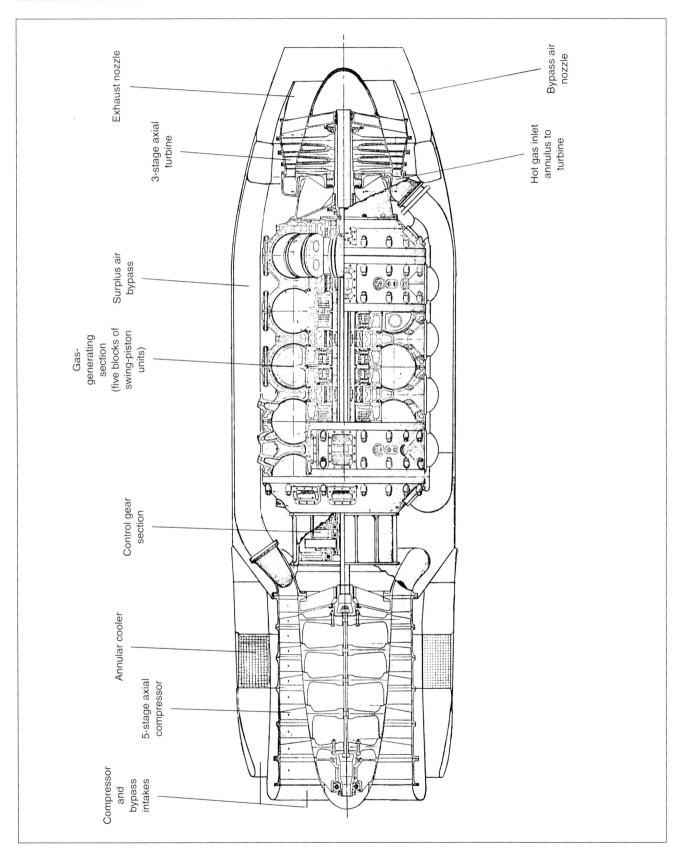

Fig 6.16 Lutz swing-piston-TL power-plant for high-speed flight. (Schnellflug-Triebwerk)

GERMAN JET ENGINES

programmes. Silica carbide composites having ceramic matrixes are showing promise, at a high cost. Also, ceramics have some specialised uses, such as in the inlet nozzles for automotive gas turbines, while air-cooled turbines were actually fitted to an American turbojet to cure over-heating troubles after it had entered airline service. Research on regenerative heat exchangers was of great interest and was assimilated in extensive research programmes after the war, but many of the hopes here were not realised. However, research on the heat exchanger continues, and some successes (such as in the field of automotive gas turbines) have been recorded.

Of fundamental importance was the AVA's research programme concerning the installation of jet units. This programme had barely begun by the end of the war, and was subsequently largely continued under Dr Küchemann at Farnborough in England. As the speed of post-war aircraft rose, so more problems of installation required solution.

As for the Lutz swing-piston gas generator, it is possible to be more specific as to the post-war direction this work took. At the end of the war, the British Admiralty had an experimental Lutz unit shipped to England, where the British Internal Combustion Engine Research Association (BICERA) took it over for tests. After completion, the Lutz unit was coupled to a starter-generator and run for about one hour as an air compressor. It was then fuelled, but with the low compression ratio of about 8.5 to 1 it refused to fire until the inlet air was heated. Following the start of combustion, the unit failed in about one minute with a torsion-type fracture of the tube shaft connecting the piston spiders to the gear couplings. In addition, there were already signs of scuffing and local distortion of the annular combustion chamber or toroid. The project was abandoned, but the lesson was learned that large forces exist in any free piston unit and these are liberated when the freedom is restricted. The 168 mm (6.6 in) diameter swinging piston on a radius arm of 195 mm (7.7 in) exerted a level of torque not met with in a normal crank motion engine, because in the latter a peak cylinder pressure occurs only near top dead centre, where there is practically no effective torque arm. On the Lutz unit, the failure was caused by excessive torque (during initial acceleration), which was increased by having to drive the heavy generator. The primary use of the Lutz unit as a high-pressure gas producer for a turbine was no longer required when the fuel consumption of gas turbines was improved. Today, two Lutz units are preserved, originally at the Cranfield Institute of Technology in England, but probably now at the Science Museum, London, and one of these is shown in Figure 6.17.

Fig 6.17 A dismantled Lutz gas generator engine at the Cranfield Institute of Technology, Bedford. Two examples were kept there at the time this photograph was taken in March 1970. (The bar on the piston assembly is a tool.)

Section 7

The Jet Helicopter

Background to the Doblhoff/WNF 342 — Brief description of the helicopter — The jet system — Flight tests — Conclusion

German interest in rotating-wing aircraft began in the 1930s, when the technical jump from the autogyro to the helicopter was about to be made. From then until the end of the Second World War, a considerable range of helicopters was designed and built in Germany, chiefly by the Anton Flettner and Focke-Achgelis companies, and a number of records and pioneering achievements were chalked up. The most common characteristic of virtually all German (and Austrian) designs was the use of rotor systems which either lacked or inherently counteracted rotor torque effects. Instead of using the classic tail rotor method of counteracting main rotor torque (first developed by Igor Sikorsky in 1939), the Germans preferred twin-rotor systems or else applied power directly at the rotor tips. Of these latter methods, the most outstanding was that experimented with by Friedrich von Doblhoff, to result in the world's first jet helicopter. His research programme was begun in October 1942, and four helicopters, representing progressive, experimental steps, were built by the Wiener Neustadter Flugzeugwerke (WNF) in the suburbs of Vienna. The RLM's designation for the machines was WNF 342. Working with Doblhoff were Dipl.-Ing. Theodor Laufer, specialising in theory, Dipl.-Ing. August Stepan, a structures engineer and test pilot, and a group of about twenty other people. They had the facilities of the WNF at their disposal, and the helicopters were designed to facilitate quick alterations as dictated by empirical experiment.

Background to the Doblhoff/WNF 342

Doblhoff was attracted to the idea of using a jet-driven helicopter rotor because it promised simplicity, a torqueless drive and a lack of weighty transmission shafts, couplings, etc. encountered with rotors driven directly from an engine. A method of applying a jet thrust to the rotor tips was to fit them with individual jet units such as ramjets or pulsejets, but this implied a concentration of weight at the rotor tips, with an undesirable increase in gyroscopic effect and rotor inertia. Another method, which Doblhoff followed, was to feed a compressed, combustible mixture out to the rotor tips for burning in simple, streamlined combustion chambers. A conventional piston engine was chosen to drive an installation for compressing air and mixing in fuel. With this system, all fuel burnt outside the piston engine was in combustion chambers remote from the fuselage installation, so that there could be no turbine to extract power for air compression and all energy extracted from this fuel had to be in the form of thrust.

In principle, the system reminds one of the method of jet propulsion patented by Dr Harris (England) in 1917 and briefly investigated by Whittle in 1929. Similar ideas were followed by Campini in Italy, who employed a piston engine to drive a ducted fan followed by a ring of fuel injectors to heat the compressed air. The jet thrust was inefficient, but to Campini's designs the Caproni N-I aircraft was built and first flew on 27 August 1940, only to be abandoned in the summer of 1942. A few months later, Doblhoff began his work. How much study he had made of previous jet work is not known, but he had certainly had no previous experience in this field.

Brief description of the helicopter

Before looking in more detail at the jet system of the WNF 342, the following notes on the helicopter may be useful. Each of the first three helicopters (V1, V2 and V3) was provided with a small rear propeller to blow air at the tail surfaces for steering, but the last machine (V4) had a second propeller mounted coaxially to provide a thrust for forward flight when clutched to the compressor engine. Thus, in the V4 machine, the rotor jets, which had a high fuel consumption, were only used for take-off, hovering and landing, and the rotor turned by autorotation during forward flight. To simplify development problems, the first three helicopters had no rotor pitch change arrangement, and vertical control was provided simply by varying the rotor speed. The ingenious pitch control method devised for the V4 will be described later, since it was connected with the jet control system.

The WNF 342 V1 was built and first flown in the spring of 1943, but was superficially damaged in 1944 during an air raid on Vienna. Following this, the test programme was moved a short distance away to

Fig 7.1 The Doblhoff WNF 342 V1 jet helicopter first flew in the spring of 1943. A small airscrew was provided at the rear merely to give an airflow over the tail surfaces at low speeds.

Fig 7.2 The Doblhoff WNF 342 V4 jet helicopter.

Obergraffendorf, where the 60 hp Walter-Mikron engine driving the compressor in the V1 machine was replaced with a 90 hp engine, and general modifications were made to the extent that the machine was redesignated V2. With the V3 and V4 machines, progressive increases in rotor diameter were made, and these two machines used the extra power of the 140 hp BMW-Bramo Sh 14A engine to drive the compressor. Only in the V4 machine did such refinements as fairings appear as semi-permanent fittings, although this was still at the experimental stage. In all machines, an Argus As 411 geared supercharger was adapted as the air compressor. The V4 had two seats (previous prototypes had only one), a rotor diameter of 10.0 m (32 ft 9 in) and a loaded weight of 640 kg (1,410 lb), which gave a rotor disc loading of 8.17 kg/sq m (1.67 lb/sq ft).

The jet system

Fuel system

A major problem which Doblhoff had to solve was that of efficient fuel/air mixing. This was particularly important because of the small size of the combustion chambers, or jets, at the tips of the three rotor blades, and the small mass of air which could be passed through the hollow blades in a given time. The conditions which had to be met were rapid combustion, thorough fuel/air mixing and an air/fuel ratio of about 14 to 1. In the first tests, pressure injection of liquid fuel into the compressed airflow was tried, but with no success, since only about half of the heat content was given up to the airflow.

However, by evaporating the fuel at 160 °C prior to injection, success was achieved. Hot exhaust gases from the piston engine provided the heat for the vaporiser, and the vaporised fuel was injected into the airflow *before* the compressor in the first three helicopter prototypes. In this system, pressure tapped off from the compressor to the fuel tank forced fuel into the vaporiser, from which it emerged at atmospheric pressure to be sucked into the compressor. The action of the compressor blades ensured thorough mixing of the fuel vapour and air. Later, many small tubes were fitted into the vaporiser to dampen a tendency for pressure oscillations. Although mixing was excellent with this system, starting was difficult, and if one rotor jet blew out the other two followed suit. Furthermore, there were inefficiencies on the compressor side since this item was drawing in a heated gas which required extra energy to compress. While, at this stage, the helicopters were flying, there was certainly no rotor thrust or lift to spare, so Doblhoff devised another fuel/air mixing system.

This second system is shown schematically in Figure 7.3, and had the essential difference that fuel vapour was injected into the airflow *after* the compressor, i.e. on the high-pressure side. Consequently, a gear-type fuel pump working at three or four atmospheres pressure was introduced into the system, together with a regulator for maintaining a constant fuel/air ratio. The areas of the membranes in the regulator were calculated to give the correct ratio of fuel pressure to boost pressure. A spring in the regulator ensured that there was always fuel available for starting and idling, and by a simple adjustment of the spring pressure the fuel/air ratio could be altered as required.

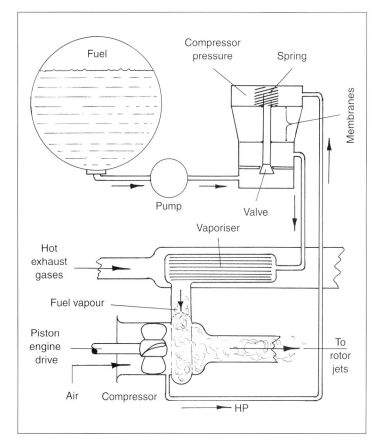

Fig 7.3 Schematic of fuel/air mixing system for Doblhoff WNF 342 (V4). *(Author)*

Rotor hub

The fuel/air mixture was ducted from the compressor outlet up to the inlet of a hollow rotor hub which is shown in Figures 7.4 and 7.5. Because the mixture was under pressure and corresponded to the throttle setting, it was utilised in an ingenious method of rotor pitch control. Each rotor blade was connected to the rotor head by means of a flexibly coupled tube flanked by steel leaf-type spring straps which were connected to an aluminium alloy casting forming the upper half of the hollow hub. This upper casting rotated in a lower, fixed casting, a seal being provided between the two. The fuel mixture flowed from inside the hollow hub out into the tubes of the three rotor blades. Passing up through the hollow hub was a hollow, fixed shaft which carried a bearing for the upper casting and was secured to the helicopter framework in a rubber mounting. Inside this hollow shaft, another shaft rotated in a spherically seated bearing to carry the blade pitch control spider at its head. Thus, angular displacement of this inner shaft tilted the spider to give appropriate cyclic pitch control. For collective pitch control, the spider was connected to the inner shaft by means of a pressure capsule connected by a pipeline to the fuel/air pressure in the hollow hub. The spider was given a vertical movement according to this pressure which was opposed by springs within the capsule. In addition, collective pitch was governed by the torsional stiffness of the centrifugally loaded spring straps. When the pilot moved his throttle control for more power, a rapid increase of fuel/air pressure and jet thrust followed, together with an increase in collective pitch, the latter maintaining the rotor speed constant. A constant rotor speed greatly simplified the controls.

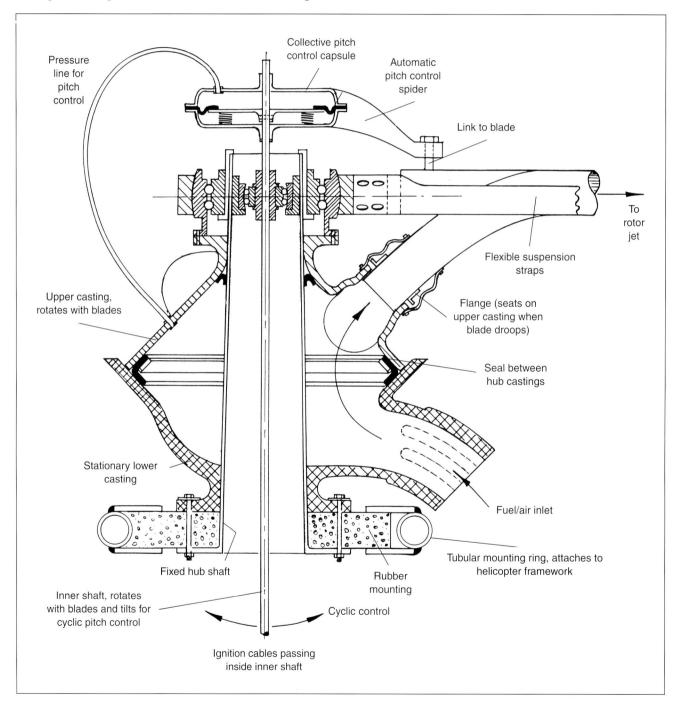

Fig 7.4 Rotor hub of Doblhoff WNF 342 (V4) jet helicopter. *(Author)*

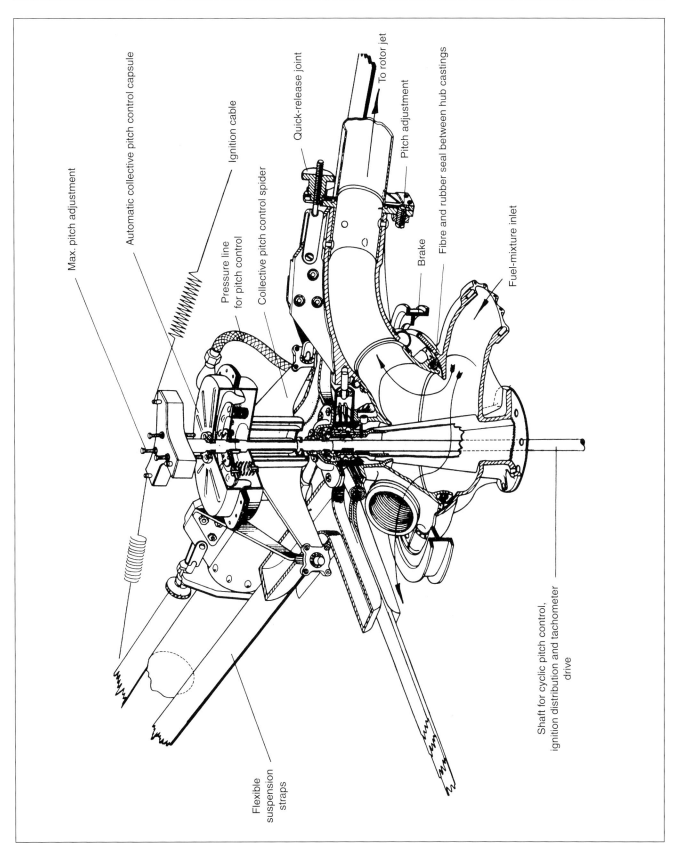

Fig 7.5 Perspective of rotor hub of Doblhoff WNF 342 (V4) jet helicopter.

Rotor blades
The rotor blades had to be aerodynamically acceptable while having internal hollow space to convey the fuel/air mixture to the jets with the minimum of pressure loss. On the third attempt, a suitable design was arrived at. A round aluminium alloy tube was drawn through dies to give a hollow, aerofoil section (NACA 23018) without its trailing edge. A wooden trailing edge with a hole for an ignition cable was then bakelite-bonded on. At the end of the rotor tube was riveted a steel fitting to which was clamped the rotor jet.

Rotor jets
Since the combustion temperature, at over 2,000 °C, was very high, the rotor jets were made in a heat-resisting steel from Gebrueder Boehler which contained 20 per cent nickel and 18 per cent chrome. Each combustion chamber was formed in two halves and welded together. To prevent combustion flashing back into the hollow rotor blade, each combustion chamber had five tubes (through and around which air could pass) to act as a filter. A small wedge at the end of each tube created sufficient turbulence to stabilise combustion. Purely for starting purposes, there was for each jet a sparking plug which was fired as the rotor blade pointed to the starboard side of the helicopter. The last length of the ignition cable leading to each jet was housed in quartz for protection. While combustion was stable enough, it was not steady and had an oscillating pressure at a frequency of 270 cycles per second.

Data for the jet system of the WNF 342 V4 are:

Maximum piston engine power	135 hp (95 hp used to hover)
Cooling power	9 hp
Steering airscrew power	3 hp
Power to compressor shaft	114 hp
Engine speed	2,250 rpm
Compressor speed	36,000 rpm
Compressor efficiency	68 per cent
Airflow (for hovering)	0.7 kg/sec (1.54 lb/sec)
Hovering { Engine fuel consumption	32.0 kg/h (70.5 lb/h)
Hovering { Rotor jet fuel consumption	135.0 kg/h (297.7 lb/h)
Combustion temperature	above 2,000 °C
Useful thrust per rotor jet	13.3 kp (29.3 lb)
Angular velocity of rotor jets	160 m/sec (525 ft/sec)
Rotor speed	305 rpm
Rotor power	525 hp

Flight tests

As stated, flying with the WNF 342 V1 began in the spring of 1943. This machine did not have mass-balanced rotor blades, even when modified into its V2 form, but the rotor was very lightly loaded and it did not reach its flutter speed. Because of the torqueless, jet-driven rotor, the helicopter was very smooth, and stick shaking was only slight.

However, trouble began with the heavier V3 prototype, which displayed severe vibration during its first tests. Twisting of the rotor blades caused lateral movements of the rotor hub which coupled with swinging of the whole machine. Below a rotor speed of 300 rpm, the situation was made worse, but above this speed flying was perfectly smooth. Unfortunately, the only way to land at this stage was to lower the rotor speed since there was no rotor pitch control. Consequently, the test pilot, August Stepan, found himself in a predicament on an early flight, since, whenever he tried to land by cutting the rotor speed, vibrations and oscillation became so severe he was forced to throttle up again. Eventually, of course, he had to land, and the violent oscillation poised the helicopter on one wheel and turned it over. The explanation of this trouble was found by calculation, and it was that the rotor blade chordwise frequency was close to rotor rpm. When the rpm was greater, all was well, but when less, violent oscillations set in (especially when the difference between the blade chordwise frequency and rpm was near to the pendular frequency of the helicopter's fuselage).

As a cure, the blade spring straps were lengthened and softened so that the natural frequency of the blades was always below the rpm. Even so, care had to be taken to keep the rotor speed as high as possible, and there was always a certain amount of resonance while the machine was on the ground. After rebuilding the rotor of the V3 machine, one of the jets was not fitted correctly, with the result that severe fluttering occurred. During tests at night, observation of the jet flames showed that the amplitude of the blade movement was as much as one metre. This error in mounting one of the jets, however, led to more calculations and theory, which indicated the need to mass-balance the rotor blades forward of the leading edge. At first, this modification, carried out on the V4 machine, resulted in very stable flying, but the controls were sluggish, and flying near the ground in gusty weather was dangerous. More-responsive control was achieved by modification of the mass-balancing weights, so that the V4 machine lacked vibration and was very smooth to fly. By the time the programme was brought to a halt, the V4 had hovered for a total of 25 hours, but had had not flown in forward flight above 40 to 48 km/h (25 to 30 mph).

Conclusion

During the three and a half years of his research programme, Doblhoff made considerable progress towards a reliable jet helicopter, but its chief drawback was the high fuel consumption of the jet rotor. Since the rotor jets used fuel at a rate more than four times greater than that for the engine driving the compressor and airscrews, rotor power was necessarily limited to vertical and hovering flight. For the rest, the machine was operated as an autogyro. To eliminate the losses involved in ducts and the system in general, Doblhoff at one stage considered turning to rotors with pulsejets and even miniature turbojets mounted directly at their tips. Towards the end of the war in 1945, in the face of the Red Army nearing Vienna, Doblhoff's team made a hasty withdrawal to Zell am See, where the WNF 342 V2 and V4 prototypes were captured by American forces. Following evaluation, the V4 machine was sent for preservation by the Smithsonian Institution in Washington.

There were many possible variations of Doblhoff's jet system, and a number were experimented with in various countries during post-war years. In 1947, the Fairey Aviation Company in England patented a jet rotor system based on Doblhoff's work. In this system, a gas turbine unit was used instead of the earlier piston engine. A centrifugal compressor supplied air via a heat exchanger to nozzles at the rotor tips, expansion of this hot air being sufficient to give thrust without burning extra fuel. A proportion of the air leaving the heat exchanger was ducted to the turbine driving the compressor, this air being first heated further in a combustion chamber. The exhaust gases from the turbine passed through the heat exchanger and flowed out either over a rudder or tail nozzles to provide steering. Refinements, in the way of automatic devices, included centrifugal governing of the rotor nozzle exit areas and fining of the rotor pitch for autorotation in the event of a turbine or compressor failure. The result of these developments was the Fairey Ultra-Light helicopter which used a Blackburn Turboméca Palouste gas turbine engine. It made many successful demonstrations, but noise and high fuel consumption forced the government to abandon its development in favour of the piston-engine-powered helicopter known as the Skeeter. The role for these helicopters was Army co-operation. After the war also, the Société Nationale de Constructions Aéronautiques du Sud-Ouest (SNCASO) became interested in Doblhoff's schemes. SNCASO soon designed and built three helicopter prototypes with rotors having burners at their tips. These were the Ariel I and II with piston engines driving the blower and Ariel II with a turbine driving the blower, and they were flown respectively in 1949, 1950 and 1951. Because of difficulties with adjusting the burners and high fuel consumption, these helicopters were supplanted by the turbine-powered Farfadet. The Farfadet only used the jet-powered rotor for vertical flight, and proceeded in forward flight in the manner of an autogiro, which was the more economical scheme that Doblhoff had originally planned. It began flying in 1953 and there were no serious problems, but development stopped after some turbine problems and a paucity of funds. Today, the most common form of jet helicopter uses the economical turbojet engine to mechanically drive the rotor, but the helicopter powered by the piston engine is still much in vogue, particularly where small helicopters are concerned.

Section 8

Pulsejets for Aviation

Paul Schmidt of Munich — The Argus Motoren Gesellschaft, Berlin — The pulsejet at Ainring — FKFS research — The H. Walter KG, Kiel — America follows the German lead — Conclusion

The history of the pulsejet (IL) is unique amongst aeronautical power plants. Its principles, known since the first decade of the last century, were more or less ignored for the next two decades until practical work began in Germany. Intensive development during the Second World War gave the pulsejet a brief but spectacular career until, soon after the war, the type, still not fully developed, became obsolete.

Basically, the pulsejet consists of a tube with a valve system at one end permitting air to flow only inwards (in some cases, more in than out), a fuel injection and mixing system, a combustion chamber with means of initiating ignition and, finally, a long exhaust tail pipe. In flight, air is forced into the combustion chamber partly by the ram air pressure and partly by a reduction in chamber pressure following combustion and exhaust. After ignition of the fuel/air mixture, the resulting hot gases are discharged through the tail pipe with a velocity greater than that of the entering air, with the net result that a propulsive force is developed in the direction of flight. Rapid and automatic repetition of the cycle results in a series of explosions and consequent fluctuations in air volume and pressure.

The aim is to achieve as near constant-volume operation as possible, and for this purpose the tail pipe is of prime importance, mainly because it supplies an air column necessary for roughly isovolumic operation. A pressure wave is reflected from the tail pipe as a rarefaction or negative wave which travels back to the combustion chamber. If the pipe is long enough, the combustion chamber pressure will already be falling off following combustion before the returning rarefaction wave arrives. The arrival of the wave reduces still further the combustion chamber pressure to a value below that of the outside atmosphere, and this facilitates the opening of the valves to admit a fresh air-charge. A final function of the tail pipe is to determine the frequency of the whole cycle, since the time taken for the pressure wave and its reflection to travel along the pipe is dependent on the pipe length.

The pulsejet, unlike its nearest relation the ramjet (dealt with in Section 9), develops thrust at zero velocity, so that theoretically it can be used for take-off, although in practice a launched or assisted take-off gives a more useful initial thrust. Although it is a simple engine, having a high power-to-weight ratio, its fuel consumption rate is very high, while other disadvantages are the pulsating thrust, noise level, vibration and destructive acoustics which it produces.

Whole volumes have been written on modern knowledge of the pulsejet, but the brief remarks above will suffice to enable the reader to appreciate in the following the labours of pioneers on the pulsejet, which, incidentally, has also been known as the resojet, buzz engine, intermittent jet engine, and aero-resonator. It should be made clear that we are concerned here with the resonating (as opposed to the non-resonating) engine. In many early designs and patent literature, the frequency of operation depended on the timing of the mixture and ignition method, so that these engines would not necessarily operate at the natural frequency of the system (non-resonating engines).

There were many early ideas for non-resonating engines, and a few will be mentioned. In 1908, René Lorin proposed to pass piston-engine exhaust gases through a regulated rotary valve and then to exhaust them through a thrust nozzle as a high-energy power jet. In the same year, Holzwarth established the fundamental type of explosion chamber, which was followed by the development of the turbine explosion chamber (see Section 5). To regulate the gas exchange processes of the non-resonating engines, many mechanical elements, such as valves and slides, were needed with all their attendant difficulties.

On the other hand, the resonating engine regulates its gas processes simply and automatically. Technically, the lineage of the resonating engine lies close to acoustic devices which, even at the beginning of the twentieth century, could produce enormous sound intensities. In the patents of Marconnet (c.1909) suggestions were made for explosion resonator operation with a view to powering aircraft. During 1910, Karavodine used such resonating explosions to operate a gas turbine for a duration of several hours, a remarkable achievement. The explosion frequency in his tube was 38 to 48 cycles per second, the turbine useful power 1.6 hp and the overall efficiency about 2.5 per cent. This low efficiency probably accounted for the subsequent neglect of resonating engines until Rheinst of Holland and Paul Schmidt of Germany began their work.

Rheinst's explosion resonator (c.1930) was a jug- or pot-shaped engine operating with self-ignition, i.e. once started, the engine did not need auxiliary aids such as sparking plugs to maintain the series of explosions. Rheinst's patents concerned the all-important process of self-ignition and detonative ignition. His resonator, in which air and exhaust gases passed through the same opening, was brought to the machine laboratory of the Technische Hochschule, Dresden, where it functioned as a steam generator.

Paul Schmidt of Munich

The name of Dipl.-Ing. Paul Schmidt is irrevocably linked with the pulsejet through his pioneering practical work. He first took an interest in the possibilities of the jet engine when, from 1928, he considered how a conventionally shaped aircraft could be given vertical take-off characteristics. While a normal airscrew would give the aircraft power for forward flight, downward-acting jets from the fuselage and wings were wanted for vertical travel, and for this a unit or units of high power-to-weight ratio were necessary. There were, at the time, proposals to augment the thrust of a rocket with external atmospheric air to give a greater mass flow, but Schmidt rejected these as an unfruitful line to pursue.

He then conceived the idea of transferring energy to the free air by means of combustion pressure in a straight tube, refilling of the tube being accomplished by means of the inertia of the departing air in the tube and through a set of check valves at its forward end. The process was to be periodic and have an almost loss-free exchange of energy resulting in large impulses.

Practical work went ahead at a slow pace, since Schmidt's main work at the time was in fluid dynamics with the consulting company Maschinen und Apparatebau. However, by 25 April 1931, Schmidt was granted a patent (523,655) entitled *Method of producing reaction forces on an aircraft*, which referred to his pulsejet-like engine, and in the same year his application for government financial support was granted. This application had been made to Adolf Baeumker, head of the Verkerhsministerium (Research Division of the Ministry of Communications), which was the predecessor of the RLM. Until the end of the war, Schmidt's work was continuously financed (first by the Verkehrsministerium and then, from 1935, by the Forschungfuhrung of the RLM), and was the first jet engine project to receive German government support.

An initial objection to the pulsejet was the ignition system, since the ignition velocities known at the time were far too slow to provide a practical thrust. From 1931, therefore, Schmidt worked on a novel device whereby ignition was produced by a shock wave, the aim being to obtain ignition velocities of up to 100 m/sec (328 ft/sec), or about ten times faster than any other then known. A 1923 paper set out theories by R. Wendlandt on the effects of shock waves, but Wendlandt had not made actual investigations of the effect of a shock wave from a hot gas on a cold fuel/air mixture, and this process was first investigated by Paul Schmidt.

Tests were initially conducted with ignition devices producing single shock waves only, whereby a fuel/air mixture was fired to rupture a diaphragm and send a shock wave along an exhaust tube. This shock wave in turn ignited various fuel/air mixtures in the tube, care being taken to eliminate any effect other than shock which might have caused ignition. Towards the end of 1934, an ignition device had been developed which gave shock waves periodically at 50 cycles/sec, the device utilising a system of valves and a piston.

The device was progressively modified until, during 1937, the first oscillographic measurements were made, and it was found that, with the ignition system operating at 50 cycles/sec, the tube itself was operating at 100 cycles/sec. Thus, every second ignition was initiated automatically, which fact did not come as a complete surprise to the engineers, since in previous tests tube ignition had sometimes been noted to continue for three or four periods after switching off the ignition system. Schmidt had, in fact, rediscovered the phenomenon of automatic ignition which Karavodine had experienced a quarter of a century earlier.

Grasping the fact that the automatic ignition witnessed was caused by the action of weak shock waves travelling back from the end of the tube after an explosion took place (thus initiating the next explosion), Schmidt decided to increase this shock wave by increasing the resistance of the tube air inlet valves a little. The effect of this was gratifying, since the tube immediately operated automatically, and the ignition device, upon which much time and effort had been lavished, was no longer needed and the road was open for the development of a practical pulsejet engine.

This important point was not reached by Schmidt without difficulty and the use of his own funds to supplement government aid. In 1934 efforts were made to obtain further financial aid so that the work could go forward at a more acceptable pace. Attempts were made to arouse interest in the original VTOL aircraft idea, and Schmidt, on a friend's advice, also worked out pulsejet applications of a more obviously military nature. Here he was assisted by Prof. Dr G. Madelung, who did the aerodynamic work for a flying bomb, interceptor fighter and a light bomber.

The flying bomb project (see Figure 8.1) was submitted to the then new RLM, which rejected it as being 'technically dubious and uninteresting from the tactical viewpoint'. This unappreciated bomb was to incorporate the pulsejet engine as an elongation of the rear fuselage and have intake valves in a flush band around the middle of the fuselage. Its calculated speed was 800 km/h (497 mph) at an altitude of 2,000 m (6,560 ft), while its span was 3.125 m (10 ft 3 in) and its length 7.15 m (23 ft 5 ᴅ in).

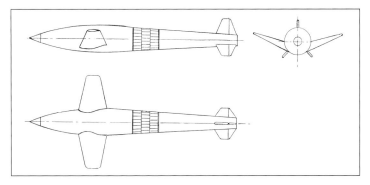

Fig 8.1 Schmidt/Madelung 1934 flying bomb project.

Financial matters improved when, in 1935, Schmidt's sponsorship by Prof. Busemann, Dr von Braun, Dr W. Dornberger, Prof. A. Nägel, and Dr Lorenz resulted in the RLM and Heereswaffenamt taking an active interest in the flying bomb by granting funds for development. Amounting to almost one million Reichsmarks (which were repaid by the end of 1944), these funds enabled Schmidt to set up in Munich a small, full-time development group with Hans Lembcke as his chief collaborator (calculations, etc.). In return for this official support (which has been described variously as 'lukewarm' and 'very strong' – depending upon which source is consulted), the Schmidt group was to develop three types of resonating engine, shown diagrammatically in Figure 8.2. After developing a straightforward pulsejet, type A, work was to proceed on a pulsejet with a second set of valves to admit further air and augment the thrust (type B). In type C, which was never built, the highest efficiency was aimed for by the addition of an oscillating-piston-compression device which was to exhaust into the tube a mass of gas which amounted to the various constituents brought together at ignition.

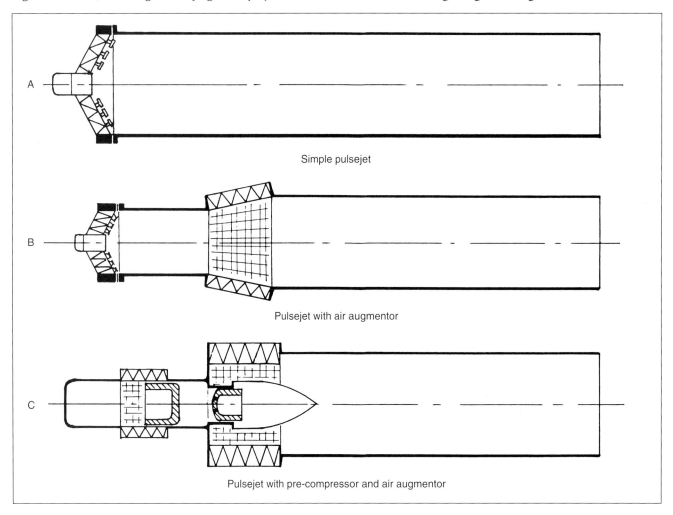

Fig 8.2 Three types of pulsejet planned for development by Paul Schmidt.

Throughout 1938, the first Schmidt pulsejet with automatic ignition was tested. Its tube measured about 2.0 m (6 ft 6^i in) long by 120 mm (4^i in) diameter and ran on an ether/air mixture sucked in from a tank. In 1939, tube diameters were increased to 200 mm (7^m in), and then up to 510 mm (1 ft 8 in), with lengths of approximately 3.50 m (11 ft 5^i in). The early months of 1940 saw a 500 mm tube, designed for a static thrust of 500 kp (1,103 lb), operating for a short time on propane or butane gas as fuel. The method of fuel/air mixing was first tested on a 200 mm tube, whereby an annular arrangement of fuel nozzles was positioned inside a short intake tube in front of the air intake valves. This system worked well but was abandoned because of the fire hazard inherent in preparing the fuel mixture outside the tube.

Development of a system for preparing the fuel mixture inside the tube was therefore begun, the first experiments with benzine being performed with a 120 mm diameter tube. By the spring of 1940, fuel atomisers (see Figure 8.3) were developed which were mounted behind the valve head, i.e. at the beginning of the combustion chamber. The flow energy of the fuel was eliminated by means of mesh guards, and a pressure head on the fuel was then provided by the intake airflow itself. Fuel was thus ejected in proportion to the airflow through the annularly spaced holes in each atomiser. Schmidt considered that the best working conditions would be obtained using an intermittent, low-pressure, mixture formation. He had in development a rhythmic atomiser which was arranged in the centre of the tube close to the valve body. The delicate mechanisms of the atomiser had a small fuel-regulating piston and a weak spring which held the atomiser closed against fuel leakage. During the intake period, the air flowing in caused a cover plate to be raised, and the correct amount of fuel was injected into the tube with a strong swirl to enhance distribution of the spray. However, this system was still under development at the end of the war.

In the meantime, development proceeded on other components, such as the air inlet valves. A tube inlet of minimum resistance was essential for high air augmentation, and at first a box-shaped construction with simple spring flaps was used. Schmidt's typical harmonica-type valves were arrived at after testing many designs, including one in 1941 which was claimed to have a life of more than twenty hours. For the harmonica type, air passageways and the flap holder were joined together to riveted, springy flaps to make a complete valve. To achieve a high mass flow into the tube, the area of the open passageways had to be as close as possible to the tube cross-sectional area. Accordingly, the conical valve was developed for the SR 500 pulsejet (Schmidt-Rohr or Schmidt tube 500). Because of the conical shape, the air passageways were made in segments and the flap seats were made with different angles of curvature. Flap valves arranged normal to the tube were also tested with success.

Figure 8.3 gives a section through the SR 500 pulsejet, while Figure 8.4 gives views inside an SR 500 without starter unit. The starter unit developed had a 24 volt electric motor which operated an axial-flow blower unit,

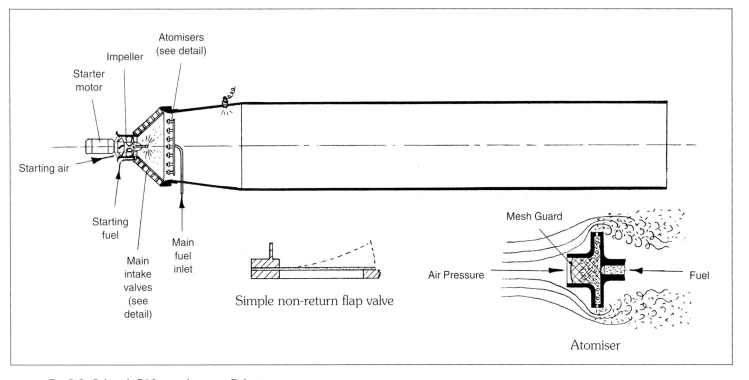

Fig 8.3 Schmidt 510 mm diameter Pulsejet.

the airstream of which was injected with fuel. The resultant fuel/air mixture was fed into the combustion chamber through the centre of the valve head. When the motor was not operating, the blower blades automatically turned to close off the central intake, leaving the main valve head for normal intake operation. Upon turning on the main fuel valve, the starting of the motor, passing of the starting and main fuels, ignition of the sparking plug and cutting out the motor and sparking plug were all automatically sequenced by an electrical relay. Delivery of fuel from the tank was performed by a pump operated by pressure pulsations from the pulsejet combustion chamber. A pilot valve was used which would not cut the fuel supply as long as combustion in the tube was regular. Should combustion fail, however, the valve closed after two or three cycles so that the tube was not flooded with fuel to create a fire hazard. Starting by the above method took about 1.5 seconds.

In 1942, an SR 500 pulsejet was uprated to give a static thrust of 750 kp (1,654 lb) by tapering the tube out to a larger diameter at the exhaust end (see photographs). This uprating was made in order to make the unit more useful in the take-off condition, although no Schmidt pulsejet was ever flown. Following further requests for more active support, the Schmidt group received an order for several pulsejet units to demonstrate the state of development. One of the subsequent demonstrations had unfortunate results when, at the turn of 1942/43, the RLM requested the testing of a 750 kp unit in the Brunswick wind tunnel of the LFA. At first, the fuel pump at Brunswick was too small, and when a second pump was obtained this delivered only about two-thirds of the normal fuel flow rate. The result was that the tube ran rather noisily at this lower operating limit and the measured thrust was only 375 kp (825 lb) at an air speed of 350 km/h (217 mph).

Soft mounting of the tube does not appear to have been used at Brunswick, with the result that the building was severely shaken and imminent collapse was feared. This fear was reasonable enough, since shocks of several tons at 50 cycles/second were acting on the building in addition to the acoustic effects of the tube. No further trials were conducted in the wind tunnel despite the fact that Schmidt had not experienced such shocks from his pulsejets with twice the thrust at Munich. There, the power peaks and vibrations produced by the periodic combustion were compensated for by sprung mountings. Tubes were connected to the mountings by means of spiral springs, resulting in power fluctuations of only about one per cent of the average thrust at the point of fixture. This method of thrust levelling was satisfactory below 1,000 kp (2,205 lb) thrust, but above this, negative power peaks occurred. To supplement the spring system, therefore, an additional hydraulic damping arrangement was provided for test-stand use.

Regarding tests, it should be realised that the intensity of sound from a pulsejet can be most uncomfortable to those nearby, and long hours of such work can cause great fatigue and exhaustion. At Munich, therefore, the Schmidt group reduced the burden on their personnel by mounting the pulsejets in vertical test stands to exhaust skywards. Thrust was measured on a hydraulic piston device, previous calibration being achieved by hanging weights on the tube. Unfortunately, this test position later proved technically inferior to other types (horizontal) of test stand, and failed to show up faulty fuel spray distribution. Once in the horizontal position and with gravity acting at right angles to the tube centre-line, fuel spray from the Schmidt atomisers was apt to distribute itself more towards the lower portion of the combustion chamber, with detrimental effects on the performance. The result was sensitivity to attitude and unstable operating characteristics.

These and other problems were gradually being solved with new auxiliary equipment under development, but by the end of the war there was no fully developed Schmidt pulsejet. Some data for the uprated version of the SR 500 pulsejet follow:

Thrust (sea level)	750 kp (1,654 lb) static
	630 kp (1,389 lb) at 350 km/h (217 mph)
	680 kp (1,500 lb) at 700 km/h (435 mph)
Pulse frequency	50 cycles/sec
Specific thrust	0.3 kg/sq cm (4.26 lb/sq in)
Specific fuel consumption	2.75
Max. tube diameter	0.565 m (1 ft 10 in)
Diameter at front of combustion chamber	0.450 m (1 ft 5^i in)
Total length	3.575 m (11 ft 8^i in)

Early in 1940, the RLM began to seek co-operation between the Schmidt group and the firm of Argus in the development of the pulsejet, and in February that year, Paul Schmidt made a single visit to the Argus company. This was followed in March by Schmidt demonstrating one of his 500 kp pulsejets, but apparently co-operation never proved very strong or fruitful. In any event, by 1941 the Argus company had forged ahead with their pulsejet development, and, in some part because of their greater resources, had taken the lead from Schmidt, despite his pioneering work.

Fig 8.4 A sectioned SR 500 Schmidt-Rohr pulsejet engine. The air inlet valve and fuel atomiser systems are clearly seen in the close-up. A wooden mandrel has been positioned where the starter unit was normally fitted. *(Donated by the Deutsches Museum, Munich)*

The Argus Motoren Gesellschaft, Berlin

The first Argus road and flight tests — Bringing the Argus pulsejet (109–014) to operational status — The 109–014 pulsejet goes to war — Further Argus pulsejet research

When, in 1939, the Technisches Amt of the RLM decided to have jet engines developed, various German companies were asked to work on different technical solutions, and this we have seen largely in Section 2. The Argus Motoren Gesellschaft of Berlin, manufacturer of small aero piston engines and super-chargers, were asked to develop a pulsejet engine. Unaware of the work of Schmidt, the task was begun in November 1939 under the leadership of Dr-Ing. Fritz Gosslau (Director of Argus) and the engineer Guenther Diedrich. Apparently a very capable engineer, Diedrich had studied at the Technisches Hochschule, Berlin, and between 1935 and 1937 had worked at the AEG steam turbine works (in Berlin) on simple steam generators. He had also studied the problems of possible steam-operated road vehicles and aircraft.

To begin with, very little information regarding the aero-resonator was available to Argus, and even the work of Rheinst was unknown. As a starting point, Diedrich was to obtain explosion resonator operation, experiment with various mixtures and obtain data on jet thrusts with a view to developing an ATO unit for aircraft. (The H. Walter KG was pursuing the same line at Kiel, but with rocket units.) It then transpired that the pulsejet was intended for flying speeds of at least 700 km/h (435 mph), and it was reasoned that air could be supplied to the pulsejet at a corresponding ram pressure, and further, if the air was admitted intermittently, fuel could be injected continuously to the combustion chamber. This early, correct analysis greatly facilitated development at Argus.

Their first test model consisted of an oscillating chamber and Borda mouth (see Figure 8.5) in which a return flow of the burning gases to the inlet was prevented by the shape of the chamber rather than mechanical valves. Airflow was arranged so that a small amount of air opposite to the inflow direction caused the formation of vortices, which caused a high flow resistance (aerodynamic choking). Continuous burning of the fuel was avoided by sinking the atomiser nozzle into a small secondary chamber and screening this chamber from the combustion chamber by a flame extinguisher on the miner's lamp principle. This model had a total length of 0.60 m (1 ft 11^1 in), including a long, narrow exhaust tube of 20 mm (^{1} in) diameter. It was first operated on 13 November 1939, and to the delight of the observers an intermittent operation with an explosion frequency close to 200 cycles/sec took place. Diedrich has stated that this first experimental model arose as a result of contact with the Schmidt group, although, officially at least, contact did not begin until early 1940. By then, Argus had built a second model on similar lines to the first, but with the essential differences that free, not compressed, air was drawn in at the front and there was no air oscillating chamber.

So far, the aerodynamic shape of the Borda mouth or vortex chamber had ensured that the combustion gases flowed essentially in the exhaust tube direction. However, contact with Paul Schmidt in February 1940 showed that his flap valve was superior to the Argus flow valve, and the adoption of simple flap valves proved advantageous for subsequent Argus development. A third model was then built with a simple leaf spring valve box, through which 90 per cent of the total airflow was sucked from the atmosphere (see Figure 8.5). The other 10 per cent of the airflow was introduced as compressed air to atomise and inject the fuel. The explosion frequency of this model was only about 10 cycles/second and its operation was non-resonating. Furthermore, it did not have automatic ignition, and could only be run with the continuous use of a sparking plug.

Development work during 1940 concentrated on fuel/air mixture formation, and by January 1941, Diedrich had found a good solution with his so-called spoiler-nozzle (or Düsenblende). By this time a pulsejet was operating which contained all the essentials of the future Argus pulsejet as used for the V1 flying bomb. Following a flap-valve grid placed at the air intake was a short venturi-shaped ring. Just past the exit of this ring, an annular region of dead airflow formed. Fuel, injected from the centre of the valve grid, spread out into the dead air and permitted even, stable combustion. The idea, whereby the incoming combustion air flows alone into the combustion chamber with *higher velocity past the fuel atomiser position* and is intensively mixed with the fuel mist, was covered in the process patent A 93713.

The first Argus road and flight tests

Immediately after the first functioning of the spoiler-nozzle mixture formation process, Diedrich began the first pulsejet road experiments (end of January 1941) for the purpose of comparing acceleration with data

PULSEJETS FOR AVIATION

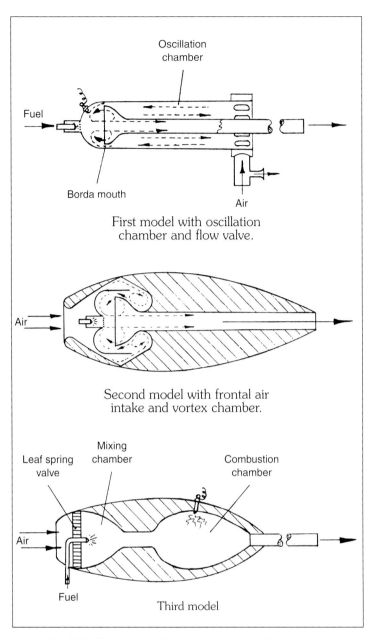

Fig 8.5 Three Argus development models for the pulsejet.

A pulsejet of 120 kp (265 lb) static thrust and 300 mm (11 13/16 in) exhaust tube diameter was expeditiously developed, and following the road tests was chosen to make the first pulsejet flight tests. This pulsejet was mounted in a balance beneath a Gotha Go 145 biplane (D-IIWS) and was swivelled to give ground clearance during take-off and landing. The first flight was made on 30 April 1941, and the aim was purely to test functions in flight (see Figure 8.6).

Fig 8.6 The first pulsejet flight tests were made with an Argus unit beneath this Gotha Go 145 biplane (D-IIWS) from Diepensee on 30 April 1941. The photograph is shot from another Go 145.

Later, a Messerschmitt Bf 109 fighter was used as a fast flying test bed with the pulsejet mounted beneath the fuselage and projecting beyond the tail surfaces to avoid the effects of heat and oscillating pressure stresses on the aircraft. The Argus flight tests were originally made at Diepensee, but later a special flight test group was formed at E-Stelle Rechlin, where Dornier Do 217 and Messerschmitt Bf 110 aircraft were used, with the pulsejet mounted above the fuselage and exhausting between the twin tail surfaces in each case. Also used were Junkers Ju 88 and Heinkel He 111 aircraft, where the pulsejet was suspended from the bomb racks between fuselage and one engine nacelle in each case. The chief Argus test pilots were Flugbaumeister Staege and Schenk.

The first aircraft to be driven solely by pulsejet power

obtained on the static test rigs. The test car was constructed on utilitarian but functional lines, and in addition to the driver's position and pulsejet under test, was equipped with a pressure tank for the fuel, a pressure-regulating valve, battery/coil ignition and pneumatic tyres. While the construction of the car and the limited road length restricted the maximum speed to 100 km/h (62 mph), the car proved to be a first-class test bed. In fact, such cars were later used as static test stands because their tyres absorbed vibration and the chassis mountings facilitated quick pulsejet changes. In this configuration, a car would be held steady and hooked to spring dynamometers to measure thrust.

Fig 8.7 An Argus pulsejet fitted above a Messerschmitt Bf 110 (GI+AZ) flying test-bed, probably at E-Stelle Rechlin. Twin-finned aircraft such as this, the Ju 88 and the Do 217, were useful in this role.

was a Gotha Go 242 cargo glider, which was tested at Rechlin in the summer of 1941. In order to improve the gliding angle and manoeuvrability during the landing approach, Argus had been asked to motorise the glider with two pulsejets, but the scheme was not pursued, and in 1942 a piston-engine-powered version of the Go 242 (the Go 244) appeared. In any event, while the pulsejet appeared suitable for speeds up to about 300 km/h (186 mph), doubts were expressed as to whether it would be suitable for high speeds. These doubts became all the more worrying for Argus since, around the end of 1941, Diedrich left the company as he believed he had proved that the pulsejet could perform no useful work at speeds in excess of 600 km/h (373 mph). By the end of 1941, the Argus pulsejet was fully developed regarding thrust on the test stand and the starting and regulation systems, and, subject to its performance in flight, a decision on its utilisation was awaited. However, matters were not improved when, in January 1942, the Heinkel He 280 V1 (prototype of a jet fighter) was moved to Rechlin to be experimentally fitted with six Argus pulsejets in lieu of its two turbojets. The aircraft made its one and only test flight in this form on 13 January. It was towed off the ground by two Messerschmitt Bf 110 Cs, but unfortunately, because of icing troubles, the pilot was forced to abandon the Heinkel in flight, resulting in no conclusion as to the characteristics of the pulsejets at high speeds.

Bringing the Argus pulsejet (109–014) to operational status

Despite the unknown factors concerning the pulsejet, the Argus company proposed, after considering other uses, that the best use for the engine would be to power a flying bomb. Eventually, the RLM decided to take a chance on the flying bomb proving technically successful, and accordingly ordered such a missile on 19 June 1942. Development of the airframe was placed in the hands of Gerhard Fieseler's aircraft firm at Kassel, an Askania autopilot was taken over for development by Siemens and the power unit was, of course, left to Argus. From this point on, the pulsejet work was completely bound up with the V1 (Fi 103) flying bomb (see Figure 8.8), Fritz Gosslau himself being a member of the V1 working staff.

In October 1942, the pulsejet came close to being written off, and with it the V1 also. That month, the LFA were requested to test the Argus 109–014 pulsejet in their wind tunnel at Brunswick. This work was carried out under the direction of Prof. Dr H. Blenk and Dr Zöbel, and the alarming result of zero thrust from the pulsejet at 620 km/h (385 mph) was obtained. At the same time, fast Argus flight tests gave very poor results. Those results were discussed by a Working Committee on Jet Propulsion Units in the autumn of 1942, this committee having been specially formed by 20 research experts on the orders of the RLM. To the urgent question, why did the pulsejet's thrust disappear at high speed, the general consensus was that its flame was

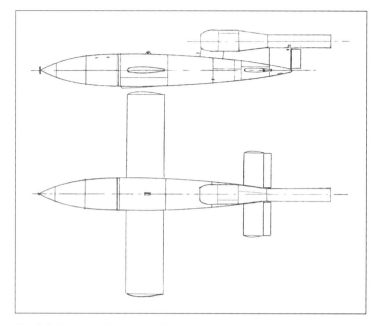

Fig 8.8 Fieseler Fi 103 (V1) flying bomb.

blown to the rear of the tube and completely upset the operation. The real answers, different for the Argus and LFA results, were found by Argus.

At the end of November 1942, Dr Volland of Argus discovered that their thrust-measuring equipment in the test aircraft was recording both positive and negative thrusts as positive readings. When the equipment was corrected, more comprehensible results were obtained, although still at variance with the LFA result. It was not until January 1943 that Argus discovered that the acoustics of the pulsejet led to false results in the enclosed space of a wind tunnel, and that, for ground tests, only test stands in the open were of any use. (The damage Schmidt's pulsejet caused to the LFA wind tunnel will be recalled.) This did not affect the *cold* aerodynamic testing in the wind tunnel, and by this means refinement of the external aerodynamic shape and the passageway of the spoiler-nozzle cross-sections for high-speed conditions was carried out. This helped to raise the thrust to an acceptable figure so that the flying bomb could at least approach its planned velocity. After only six months' work, Argus had specimens of their 109–004 pulsejet ready for testing on the V1 airframe, and while the future still held some problems, the layout of the power system was basically correct.

Figure 8.9 gives the basic details of the 109–014 pulsejet. Whereas Paul Schmidt used a pump in his fuel delivery system, the Argus method was to pressurise the fuel tank with the same compressed air supply that was used for the V1 control system. From a tank, the fuel flowed through a fuel control system and was continuously supplied to the combustion chamber through atomising nozzles. There were nine nozzles situated inside nine spoiler-nozzles (or venturis), and

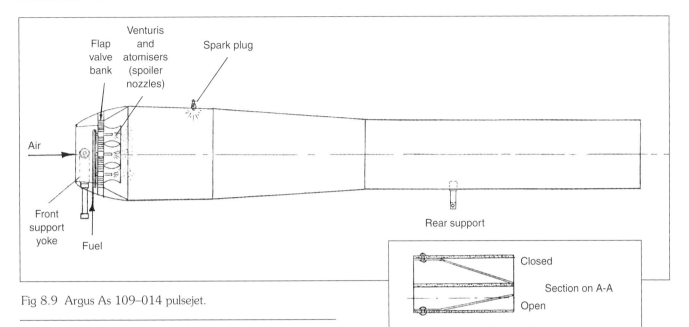

Fig 8.9 Argus As 109–014 pulsejet.

without moving parts these nozzles gave a swirl to the fuel and atomised it. The fuel control system required most careful planning and development since correct fuel pressure at the nozzles was an essential factor in successful tube operation, but it was difficult to achieve because of the fluctuating operating conditions imposed by the flying bomb. These conditions were:

	Fuel tank pressure	Pressure required at fuel nozzles
Before starting	7.0 atm.	–
At starting	6.9 atm.	1.2 atm.
Immediately after start	6.8 atm.	2.2 atm.
During launch	*9.0 atm.	2.6 atm.

(* owing to fuel inertia)

As the bomb climbed away, nozzle pressure was required to drop and then rise again as horizontal flight was assumed. Furthermore, the tank pressure slowly decreased from 7 to 6 atmospheres during the flight.

The fuel control system developed to deal with the above conditions worked as follows, and is shown diagrammatically in Figure 8.10. Fuel flowing from the tank was maintained at a constant pressure of 4 atmospheres in front of a throttle valve by means of a constant-pressure valve. The throttle valve was controlled by the difference between the airflow ram pressure and an altitude pressure capsule, the two forces acting against each other on a balance lever. Thus, with increasing altitude, the fuel flow was throttled, while with increasing ram pressure, fuel flow was increased, and the amount of fuel appropriate for the air mass flow at any given time was thereby metered.

The Argus starting system also differed from Schmidt's system in its final form. For safe starting, it was necessary to admit an amount of fuel/air mixture into the combustion chamber suddenly, so that starting commenced with a sudden and fairly violent explosion. At first, pulsejets were started with the aid of blowers, but this was not only wasteful of air, it was also uncertain with regard to an immediate start. The next two starting systems also left something to be desired. A comb of small tubes was used to press open the air inlet flap valves and to admit some compressed air. Next, small tubes were used to inject compressed air directly into the combustion chamber. Finally, three small tubes for injecting compressed air were fitted to the flap-valve grid, but in no way influenced the flaps during starting. By the use of a push-button, compressed air instantly opened up the fuel tank valve, adjusted the throttle valve and was admitted into the pulsejet.

As for mounting the pulsejet, there was a pivoted fork at the front end and a singly pinned lug at the rear end, rubber bushes and adjustable dampers being provided. Minimisation of vibration on the steel fuselage of the V1 was important because of the detrimental effects on the compass steering system, but even so, considerable vibration was transmitted which sometimes upset the guidance system.

By December 1942, Argus had solved their chief problems of providing starting and regulation systems for their 109–014 pulsejet which would suit operational conditions. Only the question of performance in flight on the V1 still remained to be proved*. Data for the 109–014 pulsejet are:

Thrust	350 kp (772 lb) static at sea level
	330 kp (728 lb) at 400 km/h (248 mph) at sea level
	240 kp (529 lb) at 645 km/h (400 mph) at 3,000 m (9,840 ft)
Weight	138 kg (304 lb)
Pulse frequency	47 cycles/sec
Specific fuel consumption on ground at 350 kp thrust	2.88
Fuel/air ratio	1 : 15
Maximum tube temperature	650 °C +
Temperature at end of tube	580 °C
Exhaust tube CSA	0.125 sq m (1.35 sq ft)
Exhaust tube diameter	400 mm (1 ft 3^i in)
Length	3.485 m (11 ft 5 in)

*The Argus pulsejet was also being developed for the Messerschmitt Me 328.

By scaling up this unit, larger pulsejets could easily have been built, such as the Argus 109–044 unit of 500 kp (1,103 lb) static thrust, but once started on the V1 programme, Argus had little time for anything that was not connected with this programme.

Early in December 1942, Gerhard Fieseler flew in a Focke-Wulf Fw 200 over Peenemünde to release the first airframe for the V1 (Fl 103) on an unpowered test flight. On Christmas Eve of the same year, the first V1 with a 109–014 pulsejet was catapulted to achieve the initial modest flight aim of about 1,000 m (1,100 yards) range. However, as more test shots were made, it became apparent that all was not well, since V1s were consistently malfunctioning. Because of the high pressure of the V1 programme, the pulsejet, control equipment, airframe and hydrogen-peroxide catapult all had to be tested simultaneously before they were fully developed, and so faults were difficult to isolate – especially as most missiles were launched out to sea and could not be retrieved. Nevertheless, haste was still urged as the Luftwaffe chiefs became concerned that the Army's A4 (V2) rocket missile would eventually take over the job of aerial bombardment!

Because of the dangers of over-supplying the fuel to the combustion chamber, fuel settings were initially kept low, and when an actual premature crash did not occur, unsatisfactory flight speeds resulted. Through gradual

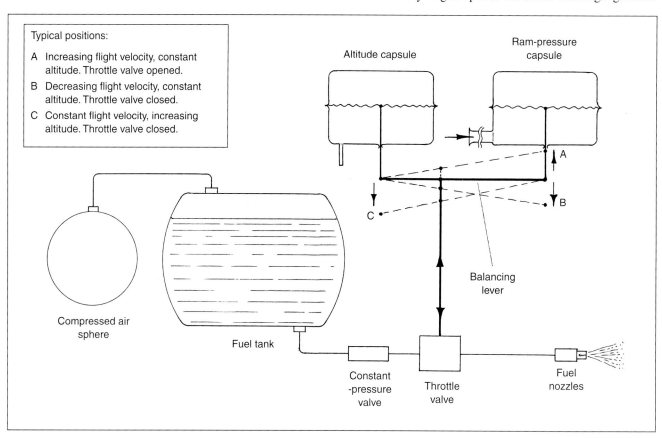

Fig 8.10 Argus pulsejet fuel-control system.

adjustment of the fuel system, speeds up to 600 km/h (375 mph) were soon attained, but then, in June 1944, a crisis arose when Peenemünde reported that flying speeds had suddenly dropped to 450 km/h (280 mph). Argus eventually discovered that a plastic diaphragm which covered the altitude capsule of the fuel control system was being replaced, unknown to them, by a material which allowed fuel to diffuse. Petrol sometimes flowed over this diaphragm during bench tests, and this gradually diffused into the capsule. Thus, the altitude capsule was caused to work incorrectly and too little fuel and power were available at low altitudes. When this error was finally rectified, the V1 speed was brought up to an operational 645 km/h (400 mph). Gradual development enabled a non-operational V1 to reach a speed of 793 km/h (493 mph) before the war ended.

The 109-014 pulsejet goes to war

The only mass-produced flying machine to be powered by the pulsejet was in fact the V1 flying bomb and its American derivatives. This ground-to-ground medium-range missile, owing to its relative inaccuracy, was a weapon of indiscriminate bombardment. On 16 August 1943, a new Luftwaffe Flak Regiment, designated 155(W), was activated under the aegis of Colonel Max Wachtel's operational research unit (the Lehrund Erprobungs Kommando) and given the task of working out V1 firing drills. Until September, this work, and also that carried out in air-launching tests, had to be performed with the under-developed V1s. As improved specimens of the bomb were received, progress was made, and soon catapult-launching sites were being constructed in occupied France.

Fig 8.11 Fi 103s (V1s) in a section of an underground tunnel production line in the Harz mountains at Nordhausen, central Germany. These tunnels were some 9 metres wide and almost as high and were dug by slave labour. *(Flight)*

Delays in production caused by incessant modifications repeatedly put off the date for opening fire on England, since sufficient stockpiles of V1s were not available for sustained initial barrages (specially demanded by Hitler). The Volkswagen and Fieseler plants were the chief centres of production, and things were made more difficult by Allied bombing of both these plants and the fixed launching sites (Operation Crossbow). Later, production went underground, notably at Nordhausen in the Harz mountains (see Figure 8.11). On 1 December 1943, a new Army Corps, the 60th, was activated under the command of Field Marshal von Rundstedt. This was a special OKW (Western Forces Command) formation with headquarters at Saint-Germain, and was staffed by both Army and Luftwaffe personnel for the control of all secret weapon formations (including, of course, those handling the V2 rocket). There were many disputes and accusations over failures and delays, and a prime intention of the V1, to disrupt the Allied preparations for an invasion of Europe (6 June 1944) was frustrated.

Finally, in the early hours of 13 June 1944, the first ten V1s were catapult-launched against London, the figure being somewhat ludicrous in view of the demanded mass launchings. Even from this small number, four crashed immediately after launching, while another two failed to make landfall. Hitler, forgetting the pressure he had brought to bear for an early launching, was furious at the premature attack, and was on the verge of cancelling the whole V1 programme. However, after seeing some exaggerated London press reports on the effects of the V1, Hitler cheered up and ordered increased production. After back-breaking efforts, the Germans managed to have over forty catapults operational by 15 June and the first mass launchings began that night.

By 29 June, the 2,000th V1 had winged on its way, although French observers were of the opinion that about one-third of bombs launched failed to fly correctly, very often because of guidance failures. On 7 July, the first air-launched V1s were used operationally when the Luftwaffe unit III./KG3 (Blitz Geschwader) attacked Southampton using He 111s. The zenith of V1 attacks on London was reached on 2 August 1944, when a landfall of 107 bombs was recorded, for which 316 bombs had been launched from 38 catapults.

The catapult method of launching the bombs at London was gradually phased out as Allied forces captured or destroyed the sites in France. Rocket-firing Typhoons often made short work of these so-called 'ski sites' during Operation Crossbow. Finally, the only method of attacking London with V1s was by air-launching from bombers based in Germany and Holland, but this method was a failure and incurred heavy German losses. By 14 January 1945, air launchings had ceased (see Figure 8.16). In the meantime, Antwerp became the prime target for catapult-launched V1s, and such attacks were intensified in December 1944 to coincide with the big German effort in the Ardennes.

ABOVE LEFT:
Fig 8.12 Fi 103 removed from storage, with wings detached.
(Bundesarchiv, Koblenz)

ABOVE RIGHT:
Fig 8.13 Fi 103 being pushed to the launching ramp.
(Bundesarchiv, Koblenz)

LEFT:
Fig 8.14 Preparing an Fi 103 for launching.

BELOW:
Fig 8.15 An Fi 103 (somewhat dented!) just launched.
(Imperial War Museum)

Fig 8.16 An Fi 103 beneath the wing of a Heinkel He 111 for air launching. Operational launches were carried out by III./KG3 'Blitz' Geschwader against England. Note the connection to the sparking plug. An Fi 103 could be carried under either wing of the He 111 H-22. *(Bundesarchiv, Koblenz)*

The constant efforts of the Nazi Party organisation to gain control of various secret weapon programmes finally met with success in January 1945, when SS General Hans Kammler assumed direction of the entire V-weapon offensive. There followed disruption of the previous chains of command, together with further mixing of Luftwaffe and Army elements, and even the SS operating V-weapons on their own.

On 29 March 1945, the last V-weapon hit London. It was a V1, and in the same month the last flying bomb landed on Antwerp. For London, 2,419 hits were recorded, and for Antwerp 2,448. Approximately four times the number of bombs had to be launched to achieve these hits. Other cities in Britain, Holland and France were also bombarded, but to a far lesser extent. Altogether, more than 30,000 V1s were manufactured during the Second World War, while, as we shall see later, copies of the V1 were also made in the USA.

A V1 weighed 2,180 kg (4,796 lb) in its operational state, which included 850 kg (1,870 lb) of explosive and 515 kg (1,133 lb) of fuel. The Walter hydrogen-peroxide catapult launched it in about one second at a speed of 105 m/sec (345 ft/sec). Climbing to a ceiling of about 3,000 m (8,840 ft), the bomb pursued its course at a maximum speed of 645 km/h (400 mph), and dived after a range of about 240 km (149 miles) was reached. Proposals to improve the V1's performance included using the Porsche 109–005 turbojet engine (see Section 2), the Argus 109–044 pulsejet or a ramjet engine. In addition, development and research continued on the pulsejet, and some of this work will be looked at.

Before doing so, however, we should backtrack on these brief notes on the V1 story to mention the curious plans to employ it as a piloted bomb. From an indiscriminate bombardment weapon, the V1 was to be transformed into a missile for attacking well-defended, pin-point targets such as warships, but with the possibility of sacrificing the pilot in the manner of the Japanese Kamikaze. By March 1944, the desperate war situation persuaded Hitler to sanction development of suicide tactics. Following various studies and experiments, work began on the adaptation of the V1 after the suggestion of Otto Skorzeny of the Waffen-SS. Piloted tests of unpowered V1s began in the summer of 1944, but accidents occurred, owing largely to high landing speeds of over 220 km/h (137 mph).

In the meantime, production of the powered, piloted V1 (designated F1 103R-IV) was begun at an assembly plant in a densely wooded area near Dannenberg, and also in a second plant known as the Pulverhof V1 assembly plant. About 175 R-IVs were produced, and some seventy pilots were selected as the first training batch from thousands of volunteers, but the piloted bomb, unlike its Japanese counterpart, never went into action. This failure was caused in part by developmental problems and lack of interest or quarrels amongst high officials. To have flown the Fi 103R-IV would have meant certain sacrifice of the pilot's life. After aiming and diving at the target, he was supposed to bale out from his cockpit just forward of the pulsejet intake. Unfortunately, it appeared that the rear edge of the hastily designed cockpit canopy interfered with the pulsejet cowling, and even if it was released, the pilot's chances of escaping injury from the cowling as he baled out were remote. Figure 8.17 shows an Fi 103R-IV at the abandoned Dannenberg assembly plant.

Further Argus pulsejet research

While the main energies of Argus were concentrated on bringing the 109–014 pulsejet to operational status and improving it for the V1 flying bomb, some time was found for general pulsejet research in addition to that carried out before the V1 was given a priority order.

Fig 8.17 An Allied soldier sits in the cockpit of a captured Fi 103 A-1/R-IV piloted bomb, without its nose-fuse attached. He is wearing the equipment intended for the pilot minus the helmet and goggles. Incredibly cramped, the cockpit only measured about 530 mm × 430 mm! *(Imperial War Museum)*

Although, by 1945, the spring flap valve was the best type found for the pulsejet, its life (about 20 minutes) was short, and so Argus conducted tests on various other forms. In 1941, a 1.50 m (4 ft 11 in) long pulsejet was tested with a rotating flap valve, but could only run with air blowing. Multi-rotating flap valves were also proposed, but the most ingenious suggestion in this direction came from Fritz Gosslau (see Figure 8.18). To give automatic rotation, Gosslau's flap had bored holes through which some of the high-pressure gases from the combustion chamber entered and exhausted in a direction tangential to the periphery. The resulting pressure impulses drove the flap around. Adjusting the flap speed to the pulsejet operating cycle, lubricating the valve and flow conditions caused by the rotation presented great problems, and much development was required before a decision on the usefulness of rotating valves could be made.

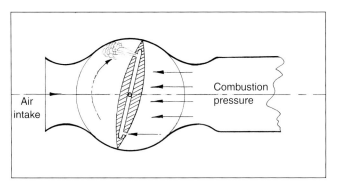

Fig 8.18 Gosslau's scheme for an automatic rotating flap valve. Tested by Argus.

On the subject of pulsejet tube shapes, a great many tests and studies were made. Diedrich proposed a hood, or outer tube, around the pulsejet tube for cooling air in order to permit mounting inside fuselages. Tests in 1941 showed that the hot tube automatically sucked cooling air through the annular space between hood and tube, and this stimulated the idea of a pulsejet/ramjet combination (ILS) in which thrust is increased by burning fuel in the annular, warmed-air flow.

Many air intake diffusers were tested after proposals by Prof. Dr Betz (AVA), Dr Zöbel (LFA), Dr Küchemann (AVA) and Argus. In May 1943 Küchemann's circular diffuser (followed by the square valve bank) was accepted as the best all round and continued to be used on the 109–014 pulsejet. However, in 1945 the design was improved by Prof. Dr Ruden by enlarging the intake to 'wet' the corners of the valve bank; the improvement offered an extra 60 kp (132 lb) of thrust at 650 km/h (404 mph). An especially interesting intake was studied in 1944 after a proposal by the gas dynamicist F. Schulz-Grunow. In this case, the aim was to reduce the effect of air ram pressure in front of the pulsejet valves and the travel of the flame front towards the rear of the tube, thereby giving, it was hoped, the same combustion process and thrust in flight as was obtained on the ground test stand. The Schulz-Grunow scheme was for a cap intake comprising a streamlined, pointed fairing in front of the valves, with a flush, annular slot after the fairing to admit air at about atmospheric pressure regardless of air speed. Owing to the streamlining, drag would also be much reduced. Other fundamental research, both on the ground and in the air, was also undertaken by Argus.

The pulsejet at Ainring

The DFS and aircraft projects

When Guenther Diedrich left the Argus company at the end of 1941, he was by no means done with the pulsejet, and appears to have spent about a year planning organisational methods for its future development. His own words, extracted from a comprehensive report he made just before the end of the war in 1945, reveal his aims: 'From observation of the structure of research, carried out while surveying the possibilities of the entire jet engine field, it was found that there existed few scientists who, upon proposing new innovations, were in a position to carry out the work. Nevertheless I succeeded in gathering a number of well-qualified workers that could travel together along a new road, so that in a short time it was possible to form the Resonator-Research Centre in Ainring.' The formation of this centre was probably begun early in 1943, and was doubtless made possible by Diedrich being appointed director of the Forschungsfuhrung (research directorate of the RLM), which was then roughly on the same level as the Technisches Amt. Head of the new research centre was a Dr Eisele, who since 1937 and at the FKFS research institute had collaborated with both Paul Schmidt and Diedrich on pulsejet questions.

From his new position, Diedrich aimed to clear up outstanding pulsejet problems and attempt to co-ordinate the work of the various companies, institutes and E-Stelle Peenemünde by assigning specific tasks to each. The FKFS institute was assigned the task of investigating the effects of altitude on the Argus pulsejet and improving the dependability of the fuel regulation.

(At an early date, the FKFS had followed up Schmidt's work with further studies, and in 1943 had built a miniature pulsejet intended as a heater to warm up aircraft engines.) Action was taken on the combination pulsejet/ramjet (ILS) idea when such a unit was ordered from Paul Schmidt's group in 1943, but it is not known if this unit materialised. At Peenemünde, various ideas of Diedrich were put to the test. One of these was for a turbo-carburettor in which a fuel pump was driven by a gas turbine receiving hot gases from the pulsejet's combustion chamber. This gave boosting with automatic boost metering from the tube's interior pressure, and was tested in February 1945. Also tested at Peenemünde was an improved flap device for the cross-circulation of air. Another method of boosting the thrust of the pulsejet was assigned to the Büssing-NAG company for investigation. The method was to inject nitrous oxide into the combustion chamber, and, the preliminary stages of the work having been reached, a 15–20 per cent increase in the thrust was anticipated. Incidentally, nitrous oxide injection for piston engines (GM1 boosting) was developed by Lutz and Winkler at the LFA for operation at high altitudes. The nitrous oxide not only supplied extra oxygen but helped to keep the combustion temperature down.

At the end of 1944, construction was still in progress on the research centre at Ainring, which was planned as the most comprehensive for pulsejet work. At this time, however, the centre was turned over to the DFS (also at Ainring) when it was still headed by Dr Eisele, but designated as Department T-2 of the DFS. The reason for this transfer is not entirely clear, but the DFS were already working on pulsejet problems. In addition, Prof. Walter Georgii, head of the DFS, was made director of the Forschungsfuhrung around November 1944, and this probably affected matters. As for Diedrich, he continued to work with Peenemünde on his projects.

The DFS and aircraft projects

Briefly, the Deutsche Forschungsanstalt für Segelflug 'Ernst Udet' (DFS), or German Research Institution for Sailplanes, grew out of the Rhön-Rossitten-Gesellschaft which was founded in 1925 for research into unpowered flight. The new DFS, under the leadership of Walter Georgii, was based first at Wasserkuppe, but transferred in 1933 to Darmstad-Griesheim in order to have airfield facilities. As the work expanded, the headquarters of most of the Institutes of the DFS were made at Ainring airfield, while the larger, more equipped airfield at Hörsching was also used when tests with heavily loaded aircraft began. The scope of DFS work gradually encompassed many fields, such as instruments, meteorology, aerodynamics, radio control, flight engineer training, and so forth, while facilities included wind tunnels, limited manufacturing equipment and assorted test aircraft. Just before the end of the war, the DFS had about 650 civilian scientists, engineers and draughtsmen at Ainring, and occupied 33 laboratory and office buildings there. Another 158 personnel moved early in

Fig 8.19 The DFS 230 glider (D-14-644) used to flight-test a pair of Argus 300 mm pulsejets. As a test bed, this assault glider had plenty of room for test equipment and various fuel system installations.

Fig 8.20 A close-up of the installation under the port wing of the DFS 230.

1944 to Prien with the aircraft construction institution (S1) to continue advanced aeronautical research.

The DFS was brought into pulsejet work towards the end of 1941, when their Department S1, under Dr Kracht, was asked to perform tests on Argus pulsejets. Development work was performed on the tube, flap valves and fuel system. The first DFS flight tests were conducted with a DFS 230 glider (coded D-14-644) fitted with two Argus pulsejets each of 300 mm ($11^{13}/_{16}$ in) tube diameter and 150 kp (331 lb) static thrust (see Figure 8.19). Unlike the earlier glider tests at Rechlin, the DFS glider had the purpose of serving as a convenient test bed for fuel system installations and of flight-testing the pulsejet components.

In addition, pulsejets were tested to destruction on test stands, and Figure 8.21 shows three views of damaged Argus pulsejet parts after 3.5 hours' running. The state of the flap valves will be particularly noted; their damage was caused not only by heat and explosions but by the constant impact with the valve seat ribs or crosspieces. (In general the Argus company only expected an efficient life of about twenty minutes from the flap valve, which was just sufficient for the time the V1 was in the air.)

Assignments given to the DFS by the Forschungsfuhrung were within the fields of fundamental research and flight testing. In connection with the latter, a project for a cheap and simple high-speed aircraft came to the DFS for attention at the beginning of 1943. The aircraft, designated Me 328, was designed by the Messerschmitt AG as a low-level bomber and possible emergency day-fighter. Development was actually handed to the Jacob Schweyer Segelflugzeugbau, which worked in co-operation with the DFS and Messerschmitt company. Only ten test aircraft were built, and the first (Me 328 V1) was mounted above a Dornier Do 217 E to measure forces acting on the Me 328 in flight. Later, the Me 328, without pulsejets, was released from the Dornier for gliding trials and proved to have barely acceptable flight characteristics.

Nevertheless, the programme was proceeded with, although early powered trials were scarcely encouraging either. In these trials, which the DFS performed from Hörsching airfield near Linz, two Argus pulsejets were mounted beneath the wings (see Figure 8.23). In this position, the pulsejet tubes ended just forward of, and below, the tailplane of the aircraft, and several accidents occurred as a result of rear airframe structural failures caused by the acoustics of the pulsejet. The usual method for starting these powered flights was to tow the aircraft off the ground using a cable-type catapult. Once airborne, the Me 328 jettisoned its main wheels, and, later, landed on its retractable skid. In the hopes of minimising the detrimental effects of the pulsejet's rhythmic exhaust explosions (which, apart from damage to the aircraft, were very uncomfortable for the pilot), tests were made with two pulsejets mounted on each side of the rear fuselage to exhaust beyond the tailplane (see Figure 8.23C). The principle of having the tail pipe extending beyond the airframe was exactly that applied in the case of the V1 flying bomb, of course. Even so, the new tests did not fully resolve the mounting of pulsejets

ABOVE AND RIGHT:
Fig. 8.21 The DFS was brought into pulsejet work towards the end of 1941. Argus pulsejets were tested to destruction on test stands, and these three views show damaged inlet flap valves, valve grid (with fuel nozzles) and tube head after 3.5 hours' running, or ten times the efficient life span expected at that time.

on the Me 328, and the type never went into production. Far from the whole project dying, however, innumerable other projects were dreamed up, but these remained on paper and cannot be discussed here. DFS use of the Me 328 for pulsejet experiments probably ended before the summer of 1943.

In 1944, after the ignition conditions of pulverised coal had been briefly tested by the DFS in an Argus pulsejet, investigations were begun at Ainring to increase the air mass flow into the unit. The investigations dealt mainly with an optimum valve designed by Diedrich. It was called optimum because the flow passageways into the pulsejet were given maximum cross-section by eliminating the ribbed crosspieces formerly used to support the valve flaps in the closed position. The necessary stiffness and support (to enable the flaps to withstand the combustion pressures) was given by simply pressing or indenting ridges into the flaps.

It was not until some time in 1944 that a fully developed pulsejet (Argus 109–014) was finally flown through the whole range of velocities and altitudes, and this work was accomplished by the DFS. With the data obtained, it was then possible to issue to the aircraft industry the pulsejet portfolio necessary to design aircraft with some measure of certainty. Then, in August 1944, as part of Germany's Emergency Fighter Programme, the RLM issued a requirement for a miniature fighter. This had to use the minimum of strategic materials and equipment such as electronics, and be the simplest possible fighter capable of bringing guns to bear on the enemy bombers. Vast production numbers were envisaged. Here, indeed, there appeared to be an opportunity for the pulsejet, and three projects to utilise it were put forward.

The three projects all dated from about November 1944. Heinkel simply offered their standard He 162 A Salamander fighter, with either one, or two less powerful, pulsejets mounted above the fuselage in lieu of the turbojet engine. From Blohm und Voss came the P.213, which had a boom to support an inverted vee tail, beneath which the single pulsejet tube passed. This pulsejet was fed from a fuselage nose intake, but the hot part of the tube was clear of the airframe. From the Junkers company came the EF 126 Elli, which largely resembled the V1 flying bomb in layout and had a single pulsejet fitted above the fuselage and cropped tail fin. In all cases, a bubble canopy was provided over the pilot's cockpit. Only the Junkers EF 126 reached the stage of a mock-up being built, and all three projects were shelved in December 1944. The EF 126 had the highest estimated speed of the three projects, 770 km/h (480 mph) at 1,000 m (3,280 ft), but its performance at moderate altitudes was estimated as poor, so that ground attack was considered as another role.

Fig 8.22 Three views of the ill-starred Messerschmitt Me 328 (RL+TY) mounted above a Dornier Do 217 E for air-launched flight tests. Here the pulsejets are mounted on the rear fuselage to exhaust beyond the tail in order to reduce the detrimental effects of the pulsejet's acoustics.

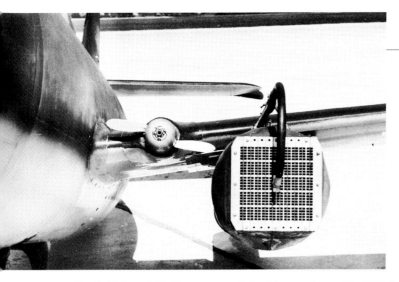

Fig 8.23A and B Two views of the Messerschmitt Me 328 low-level bomber tested with Argus pulsejets. This aircraft is at Horsching airfield near Linz, from where the DFS conducted flights. Tailpipe acoustics severely damaged the fuselage and tail unit. Since the pulsejet had no revolving parts, a small airscrew powered the aircraft's electrical generator.

Fig 8.23C Another, later, view of Me 328 (RL+TY), showing how the pulsejets were moved even further aft in an attempt to eliminate the pulsating acoustic effects on the airframe.

FKFS research

Under Prof. W.I.E. Kamm, pulsejet research work and studies were undertaken at the Forschungsinstitut für Kraftfahrwesen und Fahrzeugmotoren, Stuttgart (FKFS), at an early date. One source puts this date as early as 1931, but in any event Dr Eisele's measurement group followed up Schmidt's early work and were able with some exactness to determine the ignition process and motion of the flame front in the pulsejet. At the same time, the combustion process was studied, and a fast flight engine was designed.

In co-operation with the Forschungsinstitut Graf Zeppelin, the FKFS studied problems of pulsejet installation and tested ideas for cooling hoods. Towards the end of 1944, the institute was assigned to the investigation of the Argus pulsejet operation at altitude, and determined that it could operate up to an altitude of 8,000 m (26,240 ft), but with greatly reduced thrust. However, it was found that the thrust became practically independent of altitude when an outer duct (somewhat similar to an augmentor) was fitted concentrically around the pulsejet tube. This resulted, it was thought, in only a 15 per cent loss of thrust from sea level up to an altitude of 12,500 m (41,000 ft). For the conventional Argus pulsejet, thermal efficiencies of 7–8 per cent were reported, with a maximum static thrust of 300 kp (662 lb).

The most interesting FKFS pulsejet research was begun not long before the war ended, and concerned multi-tube pulsejets. A test stand with twin pulsejets showed that these automatically operated out of phase with each other and reduced the noise and vibration, but total thrust was less than the sum of two single pulsejets. The phase shift in frequency of combustion between ducts was an inherent characteristic of multi-tube systems.

The H. Walter KG, Kiel

The H. Walter KG at Kiel was largely noted for its work with hydrogen-peroxide-powered rockets, turbines and catapults, their rockets being used for assisted take-off, in missiles and in the world's first operational rocket fighter (Me 163). The brief pulsejet work of the company was aimed at providing a turbine drive.

A valve was designed comprising two hemispheres fitted with ports and rotating in opposite directions to provide a guide action. A frequency of about 50 cycles per second was required. One valve was constructed, tested and pronounced satisfactory, but work was stopped in the latter half of 1943 following the success of Argus pulsejets. Since the valve parts were made from scarce heat-resisting steel, they were scrapped and melted down for other uses.

Another Walter pulsejet scheme had the aim of increasing efficiency by increasing the average working pressure since, although the maximum pressure ratio in a pulsejet is about 8 to 1, the average over the whole cycle is only about 2 to 1. The scheme consisted of two interconnecting pulsejet tubes with valves at each end, the theory being that when compressed gases from one explosion reached the far end of the tube, more fuel would be injected and a new explosion would increase the pressure and specific fuel consumption. No tests were made of this somewhat complicated scheme, which, presumably, was again intended to provide hot gases to drive a turbine.

As for pulsejet flight engines, the view was erroneously held that their thrust dropped rapidly above 360 km/h (224 mph) and fell to zero at about 800 km/h (497 mph). At about 720 km/h (447 mph), it was thought that the pulsejet would have about the same thrust as a ramjet with equal fuel consumption, and as related in Section 9, the Walter KG concentrated on the ramjet for their air-breathing jet engine experiments.

America follows the German lead

Following the first mass launchings of the V1 flying bomb in June 1944, the Allies soon had enough remnants of unexploded bombs (despite their triple-fusing) to lay bare all its secrets. While the main interest in Britain was to combat the flying bomb, interest arose in the USA to copy this new weapon.

On 11 July 1944, Col. D.J. Keirn of Wright Field, Dayton, Ohio, telephoned the Ford Motor Company, Detroit, to see if the company was interested in making pulsejet engines. The Ford company was interested, although no drawings were available, only parts from unexploded V1s. Contracts for 25 experimental engines were signed on 15 July, and the first example was produced three weeks later. A contract for 1,500 engines was signed on 2 August, and later amended to increase the total to 2,550 engines on 11 January 1945. The number was subsequently increased to 3,000, with a total of 7,000 contemplated. However, in July 1945, production was cut back and the contract was completely cancelled with the victory over Japan. By then, the Ford company had built 2,401 copies of the Argus 109–014 pulsejet, which was designated PJ-31-1 by the American military. Figures 8.24 and 8.25 illustrate the production and testing of the Ford pulsejet, and because of the lack of equivalent German photographs, these illustrations are of particular interest.

While Ford were producing their pulsejets, the Republic Aviation company were building copies of the German V1 flying bomb known under the name of Thunderbug, or to the US Navy, as the KUW-1 Loon, and to the US Army, JB-2. Instead of using the elaborate Walter catapult, these missiles were launched with the aid of four, grouped, solid-fuelled rockets along a simple track, the most interesting launches in this manner being from a US Navy submarine. Launches of Thunderbugs were also made in the air from B-17 Flying Fortresses (e.g. 339 119), two missiles being carried under the wing of each aircraft. As far as is known, none of the American V1 copies saw operational service, although their planned destination was Japan. The missiles are shown in Figs 8.26 and 8.27.

Conclusion

The idea of the flying bomb gained increasing support in Germany, firstly from the Luftwaffe hierarchy, which saw the need to compete with the V2 rocket of the Army. Once development of the flying bomb was assured, the same applied to its chosen power unit, the pulsejet. Finally, high priority and the order for mass production and operational deployment was given by Hitler with his attack obsession.

Fig 8.24 Late war production of PJ-31-1 pulsejets, based closely on the Argus model, by Ford in the USA. Most of the assembly stages are shown, including the arc welding of the tubes in the bays at the top centre of the picture. *(Ford Motor Company)*

Fig 8.25 A night test of the PJ-31-1 pulsejet, using a blower, shows the exhaust flame to good effect. *(Ford Motor Company)*

Fig 8.26 A JB-2 Thunderbug (or KUW-1 Loon), clamped and weighted, is prepared for static firing of the Ford pulsejet at Farmingdale, Long Island, New York. These American copies of the German Fi 103 flying bomb were to be rocket-launched from the ground, ships or submarines, or air-launched from bombers. They were too late for WWII deployment. *(Republic Aviation)*

Fig 8.27 A KUW-1 Loon, U.S. Navy version of the JB-2 Thunderbug, being rocket-launched from the submarine USS *Carbonero* in the Pacific. During Exercise 'Miki', held in Hawaii, the Loon was guided down a line of ships as a target in an anti-aircraft exercise. *(Official US Department of Defense photo)*

While the pulsejet had shortcomings, it proved to be a good choice, under the exigencies of the time, for powering the V1. For the short range required, the valve system of the pulsejet had sufficient life to see the journey through without degrading performance too far once over enemy territory. On the economics side, the pulsejet, mechanically very simple anyway, could be made from low-alloy and mild steel, since its highest temperature was not much above 650 °C. Thus, it fulfilled the flying bomb requirements for an inexpensive, throw-away power unit. In fact, more than fifty complete flying bombs could be produced for the cost of one V2 rocket, and each bomb delivered nearly as much explosive. On the other hand, shortcomings of the V1 flying bomb, attributable to the pulsejet, were its low flying altitude and insufficient speed, so that it was vulnerable to Allied fighters and radar-controlled guns. In addition, a small fault (the remedy of which was eventually effected) caused the fuel-starved pulsejet to cut out upon commencement of the bomb's dive. This power cut both provided warning to those below to take cover and also minimised the penetration effect of the warhead.

Regarding fighters, the pulsejet appears to have had little future here because of its low power reserves and because operational ceilings would have fallen below that of the enemy's aircraft. There was also the inconvenience of the desirability of providing catapult or rocket assistance for taking-off. An alternative, which Messerschmitt put forward in several projected versions of the Me 328, was to air-launch the pulsejet fighter/s from a carrier aircraft. For low-level missions, such as ground attack and reconnaissance, the pulsejet aircraft might have held advantages for the Luftwaffe (large numbers with adequate performance), but this was never proved.

A serious objection to the pulsejet pointed to detrimental effects of its vibration and pulsating acoustics. During tests, the latter not only tore the fabric from gliders, but within a short time damaged the metal structure and panelling on more substantial aircraft. Just as serious could be the bad effects on the pilot.

Thus, in the sole case where it was married to the V1, the pulsejet served Germany quite well (omitting any argument concerning the military aspects of the V1), but this must be considered as a special technical case. Credit for the development of the pulsejet concerned goes largely to the Argus company, but without forgetting the pioneering work of Paul Schmidt. Both pioneering and energetic development were recognised officially when the RLM designated the pulsejet as an Argus-Schmidt-Rohr. This followed an inquiry in 1942/3 into pulsejet work to see if the V1 power unit could be considered as a new type. By the end of the war in 1945, the pulsejet was already being overtaken by other power-plant developments, principally the turbojet, and even before the war's end, proposals were made to supplant the pulsejet/s on the V1 and Me 328 with the turbojet (for the V1, see under Porsche in Section 2).

Mention has already been made of American wartime interest in the pulsejet. Other countries were also interested and continued to be so after the war. Then, pulsejet studies were made at a more leisurely pace, and following on from German experience, the more likely pulsejet applications, such as target drones and helicopter rotors, were centred upon.

Untypical was the continued experimentation by the Soviets of aircraft powered by the pulsejet. The Junkers EF 126 fighter or ground attack aircraft project was continued at Dessau under the Soviets with a German team led by Dipl.-Ing. Brunolf Baade. By January 1946 the EF 126 mock-up was ready, and by May the V1 prototype was ready. Flight tests began on 12 May 1946 with the V1 being towed behind a Ju 88, but the pilot, May Mathis, was killed on a subsequent flight on 21 May. Later in 1946, the V2 to V5 prototypes were fitted with more powerful Argus 109–044 pulsejets, and some gliding and powered flights were made from Stakanovo (now Zhukovskii) airfield in Russia. The EF 126 programme was cancelled in early 1948.

Other Soviet pulsejet work included, from 1945, flying captured 109–014 pulsejets above a Petlyakov Pe-2 fast, tactical bomber, and more developed pulsejets above the Tupolev Tu-2. The pulsejets exhausted between the twin fins of these aircraft. Indigenous Soviet pulsejet work had already begun under Vladimir N. Chelomey in 1942, using the units known as D-10 (or RD-13), which were flown under the wings of Lavochkin La-7 fighters in 1947.

Other post-war developments included the production of greatly improved flap valves and tests on flow-type valves. In the USA, the Marquardt Aircraft Company produced a series of pulsejets for use with helicopters and target drones. In 1948, two of their units were used to power the company's M-14 helicopter, while later on another unit powered the Globe ED5G-1 drone. An example of a Marquardt pulsejet, the MA-16, is said to have had an operational efficiency at Mach 0.45 and a service ceiling around 4,900 m (16,072 ft). In 1949, the Royal Swedish Air Board put in hand the development of a surface-to-surface ship-launched guided missile powered by a pulsejct that was an elongation of the missile's body and had air intake valves in a flush band around the pulsejet on lines similar to that proposed by Paul Schmidt in 1934; not until 1955 did this missile, the Robot 315, become operational on Royal Swedish Navy destroyers. In England, Saunders-Roe built and tested, in 1952, pulsejets for helicopters. These units were the PJ.1 of 20 kp (44.1 lb) thrust and the PJ.2 of 55 kp (121 lb) thrust. In France, V1s were launched in 1952 from a SNASE SE-1910 permanganate-powered 'rail chariot' at Cannes. The SFECMAS concern in France built, in 1953, a 180 kp (397 lb) thrust pulsejet based on the Argus units.

These are a few examples to show how the work continued. During the 1950s, applications of the pulsejet to helicopter rotors were prolific, but lack of endurance and efficiency eventually nullified the system. In the case of drones, pulsejet power became insufficient since it gave unrealistic target speeds in the face of the increasing speeds of operational jet aircraft.

Section 9

Ramjets for Aviation

The H. Walter KG, Kiel — The Trommsdorf missiles — Eugen Sänger's ramjets — The Focke-Wulf Flugzeugbau GmbH — Bayerische Motoren Works AG (BMW) — Versuchsanstalt Heerte — Conclusion

The first purposive work on ramjets was carried out in Germany just prior to and during the Second World War, and, but for various non-technical influences, the ramjet could have powered aircraft and missiles in service use. The theoretical principles of the ramjet were known in the same decade as the previously discussed pulsejet, and credit for the conception of the ramjet operating principle is generally attributed to the French engineer René Lorin, who published his theory in 1913. Patent applications for the use of a ramjet as an aircraft power-plant were made by Fono (Hungary) in 1928 and by Leduc (France) in 1933. However, up to that time, only embryonic ideas existed, and little practical work was carried out. Even the experimental work of Leduc was dogged by setbacks (attributable mainly to over-complication in design), and with the German occupation of France in 1940, his work was secreted away until after the war. Various other people trifled with the ramjet principle, but the failures and poor results obtained merely served to dampen enthusiasm and hopes for the ramjet as a power unit.

Mechanically, the ramjet is the very essence of simplicity, since it does not even have the inlet valves of its nearest relative, the pulsejet. Almost everything depends on the geometry of the unit, which should be carefully determined for a specific flight velocity for the best results. The geometry or layout of the ramjet can take various forms depending on the velocity aimed for. The first section of the unit is a diffuser for compressing the incoming air. Compression is achieved by slowing down the incoming air, which can be brought about by passing it through a divergent nozzle-shaped diffuser, a diffuser generating shock waves or a diffuser with both shock waves and a diverging inlet duct. As the air leaves the diffuser section, to some extent already heated by the compression, it is mixed with fuel and brought to a high temperature in the following combustion section. The hot gases thus produced then flow through an exhaust nozzle to expand into the atmosphere with a velocity greater than the flight speed. Because of this rate of increase in the velocity of the air passing through the engine, drag is overcome and a propulsive force is developed.

Of course, since the compression of the air in the diffuser is dependent upon velocity through the atmosphere, the ramjet cannot work from a standing start and must be brought up to operational speed by auxiliary means, such as booster rockets. The minimum air speed required is about 260 km/h (161 mph), although nobody would consider such a low speed for the ramjet today. The higher the flight speed, the higher the ram pressure and thrust, owing to the increased amount of fuel it is possible to burn efficiently; this is true up to the point where the exhaust nozzle becomes choked. In any case, air must enter the combustion chamber with a low subsonic speed, and the provision of some sort of flame stabiliser is necessary for steady combustion.

Finally, without burdening the reader with unnecessary ramjet generalities, it should be mentioned that the net thrust, (i.e. gross thrust minus drag) of a ramjet engine is often conveniently expressed in terms of a net thrust coefficient, CT, in simplified terms thus:

$$CT = \frac{T}{q\, A2}$$

where T = net thrust, q = dynamic pressure of the free-stream air and A2 = maximum cross-sectional area of the engine.

The H. Walter KG, Kiel

The Walter ramjets — Planned application of the Walter ramjet — The Walter rocket-ramjet unit (RL) — The Walter ramjet-rocket unit (LR) — The Walter single-fuel rocket-ramjet unit (Einstoff RL)

The first German ramjet work to receive government support was that performed by the H. Walter KG (HWK). This company, whose later pulsejet work was mentioned in Section 8, had its origins in 1935. The company's work received the backing of all three branches of the German armed services, but by 1937, the development section (LC II) of the RLM's Technisches Amt was the chief sponsor of HWK's work, which was then concentrating on hydrogen peroxide ATO rockets for aircraft. HWK's use of hydrogen peroxide was later seen in catapults, turbines, rocket aircraft and missiles, and was planned to power new types of U-boat and torpedoes.

First studies of the ramjet at HWK were in 1934, when an aerial torpedo with such propulsion was suggested to the Heereswaffenamt. In the same year HWK submitted a proposal for a turbojet engine with afterburning (TLS) to the RLM. However, it was probably not until the spring of 1937 that official support was forthcoming, when the Forschungsfuhrung (LCI) of the Technisches Amt provided a research contract. Work was put in hand under Dipl.-Ing. Lensch, director of research at the Walter works at Kiel. Later, when manufacture of parts was necessary, this was performed chiefly at the company's main engineering plant at Beerberg in Silesia. Testing and development was carried out at the Walter testing station at Wik near Kiel, where a Dr J. N. Schmidt was in charge of development. Schmidt had worked for Germania Werft in Kiel or the OKM before founding H. Walter KG in 1935.

At Wik, the static test rig could be supplied with air from an axial compressor driven by an 800 kW electric motor, the Kiel power station being conveniently nearby. A continuous airflow of up to 7 kg/sec (15.43 lb/sec) at 6 atmospheres and with a speed of up to 340 m/sec (1,115 ft/sec), or mach 1.0, was possible.

Interestingly enough, the compressor for this test rig came originally from a curious type of turbojet engine (TLS) investigated earlier on Walter's own initiative. Air for this engine was pre-compressed in a divergent diffuser and then further compressed by a multi-stage axial compressor up to a pressure ratio of about 5 to 1. The compressor was driven by a gas turbine which was powered by the steam and oxygen products from decomposed hydrogen peroxide (T-Stoff), these products having a temperature of about 550 °C. Unlike the conventional turbojet layout, Walter's TLS unit had the combustion chamber positioned after the turbine so that the hot gases from the combustion process gave up all their energy in the form of thrust without having to power the turbine. Also added to the hot combustion gases were the steam and oxygen, which emerged from the turbine, and the compressed airflow of course. A cooling system was provided for the unit's exhaust nozzle.

With this unit, it was hoped to surpass the turbojet in climb and speed, and surpass the rocket in endurance. High efficiency was hoped for because of the then high compression ratio. It was calculated that, if the rocket motor of the revolutionary Messerschmitt Me 163 interceptor was replaced by a TLS unit, its range would be increased threefold. The fact that the TLS unit required two separate fuels was hardly a disadvantage in this case, since the Me 163 already carried two separate propellants for its rocket motor. In any event, with the successful development of conventional turbojets by other companies, HWK's turbojet was officially abandoned. Nevertheless, the engineers concerned must have retained their interest, because Hellmuth Walter has stated that a TLS unit was operating on the test stand at the end of the war in 1945.

The Walter ramjets

The first Walter ramjet tests, made to establish the principle, were begun in 1936, when the unit shown in Figure 9.1 was employed. It had a divergent air inlet diffuser, the walls of which began parallel and then expanded into two conical sections having included angles of 4° and 8° respectively. A simple cylindrical combustion chamber and convergent exhaust nozzle followed the diffuser. Near the diffuser entrance, a ring with a large number of fuel jets was positioned to spray fuel downstream, the ring being backed by two cylindrical shields. At the beginning of the combustion chamber there was a system of baffles to stabilise combustion, and this was followed by a petrol injector and igniter to initiate combustion.

In order to minimise initial difficulties, a large combustion chamber of 500 mm (1 ft 7^1 in) diameter was chosen, together with a propane gas fuel. With the large-diameter chamber, air velocities would be low and thereby ensure easy ignition, although it was appreciated that this first unit would have a drag greater than its thrust. Easy fuel vaporisation and mixing was ensured by the use of propane gas in the first tests, but later a switch was made to petrol, and even diesel fuel in some tests. Tests were made over a range of air velocities and combustion temperatures, but the normal test condition

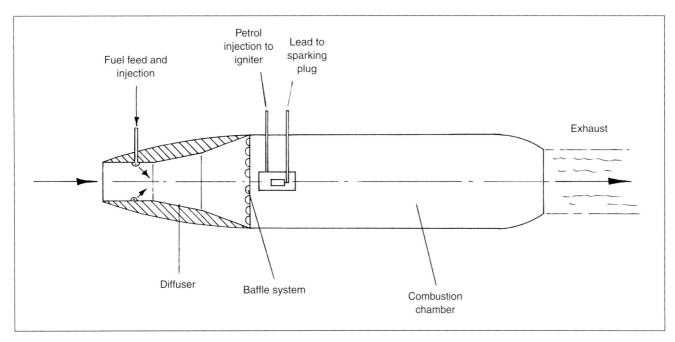

Fig 9.1 Walter Ramjet, first tested in 1936.

was a velocity of 280 m/sec (918 ft/sec) and a combustion temperature of 1,000 °C. Under this condition, the air flowing around the outside of the ramjet was sufficient to keep the temperature of its walls down to 600 °C. However, a belief existed at the time that combustion temperatures not much above 1,000 °C could be controlled in a ramjet without artificial cooling, and as a result of accepting this, low thrust coefficients of about 0.1 were recorded in the first tests. Interest in the ramjet therefore waned until 1941, when Walter's rocket experience showed that temperatures up to 1,800 °C were practical. Ramjet work therefore went forward from 1941, and temperatures up to 1,500 °C were eventually attained in the ramjets.

The results of further tests gave a compression efficiency of 86 per cent, an expansion efficiency of 92 per cent and a specific fuel consumption of 4.392 per hour under the highest velocity and temperature conditions. Assuming no losses, the theoretical specific fuel consumption had been calculated as 3.852 per hour. The dimensions of the first Walter ramjet were a diffuser entry diameter of 112 mm (4 in), a combustion chamber diameter of 500 mm (1 ft 7^1 in) × 1.20 m (3 ft 8 in) length and an exhaust nozzle exit diameter of 175 mm (6^m in).

Having completed these first tests, which were spread over a considerable time, the aim was then to reduce the dimensions of the first unit until optimum size and performance was obtained. Without altering the entry and exit orifices, the diameter of the combustion chamber was progressively reduced until 250 mm (9^m in) was reached. Overall length of the ramjet was 1.75 m (5 ft 8^m in). At this stage, using an air velocity of 300 m/sec (984 ft/sec), air entered the combustion chamber at the high speed of about 100 m/sec (328 ft/sec), which produced serious combustion difficulties and inadequate fuel mixing. To overcome this problem, two rings of fuel jets instead of one were fitted to the diffuser walls, and these jets sprayed the fuel across the airflow instead of downstream. Also, the flames behind the ring of baffles were diverted sideways by a steel ring so that they spread out to the walls. These measures gave good fuel mixing and combustion across the whole crosssection of the combustion chamber. However, if a flame-out occurred at air speeds above 185 m/sec (607 ft/sec), combustion could not be restarted, so an improved ignition device was designed. This consisted of a piece of porous material inside a tube. Petrol was injected into the porous material, behind which formed a rich mixture for igniting with a sparking plug. The resultant steady flame enabled combustion to be started at any air speed up to the maximum.

With the ramjet developed to this stage, the highest gross thrust measured on the Walter test rig was 105 kp (232 lb), from which had to be deducted an allowance for the drag of the unit. To determine the drag and thrust coefficients of the ramjet it was sent to the LFA, Völkenrode, for tests in the A9 wind tunnel under the supervision of Dr Knackstedt. For these tests, an aerodynamic fairing was fitted over the diffuser, and it was found that the ramjet gave a useful net thrust of 58–60 kp (128–132 lb) in an airflow of 1,000 km/h (621 mph). This meant that its drag was less than half its thrust. The LFA tests included combustion (using propane) at subsonic speeds and cold tests at supersonic speeds. Thrust coefficients obtained varied between 0.36 at 518 km/h (322 mph) and 0.27 at 1,000 km/h (621 mph). However, these values were considered unreliable because, when the ramjet was tested without the wind tunnel enclosure around the working section, higher thrust coefficients were obtained. The Walter ramjet was, in fact, too large for the wind tunnel, which resulted in pressures that were not uniform. The minimum specific fuel consumption measured was 7.56 per hour, considerably higher than that found by the Walter company.

Internal flow in the ramjet was not measured in tests with combustion at the LFA, but the Walter engineers realised that a parallel section was not the best for the combustion chamber internal flow. An expanding cross-section (diverging) was needed to give a constant pressure and continuous combustion. Accordingly, the ramjet shape was altered so that its greatest diameter was nearer the exit nozzle, and an increase of 5–6 per cent in net thrust resulted. Furthermore, the new external shape of the ramjet fitted in with a new discovery by Dr Zobel and Prof. Busemann of the LFA, who were working on high-speed aerodynamics. They found that, for speeds approaching the sonic region, the shape with the least drag was an inverted pear drop, with its maximum cross-section approximately one-third of its length from the tail. To further reduce the ramjet's drag, tests were made with various tail forms, and sucking away the boundary layer was also tried, but any advantage could not be measured.

The new divergent combustion chamber (also used by P. Schmidt in his pulsejets) permitted a shorter chamber length, since combustion was completed earlier owing to the maintenance of a constant pressure. To obtain a further reduction in length, experiments were begun with the injection of fuel in an upstream direction, so that mixing could begin earlier in the diffuser. Various arrangements of fuel jets and baffles were tried, and although these experiments were not completed, it was hoped to eventually reduce the combustion chamber length to 0.70 m (2 ft 3 in), which was less than 60 per cent of the length of the original ramjet's combustion chamber. Another series of tests showed that, without any serious loss of compression efficiency, the diffuser section could have its length reduced to between 350 and 400 mm (1 ft 1^i in to 1 ft 3^i in).

Planned application of the Walter ramjet

During 1944 (probably late in that year), Walter's Beerberg works began making the components for a full-sized ramjet intended as an aircraft power-plant. It was planned to mount two of these ramjets beneath the wings of the Messerschmitt Me 263 rocket fighter, the first prototype of which was completed at the Junkers Dessau factory in August 1944. The predecessor of this aircraft, the rocket-powered Me 163 B Komet, was a revolutionary, tailless aircraft which went into operational service with the Luftwaffe in June 1944.

Whilst its Walter rocket motor gave the interceptor a speed of up to mach 0.82, endurance was extremely limited. It was intended to improve endurance with the Me 263, which, with its lengthened fuselage, carried more rocket propellant.

By using the rocket motor only for boosting up to operational speed and then using the ramjets for main power, a further increase in endurance was hoped for while still obtaining a spectacular performance. One ramjet was to be mounted beneath each wing, fairly close to the fuselage, so that in the event of one unit failing, there would be the minimum of a yawing force. Accommodation for the ramjet fuel was provided by most of the space previously occupied by the rocket propellants, since only a small quantity of the latter was now required.

The design of the ramjet was basically as previously described, but scaled up. There was a diffuser and fairings in light alloy and an 840 mm (2 ft 9 in) diameter combustion chamber in 1 mm thick steel. Fuel was injected from an 18-hole streamlined ring fitted some way along and inside the diffuser. Nine of the holes directed fuel towards the axis and nine away from it, the holes being arranged in three groups which could be supplied with fuel according to a control device. At the end of the diffuser there were four or five concentric rings fitted to two perpendicular supports to form a baffle system for the combustion chamber. Mounted on the baffle system were four jets which were fed by some of the rocket propellants to ignite the ramjet's fuel/air mixture. The rocket propellants were Walter's classic T-Stoff (80 per cent hydrogen peroxide plus 20 per cent water) and C-Stoff (hydrazine hydrate, methyl alcohol and water), which reacted spontaneously when mixed. With this arrangement it was believed that the ramjet could be started at any speed up to the planned maximum.

In order to maintain the requisite ramjet combustion temperature, it was necessary to have a device which would meter the fuel according to speed and altitude. For this, it was proposed to use a device designed to maintain the speed of the Henschel Hs 117 Schmetterling (butterfly) anti-aircraft missile constant. (Development of this missile was begun by the Henschel aircraft company in 1943, and one version used a Walter rocket motor. It was nearing operational use when the war ended.) Basically, the device adjusted the ramjet's fuel throttle valve according to the difference between static, or barometric, pressure and total, or ram, pressure, these varying of course with altitude and speed. An alternative method of controlling the fuel flow was with a device operated by the actual combustion temperature, but this had to be laid aside because of the problem of obtaining suitable heat-resisting materials. A 4 hp pump working at 10 atm. supplied fuel to the ramjet, while the C-Stoff and T-Stoff for ignition had a compressed gas feed which went out when the ramjet was operating.

It was planned to test the first ramjet of this design in the new BMW wind tunnel at Oberweisenfeld, Munich (see Section 2), before beginning flight tests. Towards the end of the war in 1945, most of the components for the first ramjet had been manufactured and were sent to Lübeck, from where they were to go to Kiel. However, the components finished up near Bosau (halfway between Lübeck and Kiel), where the Walter KG had an experimental station on the shores of the Plöner lake. As soon as the war was over, Hellmuth Walter's brother, who managed the Bosau station, organised the sinking of the most important ramjet components and other secret material in the Plöner lake. This lake has some 1.5 m (5 ft) of mud on its bottom, which could protect parts from corrosion for up to a year, and by knowing the bearings of the sinkings, it was hoped to retrieve any material required at a later date.

Data available for the Walter aircraft ramjet are:

Thrusts at 800 km/h (497 mph):	450 kp (992 lb) at sea level
	200 kp (441 lb) at 10,000 m (32,800 ft) altitude
Minimum operating velocity	288 km/h (179 mph)
Maximum combustion temperature	1,200 °C
Combustion chamber diameter	840 mm (2 ft 9 in)
Diffuser length	approx. 0.95 m (3 ft 1 in)
Total ramjet length	2.50 m (8 ft 2 in)

The Walter rocket-ramjet unit (RL)

At their Hirschberg section, the Walter KG engaged in the design of combined rocket-ramjet units during 1943, although it is probable that such units were studied and the experiments carried out during the earlier phase of ramjet research. The first combination unit designed was of the RL type, shown in Figure 9.2, in which a rocket exhausted into the intake of a ramjet. The aim was to induce the necessary airflow for operation into the ramjet *while stationary*, and once operating velocity was reached, the rocket could be turned off. It is possible that this idea originated from ideas during the late 1920s, when there were proposals to augment the thrust of a rocket by adding atmospheric air to its exhaust, thereby increasing the mass flow. Engineers such as Paul Schmidt rejected such ideas (see Section 8), but the Walter scheme of using the rocket as an auxiliary, or starter, for the ramjet, and not as the prime mover of a power unit, was a different approach.

Both 'cold' and 'hot' types of Walter rockets were considered for use. In the earlier cold type, T-Stoff (80 per cent hydrogen peroxide plus 20 per cent water) was simply decomposed by passing it over a catalyst, Z-Stoff, to produce 62 per cent steam and 38 per cent oxygen at a constant temperature of about 550 °C. The hot rocket produced much higher temperatures by burning a suitable amount of fuel (e.g. C-Stoff) in the decomposed T-Stoff. During the ramjet research, mostly rockets of the cold type were used, since Walter's first hot rocket (the 109–509A–0–1) was not ready for use until the early autumn of 1942.

In the first RL units tested, the ramjet entry, where the rocket was mounted, was of a large bell-mouth shape which, even when streamlined with an external fairing, had a high drag. The bell-mouth entry, however, was progressively decreased until the entrance to the ramjet diffuser had only a rounded lip. As long as the diameter of this lip was above 20 mm (^{1} in), the rocket exhaust was able to draw a large quantity of air into the diffuser. After compression in the diffuser, this air passed with the rocket exhaust into the combustion chamber to provide oxygen for the ramjet's fuel and then the accelerated mass for thrust.

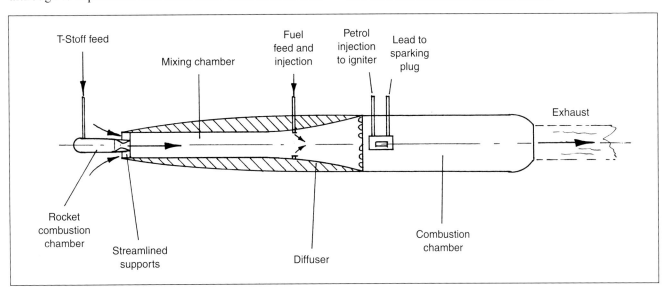

Fig 9.2 Walter rocket-ramjet unit (RL).

The ramjet diffuser was of the simple divergent type used in the previously described straightforward ramjets, but it was preceded by a parallel, tubular section some 0.70 m (2 ft 3 in) long. This section was necessary to give time for the energy of the rocket exhaust to be transferred to the incoming air and for adequate mixing of the two before entering the diffuser section. In addition, the output of the rocket had to be carefully adjusted so that the desired amount of air was drawn into the ramjet and the energy in the rocket exhaust was used up in the process.

Various sizes of rocket venturis were experimented with, ranging in size from 3 to 15 mm (to ^{1} in) diameter, until a diameter of 10 mm (in) was chosen. This size gave the lowest rocket propellant consumption of one part of T-Stoff to six or seven parts of air drawn in. The rocket produced a stream of steam and oxygen with a velocity of 1,000 m/sec (3,280 ft/sec), which imparted a velocity of 50 m/sec (164 ft/sec) to the incoming air.

Whereas Walter's 250 mm ramjet developed a net thrust of about 60 kp (132 lb) in an airflow of 1,000 km/h (621 mph), the equivalent-sized Walter RL unit gave a net thrust of up to 120 kp (264 lb) in an airflow of 840 km/h (522 mph), this result being obtained in a wind tunnel. Furthermore, the thrust of the RL unit at zero velocity was as high as 100 kp (220 lb). It therefore appeared that not only did the RL unit give a useful static thrust but its thrust at speed was consistently higher than for the simple ramjet. However, the highest RL thrusts were obtained by expending one part of rocket propellant to every four parts of air drawn in, and a thrust of about 80 kp (176 lb) only resulted when the unit was run with a minimum of rocket propellant consumption.

The Walter ramjet-rocket unit (LR)

Figure 9.3 shows the scheme of a second type of combination unit tried by the Walter KG, in which the rocket was fitted to the outlet of the ramjet. Here, the rocket exhaust was supposed to draw air through the ramjet duct with similar results as in the RL unit, but tests showed that the total thrust of the unit was only equal to the combined thrusts of the separate ramjet and rocket components. The thrusts of these two components varied according to their fuel flow quite independent of each other. In addition there were the disadvantages caused by having the rocket in the ramjet's exhaust, namely, the difficulty of designing heat-resisting supports and special cooling methods for the rocket. In fact, the LR unit was not as successful as the RL unit, and its only advantage lay in the slight reduction of skin friction and turbulence at the ramjet's exhaust nozzle because of the suction action of the rocket exhaust.

The Walter single-fuel rocket-ramjet unit (Einstoff RL)

A major disadvantage of the RL unit described earlier was that, in addition to the fuel for the ramjet component, an aircraft or missile also had to carry rocket propellant(s) such as T-Stoff and, for a 'hot' rocket, C-Stoff also. An attempt was therefore made to develop an RL unit in which both rocket and ramjet components would operate on the same fuel. The scheme of this so-called Einstoff RL unit is shown in Figure 9.4

For this scheme, the term 'rocket' is perhaps misleading, since there was no rocket component which

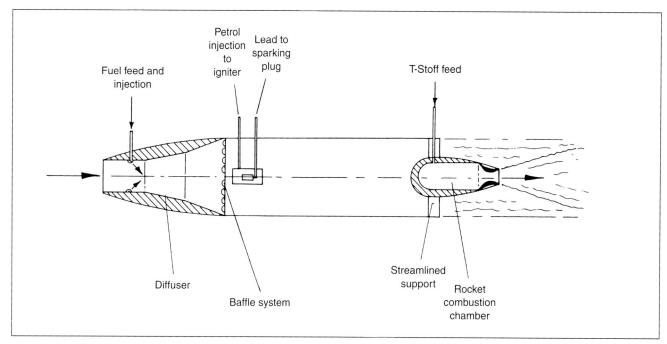

Fig 9.3 Walter ramjet-rocket unit (LR).

carried its own oxygen supply. Instead, a stream of vaporised, superheated fuel issued into the inlet of the ramjet to draw in the requisite amount of air for starting. The stream of fuel mixed with the air in a straight length of tube and compression followed in a diffuser section as before. As the mixture entered the combustion chamber, it ignited spontaneously, so that there was no need for fuel jets or ignition devices. Some kind of baffle system was still necessary to stabilise combustion.

The fuel was fed to the injection nozzle via a coil which surrounded the combustion chamber, thereby providing the heat for vaporising and superheating. During starting, when the combustion chamber was heating up, the fuel was fed via an auxiliary heater. Once the system was running on the main heating coil, the fuel vapour emerged from the nozzle with a temperature of between 500 and 700 °C, the fuel pressure used being 25 atmospheres. Under these conditions, the petrol fuel originally used was liable to crack, and so a switch was made to methyl alcohol fuel, which vaporised cleanly without residue or cracking. Furthermore, a fuel/air ratio of 1 to 7 was required to give the minimum fuel consumption while entraining the requisite amount of air, and this ratio was just that required for the complete combustion of methyl alcohol.

Whereas, in the previous RL unit, the rocket was turned off once the ramjet reached operational velocity, in the Einstoff RL scheme the 'rocket' component was left running all the time as part of the fuel system. Herein lay the pitfall, if not the downfall, of the scheme, since the temperature control of the vaporised fuel had to be precise in order to achieve spontaneous ignition at the correct point in the combustion chamber. If, for example, faulty combustion brought about a decrease in the combustion chamber temperature, the temperature of the fuel vapour would also fall and make the situation worse until combustion would probably cease. Other problems, such as combustion in the diffuser, could occur if the combustion temperature rose. The development of suitable control equipment was not taken up, and the Einstoff RL scheme did not proceed beyond the early experimental stage.

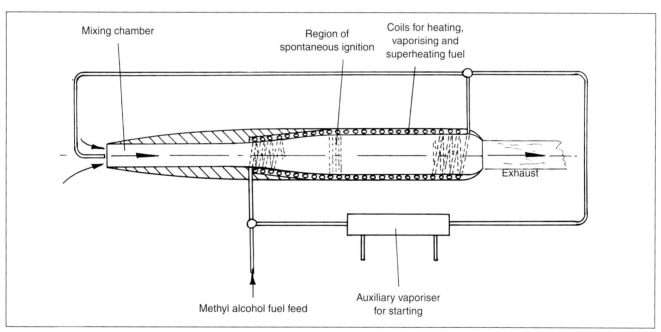

Fig 9.4 Walter single-fuel rocket-ramjet unit (Einstoff RL).

The Trommsdorf missiles

Appearance of the supersonic diffuser — Missile development with the Oswatitsch diffuser — The intercontinental ramjet missile

Perhaps surprisingly, pioneering research work with a practical end in view was carried out on supersonic ramjets under the aegis of the Heereswaffenamt (Army Ordnance Office). The story began in 1936, when Dr Wolf Trommsdorf proposed to the Heerswaffenamt the use of pulverised coal and compressed air, instead of the more expensive gunpowder (black powder), as the propellants in rocket weapons. Following successful shooting tests, Trommsdorf then conceived the idea of replacing compressed air or oxygen carried by the rocket with air taken from the surrounding atmosphere. By this means, a given size of missile could carry more explosive where there had previously been oxidant. Trommsdorf patented his idea in 1937 in Munich, and in the following year offered it to the Heereswaffenamt in the following form: 'To drive a rocket-driven missile from a gun barrel and to ram the air at the head of the fast-moving missile for introduction into the rocket combustion chamber, either as a supporting mass, or as an oxygen-carrying agent.'

Here, then, was the scheme for the missile which was to be accelerated after it had been fired from a gun, thereby increasing its speed and range. The idea was not entirely new. At about that time, for example, the H. Walter KG put forward a proposal to use a T-Stoff rocket to increase the range of a gun shell, their calculations showing that maximum range would be obtained if the rocket fired when the shell entered the upper, lower-density region of the atmosphere. Because the need to make components able to withstand high acceleration shocks from the gun appeared highly problematical, no action was taken on the Walter proposal. Notwithstanding such proposals, the novel idea in the Trommsdorf missiles, or Tr-Geschosse as they became known, was the use of atmospheric oxygen, and on 17 November 1938 Trommsdorf was offered a job by the Heereswaffenamt to develop these missiles.

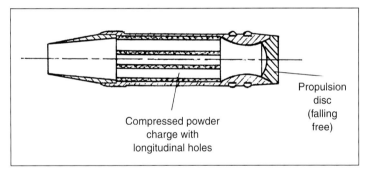

Fig 9.5 Trommsdorf 88 mm ramjet missile E1 (gun launched, fin stabilised).

He moved to the research department of the Heereswaffenamt (WAF) at Gottow, near Kummersdorf-Schiessplatz. The theory of the ramjet missiles was formulated by late 1938, when it was considered that acceptable efficiencies for the ramjet's thermodynamic process could only be attained at speeds higher than Mach 2.0. Experiments were begun late in 1939 by firing missiles of the type shown in Figure 9.5 from an 88 mm anti-aircraft gun (the famous German '88'). The missile, designated E1, had a divergent type of diffuser to compress the air, which then passed through a solid propellant charge having several longitudinal holes. The resulting combustion gases passed out through an exhaust nozzle. For firing from the gun barrel, the missile had its exhaust nozzle sealed by a propulsion disc which fell away after leaving the barrel, the firing speed being about Mach 2.5.

Firing tests with the E1 missiles were not a success. Very often the compressed powder propellant disintegrated, while the diffuser appeared unsuited to the high velocities and the propellant did not burn sufficiently. More disappointing still was the fact that no concrete proof was found that the missile could develop a useful thrust. In other words, it neither accelerated once it had left the gun barrel nor reached unusual altitudes or ranges.

Following this unpromising start, the next years were devoted to a basic research programme which was conducted on a slender budget. A small blow-down wind tunnel for ramjet combustion tests at speeds up to Mach 3.0 was built, although it was not until 1942 that the first successful burning tests with a liquid propellant were achieved. Early in the work, the use of solid propellant was rejected as ill conceived for high speeds, and liquid propellant was concentrated upon. From the start, every part and aspect of the proposed missiles constituted new problems which had to be laboriously overcome working from basic principles. Not the least of the problems was the difficulty in designing mechanical components and structures able to withstand tremendous acceleration forces, but a major stumbling block for the missiles was the lack of a satisfactory air inlet diffuser. The simple divergent diffuser used in the first test missiles provided only poor and inefficient compression of the air at the high operating speeds because of the unsatisfactory shock waves it produced. Drag was too high also.

At an early stage, the diffuser problem was put to Dr H. Ludwieg of the AVA, Göttingen, after Trommsdorf had discussed his proposed missiles with the famous armaments firm of Krupp. However, Krupp eventually lost interest, and Ludwieg could not succeed in his task through a lack of funds. On Trommsdorf's suggestion, therefore, the Heereswaffenamt enlisted the aid of two other research institutes to solve the problem.

Appearance of the supersonic diffuser

The problem of developing an air intake which could give efficient air compression, or pressure recovery, at supersonic speeds was given in 1941 to two research institutes to solve. At the LFA, Brunswick, Prof. A. Busemann and his assistant, Dr G. Guderley, produced the so-called Busemann ring. This was a divergent diffuser, similar to that already used by Trommsdorf, but with the essential difference that the intake had a sharp lip to produce a normal shock wave in front of the intake. As the supersonic airflow passed through this shock wave, it was reduced to subsonic velocity and compressed, further compression occurring in the diffuser in the normal way. The disadvantage of the Busemann ring was that it produced excessive drag above about Mach 1.85.

At the same time that the LFA produced its solution, a better one was produced by the Kaiser-Wilhelm-Institut für Strömungsforschung (KWI), Göttingen, where Dr Klaus Oswatisch and his assistant, H. Böhm, worked under Prof. L. Prandtl. The KWI solution, which was used in subsequent Trommsdorf missiles, was to slow down the supersonic airflow and bring about compression in stages by using *several* shock waves, thereby permitting operation at higher speeds than was previously possible. The device used was the multi-cone shock diffuser, which had unusually low energy losses from shock waves.

The multi-cone shock diffuser had two features of major importance. The first was a central body in the intake which was designed so that the minimum cross-section open to the incoming airflow was an annulus and not a circle. The second feature was the design of a conical nose for the central body, which consisted of a conical point with a relatively small included angle, followed by another conical portion having a larger included angle. This latter portion was followed by a third portion with a curved profile.

The elemental theory of this diffuser was fairly easily arrived at, but the determination of its exact geometry and other data by experimentation proved difficult. Oswatisch was chosen for the work, since he was the only one available who was working at the time on fundamental gas dynamic problems, Prandtl being very occupied with many different problems. Work began in a somewhat small KWI wind tunnel, which set a limit of 35 mm (1 in) diameter on the models used. Such dimensions were too small to have enough freedom from boundary layer effects, but useful preliminary results were obtained. Furtherance of the work was not possible until larger wind tunnels at the AVA, Peenemünde and Kochel were used, although, even then, the models did not have diameters exceeding 65 mm (2 in). Many calculations and tests were performed, for the most part at an airspeed of Mach 2.9. The following gives a brief outline of the course of development and the logic in determining the final shape of the diffuser.

The first basic model is shown in Figure 9.6A. This had a central body with two discontinuous increases in slope in the supersonic flow region. These slope changes produced two oblique shock waves having less total energy loss than a single shock wave (as used by Busemann). The external portion of the diffuser had its sharp leading edge at the intersection of the two oblique shock waves, and had the external slope as the forward portion of the central body. When combined with a suitably faired afterbody, this model gave a relatively low energy loss, and could give up to 75 per cent of the possible air compression, but its external drag was high. In addition, it was impractical where the Mach number of the airflow was different from that designed for (because of the different positions taken up by the shock waves), and when the airflow approached the diffuser at an angle to its centreline.

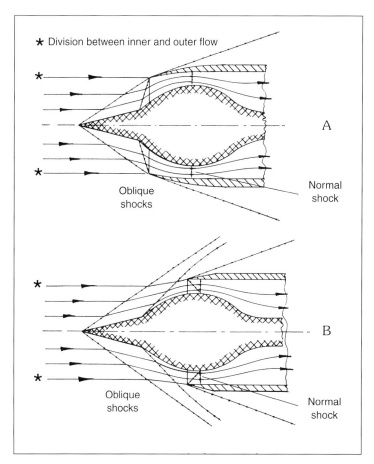

Fig 9.6 Oswatisch multi-cone supersonic diffusers –
A First basic layout
B Second basic layout (not to scale)

The second basic model is shown in Figure 9.6B. This could only produce up to 65 per cent of the possible air compression, but had a drag coefficient only 50–60 per cent of that for the first model because of the low curvature of the external portion of the diffuser. The oblique shock waves now extended into the free air stream outside the diffuser, which resulted in a reduction of air compression, but the diffuser was usable over a small range of Mach numbers and with the airflow approaching at a slight angle to the centreline. Dimensions, such as the length of the frontal cone, distance to maximum curvature of the following body and distance to the lip of the surrounding annulus (all measured from the point of the central body), were deduced to give the desired location of the shock waves relative to the surrounding annulus.

The theory indicated that, by using an infinite number of small angles on the nose of the central body (i.e. a smooth, concave curve), a large number of oblique shock waves would produce deceleration of the supersonic airflow in many stages with no energy loss. However, when Oswatisch tested a model on these lines, it actually proved inferior to his second basic model. In any case, it was easier to manufacture the central body with straight lines as far as possible, and for the Trommsdorf missiles, all-conical central bodies without curved after-bodies were used.

By 1943 Oswatitsch had sufficient data on his diffuser to publish results, and before the war ended he was able to make more accurate tests with larger models up to Mach 4.4 in the incredible wind tunnel at Kochel. Finally, in order that the diffuser could be fully developed, Oswatitsch received a commission from the Heereswaffenamt to build a wind tunnel specially for this purpose. The specification of this tunnel included a working diameter of 150 mm (5^m in) and a velocity of Mach 2.9. Operation without combustion was scheduled for the summer of 1945, and with combustion that autumn, so that the end of the war saw the tunnel incomplete.

Missile development with the Oswatitsch diffuser

Trommsdorf's team began receiving details of the new supersonic diffusers during 1942, and Oswatitsch's multi-cone diffuser was selected for use with new gun-launched, ramjet missiles. Since the missiles were to be spin stabilised, a maximum length was foreseen of five times the missile calibre. This restriction in length set, in turn, a minimum calibre for which a useful long-range gun missile could be made, since, unlike the conventional artillery shell, much of the ramjet missile is cavity and fuel, thereby leaving less room for explosive and warhead. It was therefore calculated that 280 mm (11 in) was the minimum worthwhile calibre for spin-stabilised, ramjet missiles, and these would only weigh about half as much as a conventional shell of the same calibre. Above 280 mm calibre, the ramjet missile could carry a progressively greater proportion of explosive payload.

To obtain maximum payload in a given size of missile, minimum internal cavity was striven for, and also minimum structural weight. Concerning the weight, missiles were designed with structures close to the breaking point, and in view of the considerable forces imposed during firing from the gun barrel, designs could only be proved by actual firings. Large numbers of missiles were therefore fired for such tests alone.

The central body of the diffuser provided essential storage space for the liquid fuel (and later, the explosive). In the fuel tank, paddles were later fitted so that the fuel took up the rotation of the missile; centrifugal force was thus available to supplement various arrangements for pressurising the fuel to affect its injection into the combustion chamber. Numerous small fuel nozzles were used, and fuel governing was later performed by spring-held sliding weights. In the pursuit of simplification, many fuels were tested which would ignite spontaneously in the temperature of the compressed air at the end of the diffuser section. The release of the fuel into the combustion chamber was made by the breaking away of

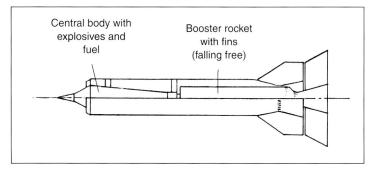

Fig 9.7 Trommsdorf ramjet missile A4 (solid-rocket launch, remote guidance).

special sealing compounds from the injection nozzles as the missile was fired.

The use of a conventional gun as the missile launcher was considered more as an expedient or necessity than as an ideal system. Other, more gradual, launching means were therefore studied in the A- and B-series missiles, both types being up to nine calibres long and stabilised by fins rather than spinning. Figure 9.7 illustrates an A-series missile which was to be launched by a solid-fuelled rocket fitted inside the combustion chamber. After launching, the rocket fell away, leaving the ramjet section to continue under tail-fin and remote control guidance. Generally similar were the B-series missiles (see Figure 9.8), designed for relatively smooth launching from long tubes called 'super-long mortars', which used medium gas pressures. The gas pressure was taken up by a propulsion collar fitted around the missile near its centre of gravity, the collar falling away after launching. Slender tail fins afforded stabilisation. The A-series missiles, at least, were not actually built, although detailed design studies were undertaken of all the components except the remote control.

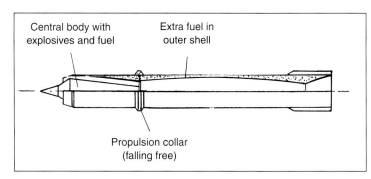

Fig 9.8 Trommsdorf ramjet missile B2 ('super-long mortar' launch, fin stabilised).

Trommsdorf's first gun-launched missiles with the Oswatitsch diffuser were in the E-series (not including the previously described E1 type), Figure 9.9 illustrating the type E4. Missiles in this series were for research purposes and carried no payload, some 260 being fired and the results compared with wind tunnel data. The range of missile calibres was 105 mm (E2), 122 mm (E3) and 150 mm (E4), but the last type was the most successful. During 1944, twenty E4 missiles were fired, followed by another twenty on 1 April 1945.

For the E4 missile, carbon disulphide fuel was used, and this was carried in the central unit and injected by means of pressure from a small powder charge and centrifugal force. The fuel emerged from four rows of nozzles (160 in all) at the beginning of the annular combustion chamber, and a step or ridge around the central unit provided a turbulent area and acted as a flame holder. A thin film of collodium sealed the fuel nozzles until destroyed upon firing, when the released fuel ignited in the compressed airflow. A combustion chamber length of only 25 cm (9^m in) proved sufficient for an efficient combustion process. For firing, a propulsion ring fitted inside the annular exhaust nozzle and later fell away. The missiles were fired with the gun barrel at an angle of 6°, which enabled the missile's trajectory to be followed by optical and acoustical equipment. To obtain data on the drag properties of the missile, dummy missiles were filled with water instead of fuel and fired from the gun. Some data for the 150 mm E4 missile are:

Net thrust	400 kp (882 lb) approx.
Total weight	28 kg (61.7 lb)
Initial (gun muzzle) velocity	3,385 km/h (2,102 mph)
Final velocity after 3.2 secs burning time	5,255 km/h (3,263 mph) or Mach 4.5
Combustion efficiency	98 per cent
Overall unit efficiency	37 per cent
Overall diameter	150 mm (5^m in)
Overall length	0.635 m (2 ft 1 in)

Once sufficient data and experience were accumulated, the development of a ramjet missile for operational use went ahead. Operational employment was to be by Army and Naval artillery and anti-aircraft guns. Figure 9.10 illustrates the type C3 in this series. This missile carried explosive in the main body of the conical diffuser and was intended for use with the 280 mm K5 gun. The remaining volume of the central unit contained a fuel comprising diesel oil and tetraline which ignited in the 700 °C temperature of the compressed air leaving the diffuser annulus. A powder charge and centrifugal force pressurised the fuel up to 180 atmospheres for injection through 480 nozzles. Compared with the E4 missile, the C3 missile had the annular portion of its diffuser much reduced in length, and this was followed by a longer combustion chamber of about 60 cm (1 ft 11^1 in). In addition, the flame-holding step on the central body was positioned before, instead of after, the injection nozzles. As with the previous missiles, a propulsion ring fitted inside the exhaust annulus, while the streamlined vanes supporting the central body inside the outer shell were

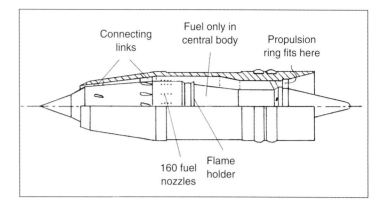

Fig 9.9 Trommsdorf 150 mm ramjet missile E4 (gun launched, spin stabilised).

set at an angle to assist in spinning the missile. Tests with C2 missiles fired from 210 mm guns showed that these were too small to carry a useful load. Data for the 280 mm C3 missile are:

Thrust (net?)	2,000 kp (4,410 lb)
Total weight	170 kg (375 lb)
Initial (gun muzzle) velocity	4,400 km/h (2,732 mph)
*Final velocity	6,700 km/h (4,160 mph), or approx. Mach 5.5
*Range	350 km (217 miles)
Overall unit efficiency	47 per cent
Overall diameter	280 mm (11 in)
Overall length	1.35 m (4 ft 5 in)

(*with fuel reserves)

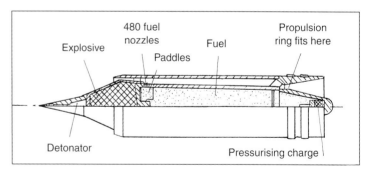

Fig 9.10 Trommsdorf 280 mm ramjet missile C3 (gun launched, spin stabilised).

The intercontinental ramjet missile

Towards the end of 1944, design studies were made by Trommsdorf's team for ambitious intercontinental ramjet missiles known as the D-series, Figure 9.11 illustrating the type D6000. This interesting missile had a cylindrical fuselage formed around the ramjet, with semi-delta wings and a high-mounted tailplane attached. Aerodynamic controls were conventional ailerons, elevator, fin and rudder. At the air intake there was a central body made up of conical sections to form an Oswatitsch multiple-shock diffuser. The central body housed the guidance and steering apparatus, followed by the warhead and some of the fuel. More fuel was contained in annular tanks which surrounded the ramjet as formers for the outer, stressed skin of the fuselage. The annular diffuser section surrounding the rear of the central body led into a divergent diffuser section which then stretched almost to the end of the fuselage. Here was located the comparatively short combustion and exhaust section.

The fuel and burner system was supported by six vanes at the end of the divergent diffuser section. Fuel was pumped from the tanks, through pipes in the supporting vanes, down to a system of fuel jets mounted in a shallow conical ring. Behind this ring was a fairing forming the inner limits of the annular exhaust nozzle. An interesting scheme was devised for both cooling the exhaust fairing and pumping the fuel. At the centre of the six supporting vanes previously mentioned was a circular inlet which admitted some of the compressed air at the end of the diffuser. This air passed through hollow spaces on the inside of the exhaust nozzle central fairing, thereby cooling it. After its journey around the fairing, during which it was further heated, the compressed air was expanded through a small, axial turbine which drove the fuel pump. The pump, turbine and compressed-air exhaust tube were all contained inside the exhaust fairing.

The height and speed initially required for launching the missile was 14,000 m (45,920 ft) and 720 km/h (447 mph) respectively, this condition being equivalent to Mach 0.67. The means of achieving this condition had not been settled, but would probably have been a carrier aircraft. From the launching height, the missile fell until, near the ground, jettisonable wing-tip booster rockets accelerated it up to Mach 2.8. At this speed, the ramjet took over and further accelerated the missile up to a speed of Mach 4.0 and a height of 24,000 m (78,720 ft). The missile then travelled about 5,000 km (3,105 miles) under ramjet power and then glided a further 300 km (186 miles) to the target. Because of its small warhead (assuming a non-nuclear device), the strategic value of this missile must have been slight, but it was remarkably more simple than the A9/A10 version of the German V2 (A4) rocket which was projected for bombardment of American cities. However, by the end of 1944, fuel was so critically short in Germany that an aircraft could not be allocated to test-launch even a model of the D6000 missile.

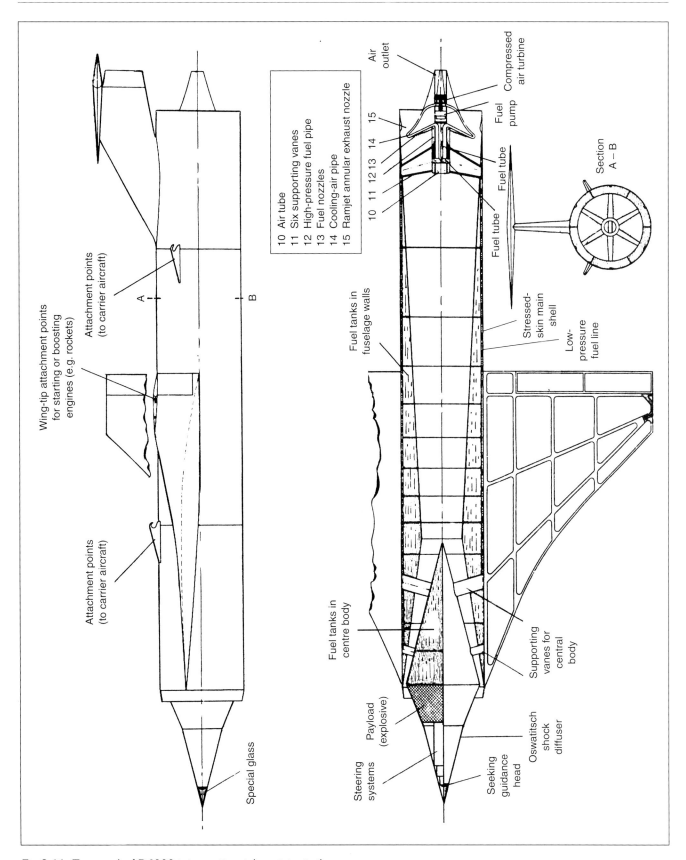

Fig 9.11 Trommsdorf D6000 intercontinental ramjet missile.

Eugen Sänger's ramjets

First ramjet work at Trauen-Fassberg — Test flights at the DFS — Foam coal fuel — Planned application of the Sänger ramjet

Dr Eugen Sänger began working in earnest on the subsonic ramjet in 1940, and the work later led to some spectacular, if inconclusive, tests. Sänger qualified in aeronautical science and as a pilot in Vienna, and by 1930 had begun rocket experiments at the Technisches Hochschule there. At the request of the RLM, he left in 1935 to assist in the establishment of a new research centre for the Luftwaffe in the depths of Luneburg Heath, near Trauen-Fassberg. Although he was given an office at the AVA, Brunswick, this was more in the nature of a cover for his real work. Around the time of 1937, Sänger, although a brilliant scientist, did not believe that the ramjet would have any surplus thrust left after deductions for drag and so forth. In this he was at variance with Helmut Schelp of the RLM's Technisches Amt (LC1) who was in charge, at that time, of ramjet and pulsejet projects and of organising funds for research on such engines. According to Schelp, who urged development of the ramjet, Sänger largely held his negative views until the results of his first ramjet flight tests were available in 1942. Nevertheless, funded by the RLM, he was obliged to investigate the ramjet, and his studies began in 1938, when he issued a report giving an extensive theoretical analysis of the subject, although his chief occupation at the Trauen centre was originally work on liquid-propellant rocket motors.

However, during 1940, some Luftwaffe General Staff officers visited the centre and voiced their opinion that revolutionary advance was needed in air defence if aircraft that the enemy were developing were to be effectively combated when the time came. To this, Sänger suggested that the ramjet would be a worthwhile form of power-plant to investigate, since it promised, in the high subsonic speed range, a superiority over rockets owing to a much lower specific fuel consumption, and a superiority over turbojets because of the higher thrusts possible. The operational envelope required for the ramjet was revealed in 1941, when the RLM specified that the anti-aircraft weapon should be able to climb to an altitude of 12,000 m (39,360 ft) in two minutes and there operate for almost one hour.

Sänger's preliminary ramjet tests were performed in collaboration with Prof. Busemann at the LFA, Volkenrode. These tests, to study internal flow, were made by diverting a wind-tunnel air stream into a tube about 3.0 m (9 ft 10 in) long, and injecting petrol from a ring of small jets positioned in the tube. Very long flames, several metres in length, resulted, which discouraged wind-tunnel tests with complete ramjets, so further work was continued at the Trauen centre, where Sänger's chief assistant was a man called Heinrich.

First ramjet work at Trauen-Fassberg

In designing his first ramjet, Sänger divided his time between this and finalising plans for a projected liquid-propellant rocket motor of 100 tonnes thrust. Comparatively large sizes were also envisaged for the ramjet, which was planned with a straightforward divergent diffuser, cylindrical combustion section and convergent exhaust nozzle. At the time, the most recent ramjet work known was that of the H. Walter KG, which had not then (1940/1) progressed far, Sänger thought, because of low combustion temperatures and lack of a wind tunnel.

From the start it was hoped to test ramjets in the air by using a suitable aircraft test bed, since ground-testing methods were considered unsatisfactory. Wind tunnels did not then exist which could reproduce the high-altitude conditions in which the ramjet was to operate, while an open-air blower would not reproduce the normal conditions met in an undisturbed atmosphere. Nevertheless, some initial ground tests were undertaken in order to obtain more data on combustion, since there was considerable doubt that the requisite high temperatures could be handled.

Ground tests began in the early autumn of 1941. Because of the haste of the programme, the first combustion tests were performed using a 500 mm (1 ft 7^1 in) diameter metal drainage pipe as a combustion chamber. This was mounted above an Opel-Blitz truck, which also carried the necessary fuel and pressure bottles, and was driven along a test course for a few minutes at speeds up to 90 km/h (56 mph). In further tests, an 800 mm (2 ft 7 in) diameter combustion chamber was used (see Figure 9.12), as well as complete ramjet units with diameters up to 500 mm and diffusers of various angles and lengths (see Figure 9.13). The combustion tests, in which temperatures up to 2,200 °C were noted without effect on the duct walls, led to the design of a grille with swirl-type nozzles for upstream injection of the fuel. Useful dimensional data, especially regarding the diffuser, were also obtained.

Sänger's first ramjet tests using the truck led to the appellation of 'flying stove pipe' for the ramjet, but not so jovial was the reaction of the Forschungsfuhrung in Berlin when the first test results were sent there. Because Sänger had no official research contract for the ramjet work already at the test stage, the Forschungsfuhrung was obliged to issue a reprimand, and only some months later could the work continue with official authorisation. Nevertheless, the scope of the work was ordered to be reduced, so flight tests of the ramjet could not be made.

RAMJETS FOR AVIATION

Fig 9.12 An 800 mm tube, mounted on an Opel-Blitz truck, used to conduct ramjet combustion tests in October 1941.

Fig 9.13 Amongst the ramjet items road-towed during Eugen Sänger's low-speed tests in 1941, was this complete 500 mm diameter ramjet engine. The appellation of Flying Stove Pipe arose at this time.

This led, early in 1942, to Sänger and one of his assistants, Irene Bredt, leaving Trauen-Fassberg and moving to the DFS at Ainring, where Prof. Walter Georgii offered them facilities to further their ramjet work.

Test flights at the DFS

At the DFS, Sänger had, at least by the end of the war, a small team of sixteen people, including non-technical staff, to help in the work. For the first ramjet flight tests, a 500 mm diameter ramjet was built and mounted above a twin-engined Dornier Do 17 Z (BC + NJ) to exhaust between the aircraft's twin fins (see Figure 9.14). The basic features of the ramjet were obtained from the previous ground tests, and remained essentially unchanged in the following years. The test aircraft was fitted with instruments to record pressures, temperatures, etc. in and around the ramjet, which was mounted to permit its gross thrust to be measured by a dynamometer. To calculate the net, or useful, thrust of the ramjet, separate flights were necessary in order to measure the duct's drag. A close-up of the installation is shown in Figure 9.15.

The first flight test with this ramjet was made on 6 March 1942, but while this was Germany's first ramjet flight test using an aircraft, it was not the world's first. (Already in 1940, a Polikarpov I-153 biplane had flown in the Soviet Union with two ramjets designed by I.A. Merkulov beneath the lower wing.) During the flights, an almost constant speed of 306 km/h (190 mph) was maintained, but altitude was varied between 100 m (328 ft) and 3,000 m (9,840 ft). Variations of duct dimensions and fuel injection arrangements were tested, and the highest thrust coefficient obtained was 0.58. Although the combustion chamber was excessively long at about 4.0 m (13 ft 1 in), it was intended to reduce this according to the results obtained. During operation, the ramjet gave a moderate, steady roar, and the combustion chamber walls did not exceed a temperature of 600 °C. The 500 mm ramjet was last flown on 12 April 1942, after 70 test flights.

Sänger next designed a huge ramjet of 1,500 mm (4 ft 11$^{1}/_{16}$ in) diameter. To flight-test this, a more powerful aircraft was needed, and Paul Spremberg, a civilian test pilot for Rheinmetall Borsig and also for the Trauen-Fassberg centre, obtained a twin-engined Dornier Do 217 E-2. Considerable apprehension prevailed over mounting and flying the large ramjet on the aircraft, despite careful wind-tunnel model studies made by the Dornier company. The ramjet, with its pumps and fuel, imposed a weight of 2,000 kg (4,410 lb) on the aircraft, and with its total length of 10.60 m (31 ft 6 in), it was akin to another fuselage. At a flight speed of 1,080 km/h (671 mph), this ramjet was calculated to develop about 20,000 hp gross. Thus, it will be easily seen, its effect on the stability and centre of gravity of the aircraft was considerable.

Figure 9.17 shows the installation of the 1,500 mm ramjet on the aircraft. To assist stability, a dorsal fin was fitted to the ramjet, while its diffuser had external, strengthening ribs to prevent buckling of the sheet metal. Such was the considered danger of the first test that the aircraft flight mechanic resigned from his post, but since Sänger always insisted on flying the tests, a new

ABOVE:
Fig 9.14 A Sänger 500 mm ramjet mounted above a Dornier Do 17 Z (BC+NJ) for flight-tests by the DFS during 1942.

RIGHT:
Fig 9.15 Close-up of Sänger 500 mm ramjet mounted on the Dornier Do 17 Z (BC+NJ). Later, a Dornier Do 217 E-2 was used to flight-test larger ramjets, but only subsonic units were tested.

BELOW:
Fig 9.16 Test flight with a Sänger 500 mm ramjet above a Dornier Do 17 Z (BC+NJ) at a speed of 313 km/h. These test flights, made in a dive, proved that the ramjet developed a net thrust.

mechanic soon stepped forward. Watched by all the personnel at Hörsching airfield, Paul Spremberg took the encumbered Dornier off the ground for the first time on a day in December 1942. Fortunately, no serious incidents occurred on any of the eighteen flights with this ramjet, the last of which was conducted on 20 January 1943.

This is not to imply that the pilot was without his problems. A flight speed of only 324 km/h (201 mph) could be reached on the aircraft's engines alone, and with the ramjet turned on, the speed was increased to only 396 km/h (246 mph). To obtain the requisite higher test speed, it was necessary to climb to about 4,000 m (13,120 ft) and then dive until a speed of 720 km/h (447 mph) was reached. In this dive, which lasted 15 to 20 seconds, the aircraft shuddered violently, and it was necessary to keep down the thrust of the ramjet by suitable adjustment of the fuel/air ratio since the nose-heaviness of the aircraft caused by the thrust would soon have been beyond the ability of the elevators to correct. As it was, the sweating flight mechanic had to enthusiastically operate the aircraft's flaps to assist in restoring control, and only at about 500 m (1,640 ft) altitude was the ramjet shut off. One of these flights was observed by the rocket-aircraft test pilot Fritz Stamer, who is alleged to have said, 'I have seen a great deal in my life but nothing so astounding as the sight of this roaring, flaming power dive . . .' As the illustration shows, a mighty flame issued from the ramjet and must have appeared to engulf the rear of the aircraft. Largely this was caused by the necessity for burning a rich mixture. Although nothing like the full power of the 1,500 mm ramjet could be demonstrated, further useful data were obtained.

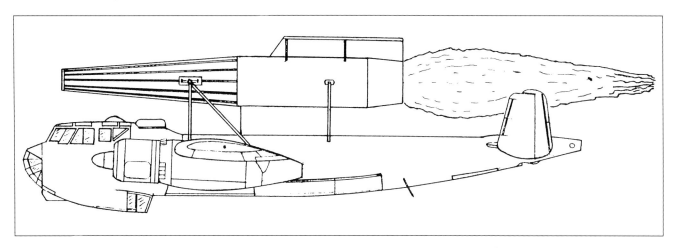

Fig 9.17 Sänger 1,500 mm diameter ramjet mounted on a Dornier Do 217 E-2. Exhaust flame shape shown at maximum speed of 720 km/h. *(Author)*

Fig 9.18 This Sänger 1,500 mm ramjet, mounted above a Dornier Do 217 E-2 (–+CD), began flight tests in December 1942. Diving with this flaming monster made for a very 'hairy ride' indeed!

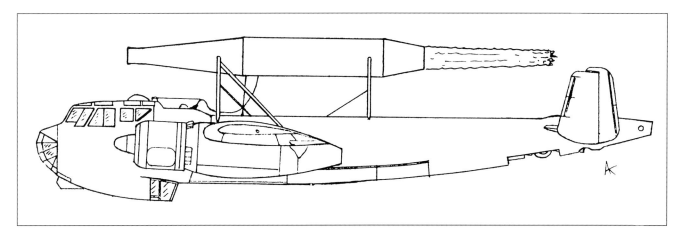

Fig 9.19 Sänger 1,000 mm diameter ramjet mounted on a Dornier Do 217 E-2. Exhaust shape shown at full power. *(Author)*

Up to this point, tests had been made to provide design data, but, this done, the size of 1,000 mm (3 ft 3 in) diameter was chosen for a ramjet to be developed as a missile or aircraft power-plant or as a booster unit. Such a ramjet was mounted above the Dornier Do 217 E-2 (see Figure 9.19) and first flown during the summer of 1944. Thirty-two flights were made at speeds up to Mach 0.37, tests at 420 km/h (261 mph) and 7,000 m (22,965 ft) corresponding to combustion chamber conditions at Mach 2.0 and 18,000 m (59,060 ft) altitude. As far as they went, these tests were considered generally satisfactory, and an average thrust coefficient of 0.55 was obtained. However, a final flight had to be made on 30 August 1944, because of severe fuel restrictions. The plan to mount two of the 1,000 mm ramjets above the turbojets of an Me 262 fighter was thereby frustrated; by this means, flights up to Mach 0.8 and altitudes of 15,000 m (49,215 ft) had been hoped for.

With the end of airborne tests, Sänger's ramjet work at the DFS had to be continued on the ground, using open-air blowers, wind and smoke tunnels and water channels. Speeds up to Mach 1.5 were simulated with models in the water channel, diffusers for such ramjets being of the sharp-lipped, divergent type with a normal shock wave as investigated by Busemann. (Previously, Sänger's subsonic divergent diffusers had featured a round lip.) Especially interesting tests were made in the DFS low-pressure chamber to investigate the effects of altitude on combustion, and it was found that, with proper fuel control, combustion could be maintained up to an altitude of 18,400 m (60,350 ft). Because of the acute fuel shortage, the possibility of using solid fuel in the ramjet was examined, and it is worth recording the curious story of one such fuel, known as Schaumkohle, or foam coal.

Foam coal fuel

It might be assumed that work on foam coal, as a substitute fuel for jet propulsion, began as a result of a desperate government order, but this in fact was not the case, or, at least, not directly. In October 1944, a Dr Heinrich Schmitt was ordered to Prague, where, along with others, he was commissioned to produce a tube of resin-impregnated wood containing coal and an oxidant sufficient for combustion of the coal. Schmitt was chosen for this job because, since 1939, his company had specialised in making products from synthetic resins, and as a physical chemist he had been active in this field since 1930. At Prague, no one was given an explanation of the purpose of the coal-filled tube, but tubes were reluctantly produced in accordance with orders. Schmitt had his doubts about combining an oxidant with coal, and when the filling for the tubes was not delivered, he turned to another idea. His idea was to produce a porous mass of coal inside the tube through which air could flow and thereby support combustion.

Almost immediate success was achieved in binding together granules of coal with a phenolic resin. The coal granules ranged in size from 0.5 to 2.5 mm and were mixed with 4 to 10 per cent (of the coal weight) of resin, moulded to shape and then dried and heated to 160–200 °C. In this way, an 80 mm (3 in) diameter tube having 6 mm (in) thick walls and a length of 550 mm (1 ft 9¦ in) was filled with foam coal. When the coal was ignited at one end and a stream of air was fed from the other end of the tube, all the coal was consumed without disintegration, while the outside of the tube remained cool enough to hold. There was no tendency for the tube to burn through, although a thin-walled metal tube would have melted.

At the Skoda company's Avia works near Prague, experiments to use foam coal as a jet propulsion fuel were begun in April 1945 under the supervision of Werner Fleck. Two tests were made using a cylinder of foam coal in a wind tunnel with an air speed of 30 m/sec (98 ft/sec). Although a fuel-burning rate of about 5 mm/min was needed to obtain a useful jet thrust, only about 1 mm/min was obtained in the Avia tests, and none of the engineers there were encouraged. In addition, there was the problem of throttling or increasing combustion and hence thrust. Similar conclusions were reached in tests made at the DFS, where 40 kg (88.2 lb) of coal lumps or foam coal units was loaded into a streamlined steel grate to fit inside a ramjet combustion chamber, a blower producing an air speed of up to 35 m/sec (115 ft/sec). Without water cooling, the grate soon melted, while using foam coal units without a grate led to less complete combustion. For reasons unknown, an iron-reinforced foam coal cylinder measuring about 1.0 m long × 500 mm diameter and having longitudinal air passages was sent to Berchtesgarten before the war ended. Coal, as a fuel for ramjets, was also investigated by the LFA and AVA research institutes.

Planned application of the Sänger ramjet

In an effort to push Sänger's ramjet work ahead, the DFS tried in 1943 to have a co-operative work programme set up with a view to having a ramjet-powered aircraft built. Little help was forthcoming from government sources, and certainly no contract to build an aircraft. During 1942, the Technisches Amt had engaged an aerodynamicist to produce an independent report on the ramjet, and this move had an unfortunate result. Apparently, the aerodynamicist was already overburdened with work, and through an error in his calculations, he arrived at only half the true theoretical thrust for a given ramjet. Not until September 1944 was this mistake fortuitously discovered at the DFS and the official false impression of the ramjet's possibilities corrected.

Prior to this, other things militated against the application of the ramjet. DFS attempts to interest the aircraft companies of Dornier, Junkers and Focke-Wulf in building a test aircraft met with no success, since these companies were already overloaded with work and were unwilling to commit resources to a dubious, if revolutionary, aircraft project. Contact with the H. Walter KG was also without result. Ramjet work barely received a mention during an important jet propulsion congress held in Berlin in September 1943, while at a meeting of the Deutschen Akademie der Luftfahrtforschung (DAL) in January 1944, even the DFS ramjet flight tests were recorded incorrectly as having not taken place.

Following the DFS correction of the report already mentioned, official interest in the ramjet greatly increased, so that a November 1944 meeting of the Chef der Technischen Luft Rüstung (Chief for Technical Air Armament which replaced the Technisches Amt in August 1944) produced a report which stated: 'At the introduction of the conference, the importance of the Lorin engine (ramjet) was pointed out . . . Because results have to be at hand as fast as possible, it will be necessary to start communication of experiences on a large scale. The development of subsonic propulsion devices shall be stressed, the supersonic speed range being encouraged only so far as the research stage is concerned.' Although this report at last officially cleared the decks for action, it should be said that, in view of the critical shortage of fuel, exaggerated hopes of using non-strategic solid fuels were held in official circles. In fact, however, any serious design of aircraft that followed discreetly made provision for carrying liquid fuel, even if a solid fuel design was put forward initially.

One of the first ramjet aircraft was designed by Prof. Alexander Lippisch. He had always been interested in tailless aircraft, his earliest work in this field dating back to 1927, while his later work on the Me 163 rocket fighter programme at the DFS and the Messerschmitt company gave him a fund of experience in high-speed aerodynamics by the time he moved to the LFA, Vienna, in May 1943. The delta wing held a special interest for Lippisch, and, probably following receipt of ramjet data from the DFS, he proceeded to his own, simplified ramjet calculations and free-flight tests with scale models. Late in 1944, work was begun on the LP-13b aircraft project, which was to consist of a pure delta wing completely enclosing a centrally mounted ramjet duct and having a triangular dorsal fin. The pilot was to sit inside the fin, which had a glazed leading-edge section. While the LP-13 version was to operate on liquid fuel, the LP-13a version was designed for solid fuel. One arrangement for the solid fuel was to have a foam coal cylinder measuring 120 mm (4^i in) diameter by about 0.40 m (1 ft 3^i in) length, with an 80 mm (3 in) diameter hole through the centre. This coal cylinder was mounted inside the ramjet, which had a divergent type of diffuser on the lines of Sänger's work. A liquid-fuelled rocket motor was envisaged for take-off and boosting to operational velocity, when the coal would be ignited by an oil burner which had, in turn, been ignited by a gas flame. For 45 minutes of flight, it was estimated that 800 kg (1,760 lb) of coal would have to be carried.

The LP-13a had a span of 5.92 m (19 ft 5 in), a loaded weight of 2,300 kg (5,072 lb) and was expected to reach a speed of 1,650 km/h (1,025 mph) at high altitude. Scale-model experiments in the Göttingen high-speed wind tunnel indicated that its structure had outstanding stability up to the maximum tested speed of Mach 2.6. In order to confirm the low-speed handling of the design, the Lippisch DM-1 glider was developed and built by Akaflieg, but had not been tested when captured by the Americans in Prien. Incidentally, Lippisch referred to the ramjet as an 'effect' rather than an 'engine', and he aimed for maximum air mass flow with the minimum temperature rise to obtain maximum thermal efficiency.

In mid-December 1944, the OKL (Luftwaffe High Command) belatedly arranged the administrative background for collaboration between the DFS and Lippisch at the LFA, although exchanges of ramjet and high-speed aerodynamic data were already being made. One exchange resulted in the projected fighter Lippisch

Delta VI being re-planned by the DFS as an unmanned aircraft carrying explosives. A single, centrally placed ramjet of 2,000 mm (6 ft 6¹ in) diameter was to be used, giving estimated thrusts for climbing at full speed of 7,300 kp (16,096 lb) at sea level and Mach 0.81, and 2,140 kp (4,719 lb) at 12,000 m (39,360 ft) altitude and Mach 0.92.

At the beginning of 1945, the Skoda-Kauba Flugzeugbau in Prague issued an order to the DFS for the design of an interceptor with ramjet propulsion. Since the Skoda Company's discouraging experiments with foam coal had not been started at that time, the ramjet was intended to use foam coal, although provision was to be made in the design for liquid fuel. The aircraft, designated SK.P14, was to have a single, central ramjet with a diameter of 1,500 mm (4 ft 11¹/₁₆ in) and a length of 9.50 m (31 ft 2 in), providing thrusts in level flight of 4,400 kp (9,680 lb) at sea level and Mach 0.83, and 1,350 kp (2,970 lb) at 10,000 m (32,800 ft) altitude and Mach 0.815. The ramjet duct formed an integral part of the fuselage construction, but the walls of the combustion chamber and the exhaust nozzle were exposed to the airstream for cooling. The pilot, who occupied a prone position beneath a glazed canopy, together with the fuel tank, tail unit and armament, were all situated along the top length of the duct. All flying surfaces were of tapered planform, the wings being on the duct centreline and the tailplane high-mounted. Jettisonable take-off rockets and undercarriage were planned with a retractable skid for landing. A maximum endurance of 43 minutes was expected with 1,200 kg (2,640 lb) of liquid fuel.

Shortly before the end of the war, the Ernst Heinkel AG received data on ramjets from the DFS, and with an order from the RLM proceeded to design a tailless, ramjet-powered fighter. This project, designated P.1080, envisaged the use of two 900 mm (2 ft 11⁷/₁₆ in) diameter Sänger ramjets mounted one on each side of the fuselage and with their outer edges faired with the wing panels. The greater proportion of the ramjet ducts was left exposed to the airstream for cooling. Swept-back wings were planned with elevon controls, and there was a single swept-back fin and rudder. The pilot's cockpit was situated well forward, leaving the rear fuselage for a fuel tank, and there were take-off rockets and a landing skid. At 500 km/h (311 mph), each ramjet was expected to give 420 kp (926 lb) of thrust, while at 1,000 km/h (611 mph) each ramjet was to give 1,560 kp (3,440 lb) of thrust. Curiously enough, the Heinkel company at Jenbach commissioned the DFS in April 1945 to design another fighter, but with a centrally mounted, liquid-fuelled ramjet. However, by the end of the war, none of the projects mentioned had been built.

The Focke-Wulf Flugzeugbau GmbH

The Pabst ramjet — Planned application of the Pabst ramjet

Formed in January 1924 by Heinrich Focke, Georg Wulf and Werner Neumann, the Focke-Wulf Flugzeugbau built a full variety of aircraft types and played a key part in Germany's aerial power before and during the Second World War. During the war, under the leadership of its chief designer, Kurt Tank, the company produced such notable aircraft as the Fw 190 Würger fighter and the Fw 200 Condor long-range aircraft. It was probably during 1941 that Eugen Sänger took a ramjet design to Kurt Tank for discussion, no doubt with a view to bringing the company's resources to bear on a ramjet aircraft project. So far as Sänger was concerned, nothing useful resulted from this contact, although, later, the DFS turned details of Sänger's ramjet experiences up to the end of 1943 over to Focke-Wulf. Nevertheless, Tank's interest in the ramjet must have been kindled by his meeting with Sänger because, at about that time, a separate ramjet programme was set up and pursued by Focke-Wulf at Bad Eilsen.

The Pabst ramjet

A 1944 Focke-Wulf ramjet programme* was conducted under the leadership of a Dr Otto Pabst of the company's gas dynamics department at Kirkhorsten (about 6 km from Bad Eilsen). After looking at the type of ramjet proposed by Sänger, Pabst believed that he could develop a better ramjet which would produce air compression and combustion within very much shorter lengths. Accordingly, he studied the basic principles of air diffusers and combustion chambers, and established new theories for both.

In the divergent nozzle type of diffuser, as used by Walter and Sänger, the pressure rise in the incoming air is brought about *inside* the diffuser by slowing the air

---

*This work was preceded by a Pabst-designed turbojet to replace the piston engine in the nose of the Fw 190 fighter but this project was cancelled in July 1942.

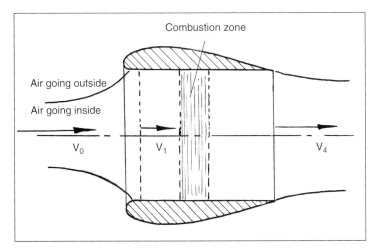

Fig 9.20 Idealised form of Pabst ramjet.

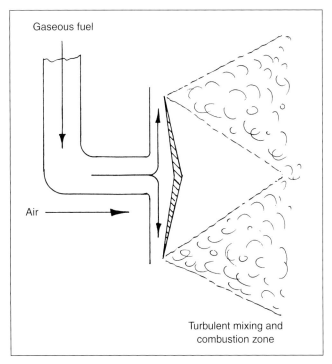

Fig 9.21 Pabst injector/flameholder.

down. This type of diffuser, known in Germany as the Einlauf-Diffuseur, requires a considerable length to avoid excessive drag. In Pabst's diffuser, known in German as the Fang-Diffuseur, the rise in pressure of the incoming air occurred *outside* the diffuser, i.e. in the free-stream air approaching the ramjet inlet. To explain how this was achieved, reference should be made to the idealised form of the Pabst ramjet sketched in Figure 9.20.

As the sketch shows, the ideal Pabst ramjet had a perfectly cylindrical form to its inside duct, a very short combustion zone and a surrounding cowl of aerofoil section. If there is no combustion, V_1 and V_4 are nearly the same as V_0, and neglecting skin friction, there is neither drag or thrust acting on the ramjet. When combustion is started, the effect is to choke the duct so that the incoming air is slowed down and V_1 becomes less than V_0. At the combustion zone, however, the added heat accelerates the gases and V_4 exceeds V_0 and a thrust results. Because the inner duct walls are parallel, the thrust must act on the *outside* of the ramjet and be caused by differences in pressure on the aerofoil-shaped cowling. As the streamlines separating the inner and outer air flows show, the cowling has a similar effect that an aerofoil has at a moderately large angle of attack, whereby there is an imbalance between the pressures on the upper and lower surfaces. It was such an imbalance which was primarily responsible for the thrust on the Pabst ramjet.

Pabst's second important ramjet contribution concerned the question of combustion, which he felt should be accomplished in the shortest distance. To begin with, fuel was to be used in a gaseous rather than a liquid form in order to achieve rapid mixing with the incoming air. For the mixing of the fuel/air gas streams, a simple but effective injector-cum-flameholder was developed, which is sketched in Figure 9.21. Fuel entered a circular chamber mounted in the airflow and was ejected downstream from a slot around the circumference of the chamber. The incoming air, flowing around the circular chamber, formed turbulent 'cones' in which the fuel could easily mix, and combustion could proceed in these lower-velocity zones. Tests with this device were made using hydrogen gas as fuel, and a diameter of 22 mm (^{m} in) was established for the chamber size. This size remained the same for all sizes of ramjet, but the number of fuel chambers used was adjusted according to the cross-sectional area of the ramjet's combustion zone. The fuel chambers were spaced one to two diameters apart so that their turbulent streams all mixed together within a short distance. The resulting short combustion flame length was independent of the ramjet size since it was governed by the fuel chamber diameter and velocity of the fuel.

Because of the importance of the ramjet's outer cowling shape, the assistance of Dr Küchemann at the AVA, Göttingen, was sought because of his extensive experience with generally similar ring cowlings for piston engines. Küchemann's assistance included the development of methods for calculating the pressure distribution and hence thrust on the Pabst ramjet. Its Fang diffuser threw up a curious fact during AVA wind-tunnel experiments which could not be fully understood: at high subsonic speeds of Mach 0.8 to 0.9, the diffuser showed no rapid rise in drag despite the high suction and local velocity around its leading edge.

After a good deal of research, Pabst designed and had constructed a ramjet model for testing in the A-9 wind tunnel of the LFA at Brunswick. A section through this ramjet is sketched in Figure 9.22. Its shell was made up from machined castings, the shape of the inner duct being somewhat removed from the theoretically ideal

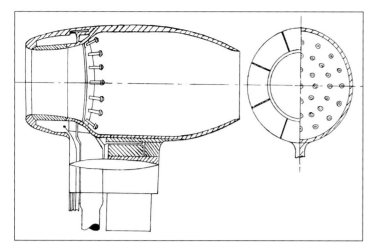

Fig 9.22 Pabst ramjet used for LFA wind tunnel tests.

parallel cylinder. Hydrogen gas was used for fuel and was fed to 59 of the previously described injectors, which were mounted on and fed from a grid of pipes.

At the LFA, a Dr Knackstedt was responsible for the ramjet tests, during which speeds up to Mach 0.9 were reached without any instability in combustion. The amount of air flowing into the ramjet was varied between two and 14 times the minimum needed to support combustion, and accordingly the exhaust jet temperature of the ramjet varied between 200 and 900 °C. Apparently, the ramjet in these tests showed a relatively high thrust coefficient and a remarkably low specific fuel consumption. However, the model was somewhat large for the wind tunnel and could have given inaccurate results because of a lack of uniformity in the pressure of the surrounding air. Also, the heating value of the hydrogen fuel was high, which gave a low fuel consumption. Nevertheless, Pabst was satisfied with the test results, which he received direct and without LFA evaluation. A report (No. 09045) written up by Pabst from the test results was sent to Knackstedt at the LFA, but he did not get a chance to read it because it was burned almost as soon as it arrived! Presumably, the LFA employed an over-enthusiastic security officer who, lacking a sense of humour, had absorbed the secrecy joke 'To be burnt before reading'. Data for the Pabst ramjet tested at the LFA are:

Maximum thrust coefficient	0.4 approx.
Specific fuel (hydrogen) consumption	1.47 at Mach 0.8 and maximum thrust
Combustion efficiency	almost 100 per cent
Max. exhaust gas temperature	900 °C
Overall diameter	237 mm (9 in)
Overall length	480 mm (1 ft 6^m in)
Intake diameter	136 mm (5 in)
Combustion chamber diameter	218 mm (8^1 in)
Exhaust nozzle diameter	138 mm (5 in)

Many combustion tests were made by Pabst using petrol instead of hydrogen as fuel. In order to achieve similar mixing and rapid combustion results to that with hydrogen, the petrol was vaporised and superheated outside the ramjet before being fed to its injectors. This was performed by passing the fuel through a heat exchanger built into the exhaust end of a miniature ramjet of similar design to the large one. Two large ramjets*, each with 50 fuel injectors, required one miniature ramjet/heat exchanger with only two fuel injectors to vaporise and superheat the fuel, i.e. about 2 per cent of the total fuel consumed was used for vaporising and superheating. Specific fuel consumption was increased by about 3 per cent if petrol instead of hydrogen was used, but the former was far more practical for aviation use.

Planned application of the Pabst ramjet

Two quite different aircraft projects were put forward by the Focke-Wulf company to utilise the Pabst ramjet. The first was a fighter, designated Focke-Wulf Ta 283 (the 'Ta' prefix being used as a tribute to Kurt Tank's excellent work). This fighter was designed for a maximum speed of Mach 1.05, or 1,125 km/h (699 mph) at its service ceiling of 10,000 m (32,800 ft). Its loaded weight was to be 5,380 kg (11,863 lb) and its range 690 km (429 miles). To avoid detrimental disturbance of the airflow, the two ramjets for the Ta 283 were mounted at the tips of the sharply swept-back tailplane. The wings, spanning 7.97 m (26 ft 1^1 in), were low-set and employed a leading edge sweep-back of 45°. Fairing back from the rear of the cockpit canopy was the fin and rudder, which had the appearance of a dorsal spine. Because the wings and cockpit were set well back, a long slim nose resulted, in which the armament was mounted. The aircraft sat very low on a retractable, tricycle undercarriage, and take-off was effected using a Walter rocket motor. While the RLM issued the designation 'Ta 283' and thereby showed official interest in this project, time ran out before a prototype could be built, although a wooden mock-up may have been finished.

Focke-Wulf's other ramjet aircraft project, designed in September 1944, was also for a fighter but was of a quite new conception. Named the Triebflügel (thrust-wing), the fighter was a tail-sitting, vertical-take-off aircraft having four similar fins forming a cruciform tail. At a point approximately one-third of the fuselage length from the nose there were three vanes or wings which rotated around the fuselage. A Pabst ramjet was mounted at the tip of each wing. For starting, the wings were set in neutral pitch and revolved up to ramjet operating speed by three 300 kp (660 lb) thrust rockets mounted inside the ramjets. With the ramjets operating, the wings were then moved into fine pitch and vertical

*Each ramjet had a diameter of 1,341 mm, a length of 2,682 mm and a projected max. power at sea level of 10,850 hp.

lift-off was attained partly by thrust from the rotating wings and partly by a vertical thrust component from the ramjets. With the aircraft turned over into level flight, the pitch of the wings was coarsened and their angular velocity reduced to maintain the ramjets at a constant Mach 0.9. Control of the aircraft was purely by movement of the tailfin trailing surfaces, which could also counteract any torque generated by the wings. During horizontal flight, the tail was slightly depressed to direct part of the thrust force into a lift force.

The main landing shock of the Triebflügel fighter was taken on its single main wheel at the base of the fuselage, each of the four fins having a small wheel for balancing purposes. All wheels were enclosed in tulip-shaped pods during flight. The pilot and armaments were positioned in the nose of the fuselage, while some 1,590 kg (3,506 lb) of rocket and ramjet fuel was carried in the fuselage behind the wings. In an emergency, the rockets could take over from the ramjets. With a rotating wing diameter of 11.5 m (37 ft 8^i in), the Triebflügel fighter was expected to attain a speed of 840 km/h (522 mph) at 11,000 m (36,080 ft), have a service ceiling of 14,000 m (45,920 ft) and a maximum cruising range of 2,400 km (1,490 miles). Each of its ramjets was 680 mm (2 ft 3 in) in diameter and 1.40 m (4 ft 7 in) long, and gave a thrust of about 840 kp (1,850 lb). Equivalent data are lacking for the Ta 283, but for both aircraft projects, the ramjets were essentially scaled-up versions of the model tested in the LFA wind tunnel. They did, of course, have detail constructional differences, and in the case of the Triebflügel fighter, a rocket chamber was fitted in the centre of the grid carrying the fuel injectors.

Bayerische Motoren Werke AG (BMW)

As we have seen in Section 2, BMW were active in the fields of turbojet, rocket and piston aero-engines, in addition to studying other types. These other types included the ramjet, studies of which were undertaken in 1944 by Dipl.-Ing. Peter Kappus and his assistant Dipl.-Ing. Huber of the company's project study department EZS. No practical work was conducted, and the value of the calculations was limited by lack of data on supersonic diffusers. The following notes on the BMW work are based on the company's document EZS 55 of 14 September 1944: 'The athodyd as an artillery weapon'.

For the purposes of calculation and comparison, the various ramjet missile arrangements studied were given the following specification:

Speed	950 km/h (590 mph)
Combustion chamber temperature (ideal)	1,500 °C
Maximum combustion flame velocity	18 m/sec (59 ft/sec)
Net weight (without boosters)	500 kg (1,102 lb)
Explosive weight	1,000 kg (2,205 lb)
Fuel weight	1,000 kg (2,205 lb)
Booster fuel	440 kg (970 lb)
Range at 1,000 m (3,280 ft)	300 km (186 miles)
Range at 3,500 m (11,480 ft)	400 km (248 miles)

Figure 9.23 gives the schematic layout of a ramjet missile proposed. Its air intake had a rounded lip, followed by a divergent diffuser at the end of which there were fuel nozzles to evenly atomise the fuel across the cross-section. A propeller-driven fuel pump was positioned inside the diffuser where it would not affect ram exploitation. In the fairing surrounding the diffuser

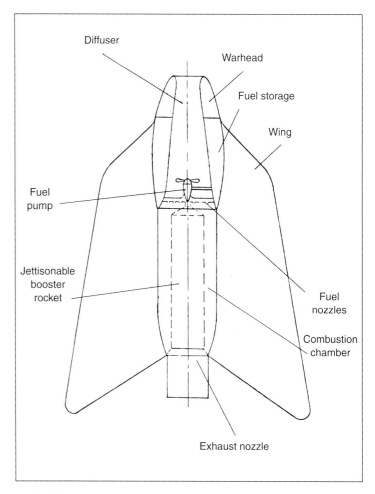

Fig 9.23 BMW ramjet missile with continuous airflow.

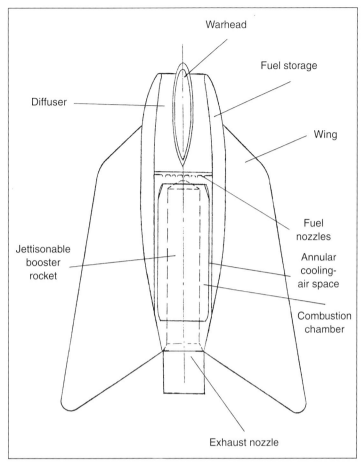

Fig 9.24 BMW ramjet missile with central warhead and internally cooled combustion chamber.

was carried the explosive and fuel leaving the combustion chamber wall to receive an external cooling airflow. Any control equipment necessary to operate surfaces on the wings was to be installed inside the diffuser fairing. For take-off, a solid-fuelled booster rocket fitted inside the cylindrical combustion chamber and later fell away.

In Figure 9.24 we see a scheme for a different arrangement, in which the explosive is concentrated inside a fairing inside the diffuser. Furthermore, the fuel is carried in a fairing which extends over the combustion section. To air-cool the combustion chamber, part of the airflow passed along the annular space between the combustion chamber and the inner wall of the missile. Combustion heat tended to vaporise the fuel, which could then be utilised in a similar manner to that in a camping stove. To avoid an undesirable shift in the missile's centre of gravity through movement of fuel in the tanks, it was proposed to fill the tanks with a suitable sponge or foam mass. Fuel was forced out of this mass by the pressure of the vaporised fuel. Again, a jettisonable rocket, fitted inside the combustion chamber, provided the means for take-off, which was to be at an angle of 45° from a modified gun carriage.

To be brief, the BMW deliberations are best summed up by stating their conclusions as follows. The ramjet missile was considered unsuitable for high supersonic speeds (obviously because of a lack of diffuser knowledge) and unpractical as a substitute for artillery weapons. For fuel, alcohol was considered because of its high total energy content and availability, although a reduction in range would follow from its smaller heating value. Because of the range anticipated and the missile's comparatively low speed, the use of accurate course control or guidance was considered imperative but worthwhile for large missiles. Compared with the V1 flying bomb (pulsejet powered), the ramjet missile offered 50 per cent more speed and therefore less susceptibility to interception, but a far higher fuel consumption was involved, together with a more energetic means of take-off. Finally, the ramjet missile promised ranges far above rocket and artillery ranges of the day, and only at very high range was the rocket weapon considered at an advantage once again. It is believed that other engine companies were asked to examine the ramjet, but they do not appear to have done so. In the case of Junkers, at least, full commitment on other tasks was given as the reason for not tackling ramjet questions.

Versuchsanstalt Heerte

At some time around 1935, engineers at the Hermann Goering steel works had some ideas on rockets, but little or nothing was done on the subject until 1944, when Chief Director Pleiger applied to the Speer Ministry for permission to begin work. Permission granted, the company set up the Versuchsanstalt (research station) Heerte near the LFA, Brunswick, from which Otto Lutz was to act as adviser. Work on both rockets and ramjets was begun, but according to Lutz and Busemann of the LFA, this work was of a low calibre and the prospect of anything useful being produced was bleak. Nevertheless, a few notes on the ramjet work at Heerte, which was headed by a Dr Rheinlander, are of interest, if only to complete our picture of the German ramjet effort.

A straightforward, subsonic ramjet, with divergent diffuser and convergent exhaust nozzle, was tested at low air speeds and pressures. Because of the low air speed, no baffles or flame holders were needed, and the petrol fuel was injected through various types and combinations of nozzles set into the wall of the diffuser.

The fuel was fed from a large settling tank, which tapered down smoothly to the diffuser entrance. A series of tests was made between 19 and 23 March 1945, average data from which are included in the following details of this small ramjet:

Thrust	0.70 kp (1.54 lb)
Thrust coefficient	0.30 to 0.33
Fuel consumption	0.74 gm/sec (0.026 oz/sec)
Combustion chamber diameter	115 mm (4 in)
Diffuser length	376 mm (1 ft 2¹ in)
Total ramjet length	1.476 m (4 ft 10 in)

In a more interesting experiment, it was planned to mount two ramjets on the ends of two rotating arms, the whole assembly being driven up to operating speed by an electric motor. Each ramjet had the usual divergent diffuser and convergent exhaust nozzle, but the constant-diameter combustion section was curved to match a radius of 2.716 m (8 ft 10^m in) from the point of rotation. The diameter of the combustion section was 522 mm (1 ft 8 in), and total length of the ramjet was 2.882 m (9 ft 5⁹⁄₁₆ in). Fuel was piped along the arms and fed into each ramjet by a single nozzle at the end of the diffuser. The incentive for this experiment undoubtedly came from Lutz, who had proposed a series of ramjets mounted on the circumference of a wheel, the spokes of which formed the blades of an airscrew. Its purpose was to provide take-off assistance to aircraft, the wheel fitting on the trailing edge of a wing and being spun up to operating velocity.

Conclusion

Although work on the ramjet in Germany began in the 1930s, initial progress was slow because of the small amount of resources and numbers of personnel assigned to development. In addition, first results were unpromising owing largely to the relatively low combustion temperatures and air speeds worked with. Not until late 1944 was an official priority raised for ramjet development and a belated attempt made to co-ordinate the various individual groups. For the most part, the ramjets investigated were of the subsonic type with conical, divergent diffusers and combustion temperatures around 1,500 °C, and this type received most attention from Walter and Sänger.

The Walter company was the only German *engine* company to make actual ramjet tests. Its work was slow but diverse, and sufficient confidence had been gained near to the end of the war to begin construction of experimental ramjets for an interceptor. Combustion development was extensive, and some effort was made to obtain efficient thrust at zero speed using hybrid rocket/ramjet units. However, these units turned out to have excessive fuel consumption or an unstable operating cycle. When, after the war, Helmut Walter worked in England, his attention was directed to rocket and turbine problems far more than to ramjets, since the latter had formed a minor section of his company's activities. Subsequently, Walter moved to the USA and eventually become Vice-President of the giant Worthington Corporation, while his son served with the US Army in Germany.

The results of Sänger's work are more difficult to evaluate, as investigators found at the end of the war. This was mainly because of the manner in which data had been collected, and net thrust was difficult to determine. In terms of thrust coefficient, Sänger's ramjets appear to have had the edge on Walter's, but this was probably because of the use of higher combustion temperatures. Diversity of design is not noted in Sänger's work, and the spectacular flight tests concentrated on a more or less set design to provide improvements in detail and a mass of data. By favouring actual flight tests (using slow aircraft) over wind-tunnel tests, rather low air speeds were used, with consequent inefficiency. Nevertheless, data based on this work went into most of the ramjet aircraft projects worked on in Germany. After the war, Sänger turned his mind to other aeronautical matters. In 1946, he married his former assistant Irene Bredt in Paris, and the two then worked for a time at the French Air Force Aeronautical Arsenal at Chatillon.

Still dealing with subsonic ramjets, Pabst of Focke-Wulf propounded quite new ideas, which resulted in a ramjet of remarkably short length in comparison with its diameter. The Pabst ramjet had a length of only two diameters, compared with six or seven diameters for the ramjets of other workers, although Walter hoped to achieve a length of only three or four diameters eventually. Because it was short, the Pabst ramjet was particularly valuable for rotating, as at the tips of rotating wings. In post-war years, such ramjets were used experimentally to power helicopter rotors, but because of high fuel consumption, the ramjet helicopter was abandoned in due course.

In fact, high fuel consumption attended all subsonic ramjet units, whether used for aircraft or missiles. Only the magnet of extreme simplicity encouraged post-war attention to the ramjet, but there were still the disadvantages of expensive take-offs and a limited speed range at which efficient operation could be achieved.

Part of the answer lay in using the ramjet, not for aircraft, but for missiles which could operate not only at high supersonic speeds but at a constant speed for which the ramjet could be specifically designed. In addition, the missile had to take off only once, so that the expense of auxiliary boosters was a minor consideration. Thus, the most valuable post-war contribution arising from Germany's ramjet work came from the supersonic missiles of Trommsdorf. These remarkable missiles owed their success largely to the multi-cone shock diffuser developed by Oswatitsch. Tests with Trommsdorf missiles continued in the Soviet Union over the firing range at Putilowa (near Moscow) from 1946 to 1952. Successful firings were performed with E5 missiles while C3 missiles with payloads covered large distances. The multi-cone diffuser also became well known in the West, since Oswatitsch moved to England and, from 1947, worked with models in the high-speed wind tunnel at Farnborough.

Trommsdorf-type gun-launched missiles became obsolescent as the world became increasingly more interested in missiles than in long-range artillery, but the Oswatitsch diffuser found many applications where a high-speed air intake is required. Therefore, while the ramjet found application in missiles, the Oswatitsch diffuser was used with both ramjet and turbojet missiles and turbojet aircraft where their speed approached and exceeded Mach 2.0. Of special value to supersonic aircraft was the fact that the central body could be moved longitudinally in the intake, so that the shock waves were adjusted, within limits, to suit varying speeds and altitudes. Whether fixed or variable, the diffuser was the subject of world-wide intensive research. Examples of its use appeared in missiles such as the Bristol Bloodhound (Great Britain), Bomarc IM-99 (USA), Hound Dog GAM-77 (USA) and Talos SAM-N-6 (USA), and in military aircraft such as the Mig 21 (USSR), EE Lightning (Great Britain), Dassault Mirage III (France), Lockheed Starfighter (USA) and Lockheed SR-71 (USA). The vital principle of intake pressure recovery using shock waves will always be used in fast aircraft.

As for Lippisch, his ramjet studies had little post-war value, but his work on high-speed delta aircraft forms had great value, and led in the USA to such aircraft as the experimental Convair XF-92 and the F-102 fighter. The ideas of Lippisch, Sänger and others on using coal as a fuel for the ramjet were more or less written off as curious German wartime 'whistling in the dark'. It was left to France's René Leduc to design the world's first true ramjet aircraft, the Leduc 0.10, which made its first flight on 21 April 1949 and led to other ambitious and remarkable research aircraft. Today, there are still projected schemes to use the ramjet for aircraft which might, in high-speed, high-altitude cruising flight, switch to auxiliary ramjet units. For the short-time boosting of military aircraft, the vogue is to use turbojet afterburning or reheat, which in essence adds a ramjet to the end of a turbojet engine.

Chronology of German Jet Engine and Gas Turbine Development

To give the reader an overall picture of the German history, c.1930–45, the following chronology lists some of the more important events to the nearest month, although more exact dating can often be found in the preceding text. Where the month is unknown (and in a few cases even the year is uncertain), the event is listed in an approximate or surmised relative position to other events.

1929
— H. Oestrich (at Bramo) makes jet propulsion studies but rejects for speeds then in view.

1930
Feb H. Holzwarth proposes his constant-volume gas turbine as a vehicle drive.

1931
Apr P. Schmidt granted patent for a pulsejet design.
— P. Schmidt obtains government support for his pulsejet work.

1932
Apr Patents filed for Brown Boveri (BBC) Velox steam generator and gas turbine schemes.

1933
— H. Holzwarth's fourth constant-volume gas turbine goes into steelworks service.

1934
Feb K. Leist files a patent for a turboprop scheme using a partially impinged turbine.
— H. Walter begins ramjet studies.
— H. Walter suggests a turbojet with afterburning to the RLM.
— Junkers Motorenbau make studies of free-piston gas generator to power a turbine driving an airscrew.

1935
— von Ohain's ideas on jet propulsion formed and patented.
— von Ohain's 'garage engine' turbojet built and running attempted.

1936
Apr von Ohain begins work at Heinkel on first demonstration turbojet (HeS 1).
— H. Wagner of Junkers Flugzeugwerk suggests turboprop study worthwhile, so special facilities set up.
— H. Weinrich submits a contra-rotating turboprop design to RLM.
— H. Walter begins ramjet experiments and receives research contract from RLM.

1937
Apr Bench-running begun with first demonstration turbojet (von Ohain's HeS 1).
— P. Schmidt rediscovers automatic ignition.
— W. Trommsdorf patents scheme for a ramjet missile.
— H. Weinrich submits plans for a marine, contra-rotating, gas turbine to OKM, which later supports building of an experimental unit.
— R. Friedrich designs a 50 per cent reaction axial compressor for a turbojet.

1938
Mar First Heinkel flight turbojet, HeS 3, bench-run by von Ohain.
— BMW begin work on a centrifugal turbojet (K. Loehner's P.3303).
— E. Schmidt begins studies leading to water-cooled turbine.
— P. Schmidt's first pulsejet with automatic ignition tested.
— E. Sänger expounds a theory of ramjets.
Aug Junkers Motorenbau begin jet propulsion study under A. Franz.
Nov W. Trommsdorf begins ramjet missile research for Heereswaffenamt.
— M.A. Mueller's axial turbojet with R. Friedrich's 50 per cent reaction compressor begins bench-running at Junkers.
— Bramo build a piston-driven ducted fan (ML).
— Bramo begin design of an axial turbojet.
— Official consultations with aero-engine companies to begin turbojet development.

1939
— Bramo test a piston-driven ducted fan with afterburning (MLS).
— Bramo taken over by BMW.
July First flight-tests with a turbojet (Heinkel HeS 3 beneath a test bed).
Aug First all-turbojet aircraft (He 178) makes its first flight.
— Junkers receive contract to develop an axial turbojet (109–004).
— BMW begin work on a contra-rotating axial turbojet (109–002).
— AEG and MAN begin studies for industrial gas turbines.
Nov First Argus pulsejet test model operates with intermittent combustion.
Nov First official demonstration of Heinkel He 178 turbojet aircraft.
— BMW begin design of axial turbojet (P.3302).
Dec Work begun on Junkers 109–004 A following failure of an earlier model.
— Daimler-Benz decide to begin turbojet work.
— MAN make investigations into ceramics for gas turbines.
— First, unsuccessful, firings of Trommsdorf ramjet missiles.

1940
— Work begun on piston-engined ducted fans (ML, MLS, MTL) at Heinkel.
— BMW make studies for a large turboprop.
— Brückner-Kanis begin work on rotating-boiler gas turbines.
— BBC begin construction of furnace gas turbine blower.
— LFA make investigations into ceramics for gas turbines.
Aug First run of BMW P.3302 turbojet
Oct First run of Junkers 109–004 A turbojet.

1941
Jan First Argus pulsejet road tests begun.
— O. Pabst at Focke-Wulf begins ramjet experiments.
Apr First jet fighter prototype (Heinkel He 280 V1) makes its maiden flight.
Apr Argus make first pulsejet flight tests.
— BBC set up a turbojet study department.
— First aircraft flown solely on pulsejet power (glider with Argus units).
Nov Abortive attempt to fly Me 262 prototype on two BMW P.3302 turbojets.
— Heinkel begin work on a diagonal-flow turbojet (HeS 11).
— E. Sänger begins ramjet tests.

1942

Jan	Only test flight of He 280 fighter prototype with Argus pulsejets.
—	BMW abandon contra-rotating turbojet (109–002).
—	MAN offer a marine gas turbine design to OKM, but rejected.
Mar	First flight-test of Sänger ramjet.
Mar	First flight-test of Junkers 109–004 A turbojet.
Mar	Me 262 fighter prototype takes off with two BMW P.3302 turbojets which fail in flight.
Mar	Blohm und Voss begin design of a marine gas turbine backed by OKM.
Jun	RLM order development of V1 flying bomb with pulsejet unit.
Jul	First all-jet flight of Me 262 fighter prototype (with two Junkers 109–004 A turbojets).
Jul	BMW begin work on an 8,000 eshp turboprop.
Oct	Heinkel's axial turbojet, HeS 30, running with success.
Oct	Doblhoff's jet helicopter research programme begun.
Oct	F. Neugebauer lectures on high-pressure combustion chambers for turbines and jet units.
Nov	LFA begin using interferometer for turbine blade study.
Dec	First runs of Junkers 109–004 B turbojet.
Dec	Argus 109–014 pulsejet developed.
Dec	First V1 flying bomb with Argus pulsejet launched.
—	O. Lutz begins development of high-pressure gas generator for turbines.
—	H. Walter begins work on pulsejet for turbine drive.
—	K. Oswatitsch develops the multi-cone supersonic shock diffuser (for ramjets).

1943

Jan	Pre-production Junkers 109–004 B turbojets ready.
May	Manufacturers meet to discuss mass production of hollow turbine blades.
May	First hot tests with Daimler-Benz 109–007 ducted-fan turbojet.
—	First jet helicopter (Doblhoff WNF 342 V1) flies.
Jun	Prototype of first jet bomber, Ar 234 V1, flies (using Junkers 109–004 A turbojets).
—	A. Müller tries to interest the military in a gas turbine for tanks. Unsuccessful.
—	AEG and BBC companies begin design of turboprops using the Ritz regenerative heat exchanger.
—	Pulsejet/ramjet combination (ILS) unit ordered from P. Schmidt.
—	First pulsejet-powered fighter (Me 328) tested.
—	BMW studies for turbojet/rocket (TLR) combination unit 109–003 R begin.
Aug	First pure-jet flights with BMW 109–003 turbojet begin (using He 280 V4).
Sep	Ar 234 V13 reaches a record 13,000 m (42,640 ft) altitude with BMW 109–003s.
Sep	Contract issued for first turbojet/rocket unit (BMW 109–003 R).

1944

—	OKM orders Brückner-Kanis to develop a large marine gas turbine with four gas producers.
—	von Ohain begins work on Tuttlingen engine with single compressor/turbine element.
—	Junkers begin work on a larger turbojet (109–012) and a turboprop (109–022).
—	Construction of a larger turbojet (109–018) begun by BMW.
—	AVA begins extensive studies concerning turbojet installation.
—	LFA begin work on experimental gas turbine with ceramic blades.
—	Büssing company build and test experimental Lutz gas generator (for turbines).
—	First pre-production jet fighters and bombers (Me 262 A–0 and Ar 234 B–0) go for official tests at E-Stelle Rechlin.
Mar	MAN begin design of a gas turbine installation to drive a 12,000 kW generator.
May	Daimler-Benz 109–007 turbojet development stopped, and company turns to a turboprop project (109–021).
Jun	First V1 flying bombs launched at London, 2,000 launched that month.
Jun	Heereswaffenamt make plans for tank gas turbine development to start.
—	Technisches Amt orders investigations into possibilities for ceramics in the gas turbine.
Jul	First operational *air-launching* of V1 flying bomb.
—	Successful firings of Trommsdorf supersonic ramjet missiles.
Jul	First experimental sorties with jet aircraft (Me 262 and Ar 234).
Jul	A. Müller begins work on first gas turbine design for tanks.
Aug	Ford in USA begin production of Argus pulsejet copies, and Republic Aviation build copies of V1 flying bomb.
Aug	Ju 287 bomber prototype (with swept-forward wings) makes maiden flight. Uses four Junkers 109–004 B turbojets.
Aug	Last air-test of Sänger ramjet.
Sep	BMW report on ramjet as an artillery weapon.
Sep	Special reconnaissance squadron formed with Ar 234 jets.
Sep	First German turbojets (Junkers 109–004) captured.
Sep	Most BMW 109–003 turbojets earmarked for He 162 (Volksjäger) fighter.
Sep	Panther V tank chosen for gas turbine installation.
Sep	Ar 234 flies with four BMW 109–003 turbojets.
Oct	First fully-operational jet fighter unit begins operations (using Me 262).
Oct	First early production BMW 109–003 turbojets appear.
Oct	Official order for development of expendable turbojet for missiles.
Nov	Official priority raised for aircraft ramjet development.
Dec	Prototype of He 162 (Volksjäger) fighter flies (using a BMW 109–003 turbojet.)
Dec	First jet bomber sorties (with Ar 234s).
—	Tank gas turbine with heat exchanger put forward.

1945

Mar	Last operational V1 flying bomb launched.
Mar	First flight of turbojet/rocket unit (two BMW 109–003 Rs with Me 262).
Apr	First jet night-fighter sorties.
Apr	Last jet reconnaissance sortie (by Ar 234).
Apr	Avia company make preliminary tests on foam coal as a ramjet fuel.
—	Installation of BBC blast furnace gas turbine blowers begun.
May	Surrender of Me 262 jet fighter unit.
May	Unit formed with He 162 jet fighters but surrendered same month.

In the immediate post-1945 years, the following technical events occurred:

1947

The first gas-turbine-powered vessel puts to sea (Metropolitan-Vickers).

1952

Atlantic crossing by a ship propelled solely by gas turbine power (British-Thomson-Houston).

1959

Royal Navy frigate commissioned using combined steam and gas turbine machinery (COSAG) by AEI.
Gas turbines provide electricity and compressed air by consuming sewerage gas (Ruston & Hornsby).
US Navy test gas-turbine-powered hydrofoil (Lycoming).

And so on . . .

Sources and Bibliography

Abbreviations for Allied reports are as follows. Dates given, especially when multiple dates, are often the dates of visits to targets. Minor mistakes in report titles have been left in to assist searching.

AI(2)g: Allied Intelligence, Technical
BIOS: British Intelligence Objectives Sub-Committee
CIOS: Combined Intelligence Objectives Sub-Committee
FIAT: Field Information Agency, Technical (United States Group Control Council for Germany)
TIR: Technical Intelligence Report (US Army, etc.)

Section 1

CIOS XXXII-1, 8 Items 5, 26. Interrogation of Dipl.-Ing. Helmut Schelp (August 1945)
Kay, Antony L. 'Some origins of German jet power'. *Air Extra* No. 1
Meyer, Adolf. 'The combustion gas turbine: its history, development and prospects'. *Institute of Mechanical Engines Proceedings*, **141**, 1939

Section 2

A12(g) Report 2339. Heinkel 109–011 turbojet propulsion unit (24 April 1945)
A12(g) Report 2372. BMW jet propulsion units (14 May 1946)
A12(g) Report 2373. Data tables for German jet and rocket units (5 September 1945)
Bentele, Max, *Engine Revolutions* (Autobiography). Society of Automotive Engineers, Warrendale, Pa. 1991.
BIOS Evaluation Report 12. Interrogation of Dr-Ing. Oestrich (20 September 1945)
BIOS Overall Report 12. German gas turbine developments 1939–45 (1949)
BIOS Final Report 35. Item 26. Report on visit to Daimler-Benz AG at Stuttgart-Unterturkheim) (August/September 1945)
BIOS Final Report 255. Gas turbine and reciprocating engine activities (1945)
BIOS (–?). Report on German aero-engine industry (15 November 1945)
BIOS/FIAT Final Report 1148. The use of heat-resisting steels in the manufacture of gas turbine blades in Germany (3 June 1947)
BIOS/FIAT Final Report 1152. Design practices and construction of centrifugal compressors by leading German manufacturers (30 May 1947)
BMW Report (via Peter G. Kappus). Das TL-Gerät mit R-Schubhilfe als Antrieb fur Jäger (16 January 1943)
CIOS IV-1, 8 Items 25, 26. Investigation of gas turbine and jet propulsion work in Paris (1945)
CIOS VI-28 & 29 and VIII-13 Item 5. William Prym, Stolberg and Zweifall
CIOS XI-6, XII-9 and XIV-4 Items 5, 26. Description of Junkers 004 jet propulsion engines. Three reports (5 Dec 1944)
CIOS XIX-2 Item 25 The Horten tailless aircraft (1945)
CIOS XXIII-6 Item 25. The Horten tailless aircraft (May 1945)
CIOS XXIII-14 Item 5. Turbine engine activity at Ernst Heinkel Aktiengesellschaft
CIOS XXIV-6 Items 5, 26. Gas turbine development. BMW, Junkers, Daimler-Benz (1945)
CIOS XXV-23 Items 19, 5. Junkers Flugzeug-und Motorenwerke AG Ausbildung, Dessau (1945)
CIOS XXVI-29 Item 5. Research and development on gas turbines at Junkers Motorenwerke, Dessau. Plus: Appendix on AVA, Göttingen (30 and 31 May 1945)
CIOS XXVI-30 Item 5. Gas turbine development by BMW (1945)
CIOS XXX-59 Item 5. Manufacturing process for fabrication of turbine blades used in 109–003 BMW jet engine (undated)
CIOS XXX-80 Items 5, 26. BMW. A production survey (May 1945)
CIOS XXXI-36 Items 5, 25, 26. Junkers aircraft and engine facilities, Dessau, etc. (1945)
CIOS XXXI-66 Item 5. Notes on aircraft gas turbine engine developments at Junkers, Dessau, etc. (15 and 17 June 1945)
CIOS Evaluation Report 43. Interrogation of Bruckmann and Hagen of BMW jet engines (31 May 1945)
CIOS Evaluation Report 69. Junkers Werke, Dessau (8 June 1945)
CIOS Evaluation Report 149. Junkers aircraft targets at Dessau, Aschersleben, Bernberg, etc. (27 June 1945)
CIOS Evaluation Report 323. Interrogation of Gen. Dir. K. Frydag and Prof. E. Heinkel (14 August 1945)
Daimler Benz letters from Max Kuhl (re 021 PTL) (July–September 1944)
Daimler-Benz letter with details of 109–021 and 109–007 engines (3 April 1968)
Daimler-Benz report on the 109–021 and 109–007 engines (February 1945)
Muller, M.A., DAL Report. Translation GDC. 10/5006T. Axial TL-jet units and engine-jet units (31 January 1941)
FIAT Final Report 441. Investigation of the BMW 003 turbine and compressor blading (31 October 1945)
Gerler, Capt. Warren C. 'The German Jumo 004 Engine'. Society of Automotive Engines, New York, 7/11 January 1946
'German Jet Developments'. *Flight* magazine, 13 December 1945
Junkers 109–004B Triebwerk-Handbuch. Oberkommando der Luftwaffe Chef der Technischen Luftrustung (Berlin, 26 January 1945)
Kay, Antony L. 'Facets of BMW liquid-propellant rocket development 1939–45'. *Spaceflight* **9** No. 12, December 1967
Lloyd, P.L., Wright-Patterson Air Force Base report: Note on gas turbine development by Daimler-Benz (File 45213, 9 June 1945)
Oestrich, Dr H., Report (via SNECMA): Die Entwicklung der Flug-Gasturbinen bei den Bayerische Motorenwerken wahrend des Krieges (4 April 1950)
Power Jets Report No. R1089. The Jumo 004 jet engine (April 1945)
Report: Le turboreacteur ATAR 101B (SNECMA)
Report (from MAN Turbo): 'der TL-Triebwerke 003A, C, D, E bei der Bayerische Motorenwerke A.G.' by Dipl.-Ing. Hagen, Sawert, Muller, Ziegler and Kappus. (May/June 1945)
Schlaifer, R. and Heron, D. *Development of Aircraft Engines and Fuels*. Harvard University 1950
TIR No. A-411. German BMW jet propulsion unit (3 June 1945)
TIR No. I-67. Interrogation of Dipl.-Ing. Max Mueller (5 July 1945)
Whittle, Heinkel, etc. 'The jet controversy'. *Inter Avia* **VII** No. 9 (1952)

Section 3

BIOS Final Report 98. Items 18, 26. German development of gas turbines for armoured fighting vehicles (1945)
Bright, Col. R.H. 'Development of gas turbine powerplants for traction purposes in Germany'. *IME War Emergency Proceedings*, **157** 375–85 (16 November 1945)
Feist, Uwe. *The German Panzers from Mark I to Mark V 'Panther'*. Aero Publishers Inc. 1966

Section 4

BIOS Final Report 931. Item 29. Vorkauf rotating boiler (Drehkessel) and rotating boiler gas turbine (Drehkessel turbine) (April/May 1946)

BIOS Final Report 98. Items 18, 26 (see Section 3)
CIOS XXV-15. Items 26, 29. A) MAN Augsburg and Harburg, B) The Franciskaner Keller, Munich (4 May 1945)
FIAT Final Report 291. Gas turbine project for a schnell boat developed by Blohm und Voss, Hamburg (3 October 1945)

Section 5

BIOS Evaluation Report 257. Brown Boveri & Cie, AG, Mannheim (1946)

Section 6

BIOS Final Report 16 Item 26. Interference method of studying air flow in turbine and compressor blade cascades as used at LFA, Volkenrode (1945)
BIOS Final Report 231. Prototype manufacture of German jet engines (undated)
BIOS Final Report 298 Items 5, 26. Regenerative-type heat exchanger. AVA, Göttingen, 14 September and 9 October 1945
BIOS Final Report 470 Items 21, 22. Specialised ceramic materials with particular reference to ceramic gas turbine blades (August 1945)
Board of Trade. FD. 1300/47. Friedrich Krupp AG Research Institute (January 1947)
Burkhardt, Dr Arthur. Correspondence/lecture notes on WMF blades, etc. (October 1969)
CIOS XXV-2 Item 25. Luftfahrtforschungsanstalt (LFA) Herman Goering, Volkenrode, Brunswick (1945)
CIOS (–?) Item 26 (Target 26/228). Research on design of axial flow compressors at AVA, Göttingen (1945)
CIOS XXV-9 Item 5. Development of ceramic materials for use in turbine engines (1945)
CIOS XXV-10 Item 5. Research and development on gas turbines – Herman Goering Institute, LFA Volkenrode (1945)
CIOS XXVII-20 Item 5. Prof. Dr-Ing. Emil Sorensen. MAN AG (June 1945)
CIOS Final Report XXVII-22. Brown Boveri & Cie (June 1945)
CIOS XXVIII-47 Item 25. High-speed tunnels and other research in Germany (17–29 June 1945)
CIOS XXXI-2 Items 1, 7. Research work undertaken by the German universities and technical high schools (8 June 1945)
CIOS XXXI-51 Items 2, 4, 25, 26. LFA, Volkenrode (1945)
CIOS XXXII-31 Item 19. Axial flow compressor development at the Stuttgart Research Institute (11 August 1945)
CIOS XXXII-45 Item 25. Gas turbine developments (July 1945)
Kuchemann, D. (ed.) *The Installation of Jet Propulsion Units*. AVA Monograph K3 (Reports and Translations No. 1937) (1 October 1947)
Ritz, L. 'Summary of the theory and construction of a rational gas turbine'. AVA, Göttingen, July 1945
TIR No. I-50. Interrogations at AVA, Göttingen (7 June 1945)
TIR No. N-29. LFA at Braunsweig (1945?)

Section 7

CIOS XXXI-5 Item 5. Doblhoff Jet-Propelled Helicopter (August 1945)
Lambermont, P. and Pirie, A. *Helicopters and Autogyros of the World*
Smith, G. Geoffrey. *Gas Turbine and Jet Propulsion*. Iliffe 1955

Section 8

CIOS XXVIII-53 Item 5. Walter Werke, Kiel (6 June 1945)
Consolidated Vultee Aircraft Corp. (Convair) Report GT-125: Interrogation of German scientists at Wright Field, Dayton, Ohio (5 June 1946)
Deutsche Forschungsanstalt fur Segelflug (DFS) Report NR. 58 (2458) (17 December 1942)
Edelman, L.B. The pulsating jet engine – its evolution and future prospects. *SAE Quarterly Transactions* 1 204–16 (1947)
Kay, Antony L. *Buzz Bomb*. Monogram 1977
TIR No. A-446. Target inspection of DFS (29 May 1945)
TIR No. E-21. Visit to DFS at Prien (1 July 1945)

Section 9

BIOS/FIAT Final Report No. 508. Interrogation regarding use of coal for firing gas turbines (19 November 1945)
CIOS XXVII-53 Item 5 (see Section 8)
CIOS XXVII-67 Items 4, 6. Aerodynamics of rockets and ramjets research and development at LFA Hermann Goering, Volkenrode.
CIOS XXVII-86 Item 4 & 6. Research activities at Göttingen on aerodynamics of projectiles, missiles and ramjets
CIOS XXX-12 Item 30. Schaumkohle and Dr Heinrich Schmitt-Werke KG (15 August 1945)
CIOS XXXI-13 Items 4, 6. Ramjet and rocket work. Heerte (undated)
CIOS XXX-81 (Final Report). Survey of German ramjets (July 1945)
Convair Report GT-125 (see Section 8)
Intelligence T-2 Report W.61912–85. Discussion by Dr W. Noeggerath and Dr R. Edse of possibilities for Athodyd propulsion of a missile in connection with basic suggestions presented in BMW document No. BES (?) 55 (2 November 1945)
Letter from Dr Wolf Trommsdorf dated 4 March 1968
TIR No. I-82. Interrogation of Dr Alexander Lippisch re Athodyd form of propulsion (2 July 1945)
TIR No. I-83 (V-17937). Interrogation of Dr Oswatitsch (re Supersonic Diffusers) (1 May 1946)

General

Barker, Nonweiler and Smelt. *Jets and Rockets*. Chapman & Hall 1959
BIOS Final Report 195 Item 4. Interrogation of personnel of Elektro-Mekanische Werke (EMW) (17 October 1945)
Francillon, René J. *Japanese Aircraft of the Pacific War*. Putnam 1979
Galland, Adolf. *The First and the Last*. Methuen 1955
Gartmann, Heinz. *The Men Behind the Space Rockets*
Gordon, Yefim and Gunston, Bill. *Soviet X-Planes*. Midland Publishing 2000
Gunston, Bill. *World Encyclopaedia of Aero Engines*. Airlife 1986
Jones, R.V. *Most Secret War*. Hamish Hamilton 1978
Killen, John. *The Luftwaffe*. Frederick Muller 1967
Scaife, W. Garrett. *From Galaxies to Turbines*. Institute of Physics Publishing 2000
Simmons, C.R. *Gas Turbine Manual*
Smith, G. Geoffrey. Gas Turbine and Jet Propulsion. *Flight*, 1942, 1944, etc.
Smith, J.R. and Kay, Antony L. *German Aircraft of the Second World War*. Putnam 1972
SOCEMA Aircraft Turbines. *Flight*, 18 November 1948
Späte, Wolfgang. *Test Pilots*. Independent Books 1995

'Let my son often read and reflect on history: this is the only true philosophy.'

Napoleon Bonaparte

Appendix

Preserved examples of engines in the world

This list is chiefly concerned with some of the aeronautical engines dealt with in this book because those are the main items which have been preserved. Tanks and so forth can also be seen in museums, of course. Before travelling to a museum, readers should always check that a specific item is on display because it may be in storage, may be undergoing work or may have been moved to another location.

This list is not definitive and newly acquired items sometimes appear.

Engine	Keeper	Location
HeS 3b replica	NASM, Washington, DC	USA
H-H 109–011	Imperial War Museum	London
H-H 109–011	NASM, Washington	USA
H-H 109–011	Crawford Auto-Aviation Museum, Cleveland	USA
Junkers 109–004	Moorabbin Air Museum, Melbourne	Australia
Junkers 109–004	Treolar Centre, Canberra	Australia
Junkers 109–004	RAF Museum, Hendon, London	England
Junkers 109–004	Science Museum, London	England
Junkers 109–004	Imperial War Museum, Duxford	England
Junkers 109–004	Vigna di Valle, Rome	Italy
Junkers 109–004 (V34?)	Luftfahrtmuseum, Hannover	Germany
Junkers 109–004	Wehrtechnische Studiensammlung, Koblenz	Germany
Junkers 109–004	Iowa State University, Ames	USA
Junkers 109–004	San Diego Aerospace Museum, CA	USA
Junkers 109–004	USAF Museum, Dayton, Ohio	USA
Junkers 109–004	Massachusetts Institute of Technology	USA
RD–10A (Soviet version)	Muzeum Lotnictwa Polskiego, Krakow	Poland
BMW 109–003	The Camden Museum of Aviation, Narallan	Australia
BMW 109–003	Narodni Technicke Muzeij, Prague	Czech Republic
BMW 109–003	Science Museum, London	England
BMW 109–003	Luftwaffe Museum, Gatow	Germany
BMW 109–003	NASM, Washington	USA
BMW 109–003 E-1	Planes of Fame Museum, Chino	USA
Ne-20 (Japanese version of 109–003)	NASM, Washington	USA
Lutz swing piston	Imperial War Museum, London	England
Lutz swing piston	Cranfield Inst. Of Technology, Bedford	England
Lutz swing piston	NASM, Washington	USA
Schmidt-Rohr 500 mm	Deutsches Museum, Munich	Germany

For the Argus 109–014 pulsejet, seek out the Fi 103 flying bombs:

Missile or Aircraft	Keeper	Country
Fi-103A-1 (443313)	Australian War Memorial, Treloar Centre	Australia
Fi-103A-1 (partial replica)	Musée Royal de l'Armée, Brussels	Belgium
Fi 103A-1	Wilrijk, Antwerpen	Belgium
Fi 103A-1/Rre 4	Wilrijk, Antwerpen	Belgium
Fi 103A-1	The National Aeronautical Collection	Canada
Fi 103A-1	Atlantic Aviation Museum, Nova Scotia	Canada
Fi 103A-1/Re 4	Canadian War Museum, Ottawa, Ontario	Canada
Fi 103A-1	Tojhusmuseet, Copenhagen	Denmark
Fi 103A-1 (477663)	Imperial War Museum, London	England
Fi 103A-1	RAF Museum, Cosford	England
Fi 103A-1 (442795)	Science Museum, London	England
Fi 103-1 (418947)	Ministry of Technology Rocket Propulsion	England
Fi 103A-1	Fort Clarence, Horsham	England
Fi 103A-1	RAF Museum, Cardington	England
Fi 103F-1 (477663)	Imperial War Museum, Duxford	England
Fi 103A-1	RAF Museum, Hendon	England
Fi 103A-1	Musée de l'Air, Paris	France
Fi 103	Musée de l'Air, Paris	France
Fi 103A	Musée du Août 19, Pourveille	France
Fi 103A-1/Re4	La Coupole, St Omer, Wizernes	France
Fi 103A	La Coupole, St Omer, Wizernes	France
Fi 103A-1	Militaer Historisches Museum, Dresden	Germany
Fi 103F-1 (478374)	Deutsches Museum, Munich	Germany
Fi 103A-1	Museum in Köln, Butzweilerhof	Germany
Fi 103A-1	Museum fur Verkher und Technik, Berlin	Germany
Fi 103	Wehrtechnische Studien-Sammlung, Koblenz	Germany
Fi 103 (replica?)	Hist.-Tech. Info., Peenemünde	Germany
Fi 103	Luftwaffen Museum, Gatow, Berlin	Germany
Fi 103A-1	NNOV, Overloon	Netherlands

Missile or Aircraft	Keeper	Country
Fi 103A-1	Aviodom, Schiphol, Amsterdam	Netherlands
Fi 103A-1	Leger en Wapen Museum, Delft	Netherlands
Fi 103A-1/Re4	Leger en Wapen Museum, Delft	Netherlands
Fi 103	Museum of Trans., Tech, and Social Hist.	New Zealand
Fi 103A-1	The Domain War Memorial, Auckland	New Zealand
Fi 103A-1	Forsvarsmuseet, Oslo	Norway
Fi 103A-1	Flygvapenmuseum, Malmslätt, Linköping	Sweden
Fi 103	Kansas Cosmophere and Space Center	USA
Fi 103	Kermit Weeks Fantasy of Flight	USA
Fi 103A-1	Greencastle, Indiana	USA
Fi 103A-1	Planes of Fame, Chino, California	USA
Fi 103A-1 (477937)	US Army Ordnance School, Aberdeen	USA
Fi 103A-1	NASM, Washington	USA
Ford JB-2	Alabama Space and Rocket Center	USA
Ford JB-2	Cradle of Aviation Museum, Garden City	USA
Ford JB-2	Holloman AFB, New Mexico	USA
Ford JB-2	Keesler AFB, Mississippi	USA
Ford JB-2	Lackland AFB, Texas	USA
Ford JB-2	NASM, Washington	USA
Ford JB-2	New England Air Museum, Windsor Locks	USA
Ford JB-2	USAFM, Dayton Ohio	USA
Ford JB-2	San Diego Aerospace Museum	USA
Ford JB-2	Travel Town Museum, Griffith Park	USA
Ford JB-2	Cape Canaveral Florida Space Museum	USA
Ford JB-2	Roswell Museum, New Mexico	USA
LTV-N-2	Aerospace Park, North Carolina	USA
LTV-N-2	Hickory Airport, North Carolina	USA
LTV-N-2	Missile Park, White Sands, New Mexico	USA

Also, the following turbojet aircraft are preserved:

Aircraft	Keeper	Country
Me-262A-2a (112372)	RAF Museum, Cosford	England
Me-262A-1a/U3(111617/9)	Planes of Fame, Chino	USA
Me-262A-1b (500071)	Deutches Museum, Munich	Germany
Me-262A-1b (500491)	NASM, Washington	USA
Me-262A-1a	USAFM, Dayton, Ohio	USA
Me-262B-1a/U1(110305)	South African Nat. Museum of Military History	South Africa
Me-262B-1a (110639)	NASM, Willow Grove	USA
Avia S.92A-1a (4)	Vojenske Museum, Kbely AB	Czech Rep
Avia CS.92B-1a (51104)	Vojenske Museum, Kbely AB	Czech Rep
Ar 234B-2 (3673)	NASM, Washington	USA
He-162A-2 (120076)	Canadian National Aviation Museum	Canada
He-162A-2 (120086)	Canadian National Aviation Museum	Canada
He-162A-2 (120223)	Musée de l'Air, Paris	France
He-162A-2 (120227)	RAF Museum, Hendon	England
He-162A-2 (120230)	NASM, Washington	USA
He-162A-1 (120235)	Imperial War Museum, London	England
Lippisch DM-1	NASM, Washington	USA
Doblhoff WNF 342 V4	NASM, Washington	USA

Index

Aerodynamische Versuchsanstalt (AVA), 16, 96, 100, 171, 186, 209, 220, 230, 270, 281, 283
Allgemeine Elektricitäts-Gesellschaft (AEG), 207
Amman, Rolf, 98
Antz, Hans M. 17
Anxionnaz, René, 11
Arado Ar 234 jet bomber, 82
Argus Motoren Gesellschaft, 244
Armengaud, René, 10, 196

Baeumker, Adolf, 239
Bates, Dr, 75
Baur, Karl, 129
Bayerische Motoren Werke AG (BMW), 96
BBC blast furnace schemes, 196, *et seq.*
Bentele, Max, 42, 44, 49, 55, 60, 63, 104
Bergar, Dipl.-Ing., 142
Betz, Albert, 16, 102, 210
Biefang, Dr, 98
Blade production, 75, 76
Blade, turbine, 'Topschaufel', 50
Blast furnace schemes, BBC, 196, *et seq.*
Blenk, H. 246
Blohm und Voss Schiffswerft, 186
Bock, Guenther, 11
Bosch, 214
Bredt, Irene, 277
Brisken, Walter, 17
Broeker, Christian, 10
Brown Boveri & Cie, (BBC), 125, 166, 170, 172, 192, 193, 194, 196, 220
Bruckmann, Bruno, 98, 141
Brückner-Kanis GmbH, 171, 176
Büchi, Alfred J. 196
Burkhardt, Arthur, 105
Busemann, A. 270, 276, 280
Büssing-NAG, 226, 254

Carl Zeiss, 218
Compound engine, HeS 60, 37
Compressor, Hermso I and II, 125

Daimler, Gottlieb, 142
Daimler-Benz AG, 142
Deutsche Forschungsanstalt für Segelflug 'Ernst Udet' (DFS), 254, 277, 281
Deutsche Versuchsanstalt fur Luftfahrt (DVL), 143, 186, 209, 217, 218, 220
Deutschen Akademie der Luftfahrtforschung (DAL), 209, 224
Diedrich, Guenther, 244, 253
Diffuser, Fang, 282
Diffuser, supersonic, 270
Doblhoff, Friedrich von, 231
Dornier Do 17 bomber, 277
Do 217 bomber, 245, 277
Dreher, Georg, 10

Ebert, Hans J. 137
Eckert, B. 210

Eisenlohr, Wolfram, 16, 32, 209
Encke, W. 16, 25, 49, 89, 209
Enders, Karl, 10
Ernst Heinkel AG, 18
Escher-Wyss, 208

Farrow, Lilian, 10
Fattler, Dr, 107
Fieseler, Gerhard, 246, 248
Flying bomb, piloted,
V1 (Fi 103R-IV), 252
V1 (Fi 103), 246
V1, American copies, 259
Focke-Wulf Flugzeugbau, 282
Fw 200, 248
Ford Motor Company, 259
Forschungsinstitut für Kraftfahrwesen und Fahrzeugmotoren, Stuttgart (FKFS), 35, 210, 253, 258
Forschungsinstitut Graf Zeppelin, 258
Franz, Anselm, 57, 173
Friedrich, Rudolph, 126, 171, 211
Fuel injector, Duplex, 50, 150, 214
Fuel, foam coal, 280
J2, 108

Galland, Adolf, 58, 121
Gas generator, Lutz swing-piston type, 224
Gas producer, Lurgi, 204
Gas turbine, armoured fighting vehicle, 156
compound, MAN L760, 204
GT 101, 157
GT 102 Ausf. 2, 171
GT 102, 163
GT 103, 170
K229, 176
K236, 179
K290, 178
MAN 5,000 kW generator, 206
marine, Auftrag 353, 186
Brückner-Kanis 10,000 shp unit, 179
MAN 7,500 shp unit, 190
Vorkhauf's Drehkessel, 176
Generalluftzeugmeister (GL), 11
Georgii, Walter, 254, 277
Goering, Hermann, 11
Gosslau, Fritz, 244, 253
Gotha Go 145 biplane, 245
Go 229 jet fighter, 85
Go 242 glider, 246
Grothey, Karl, 211
Guillaume, 11
Gundermann, Wilhelm, 19
Gutehoffnungshütte (GHH), 208

H. Walter KG, 259, 264
Hagen, Dipl.-Ing., 130
Hahn, Max, 18
Hart, Oscar, 10
Heatt exchanger, 220
Heereswaffenamt (HWA), 12, 156, 269, 272
Heinkel He 111 bomber, 245
He 162 Salamander jet fighter, 117 *et seq.*

He 178 first turbojet aircraft, 21, 219
He 280 jet fighter, 25, 246
Helicopter, jet, WNF 342, 231
Hermann, Walter, 210
Hetterick, Joseph, 10
Hille, Fritz, 98
Hoehere Luftwaffenschule, 11
Hoffman, Ludwig, 87
Hofmann, F. 193
Holzapfel, Heinrich, 179
Holzwarth, Hans, 10, 193, 238
Hoss, Dr, 196
Houdry process, 196
Hyrniszak, W. 170, 220

Instituto Nacional de Industria (INI) of Spain, 55

J2 fuel, 108
Jendrasssik, György, 10
Jung, Dr 213
Junkers Flugzeug- und Motorenwerke AG, 57
Ju 287 jet bomber, 84
early compound engine (MLS), 57
Ju 88 bomber, 245
Junkers, Hugo, 57
Junkers, Klaus, 57

Kaiser-Wilhelm-Institut (KWI), 271
Kamm, Wunibald I, 35, 156, 172, 210, 258
Kammler, Hans, 17, 251
Kappus, Peter G. 98, 120, 127, 141, 285
Karavodine, Victor, 10, 238
Keirn, D.J. 259
Knemeyer, Siegfried, 17, 52
Köhl, Hermann, 75
Kraft, Ernst, 58, 207
Kraftfahr Technische Versuchsanstalt der SS, 156, 174
Kriegsmarine Technischesamt (KTA), 12
Küchemann, D. 210, 220, 230, 253, 283
Kuehne, Willy, 207
Kunze, N, 120
Kunzel (flight engineer), 21

L'Orange, 148, 150, 214
Laderriere, Marcel de, 10
Lasley, Robert E. 10
Laufer, Theodor, 231
Leduc, René, 263
Leich, Karl F. 10
Leist, Karl, 142
Lemale, C, 10, 196
Lembcke, Hans, 240
Lichter, Richard, 58, 59
Lippisch, Alexander, 281
Ljunstrom, Frederick, 10
Loehner, Kurt, 97
Ludwieg, H. 270
Luftfahrtforschungsanstalt Hermann Goering (LFA), 210, 215, 218, 226, 246, 254, 266, 276, 281, 283
Luftkriegsakademie (LKA), 213

Luftwaffe, 11
Lurgi gas producer, 204
Lutz, Otto, 224
Lysholm, Alfred, 10, 18

Madelung, G. 239
Mader, Otto, 58
Marconnet, 238
Maschinenfabrik Augsburg-Nürnberg AG (MAN), 190, 204, 215
Mauch, Han. A. 16, 21, 58, 142
Messerschmitt Bf 109 fighter, 245
  P.1101 jet fighter, 52
  Bf 110 fighter, 245, 246
  Me 163 rocket interceptor, 266
  Me 262 jet fighter, 60, et seq.
  Me 264 bomber, 221
  Me 328 jet, 255
Milch, Erhard, 17, 40
Mueller, Max A. 25, 30, 32, 33, 37, 40, 58, 153, 155, 172
Müller, Alfred, 97, 138, 153, 156, 168, 172

Nagel, Dr, 100
Nakajima J8N1 Kikka jet fighter (Japanese), 85
Nallinger, Fritz, 142
Neugebauer, Franz, 224
Neumann, Gerhard, 141

Oberkommando der Kriegsmarine (OKM), 175
Oberkommando der Wehrmacht (OKW), 13
Oberth, Hermann, 10
Oestrich, H. 98
Ohain, Hans von, 10, 18, 40, 52, 55, 56
Oswatitsch, Klaus, 271

Pabst, Otto, 282
Pape, Wilhelm, 10
Pescara, Raul P. 57
Peters, Heinrich, 176
Pohl, Robert W. 18
Porsche KG, 153
Porsche, Ferdinand, 153
Prandtl, Ludwig, 16, 210, 271
Prym, William (blade production), 75, 76
Puffer, S.R. 207
Pulsejet, Argus 109–014, 249
  Schmidt, 241
  SR500, 241

Quick, Prof. 15

Ramjet missile,
  general, 263
  BMW, 285
  intercontinental, 273
  Pabst, 282
  Trommsdorf, 269
  Walter, 265
Ramjet-rocket, Walter, 268

Rateau, Auguste and Lucie, 10
Reichsluftfahrtministerium (RLM), 11
Republic Aviation, 259
Reuter, Hermann, 212
Riedel starter motor, 51, 71
Ritz, Ludolf, 220
Rocket-ramjet, Walter, 267
  single-fuel type, Walter, 268
Roeder, Karl, 10, 181
Rohrbach, August, 10

Saechsicher Metallwarenfabrik Wellner (blade production), 76
Sänger, Eugen, 276
Schaaf, Wilhelm, 98
Schäfer, Fritz, 25
Schafer, G, 193
Schattschneider, Max, 196
Schelp, Helmut, 15, 54, 56, 133, 142, 209, 212, 226, 276
Schepler, Hermann, 186
Scheutte, Alfred, 190, 204
Schmidt, Dr (at H. Walter KG), 264
Schmidt, Ernst, 206, 210, 214
Schmidt, F.A.F. 209 217
Schmidt, Paul, 11, 239, 262
Schmitt, Heinrich, 280
Schulz-Grunow, F. 253
Skoda-Kauba Flugzeugbau, 280, 282
Skorzeny, Otto, 252
Soestmeyer, Dr, 98
Sorensen, Emil, 204, 215
Speiser, Viktor, 170
Starter motor, Riedel, 51, 71
Steam generator, Velox, 196
Stein, Dr, 89
Stepan, August, 231
Stoekicht, Dipl.-Ing., 137
Stroehlen, Richard, 207

Tank, Jagdpanzer VI Jagdtiger, 157
  Panzerkampfwagen VI Tiger, 157
Tank, Kurt, 284
Technischen Akademie der Luftwaffe (TAL), 210
Technisches Amt, 11, 244
Tecklenburg, 10
Test chamber, high altitude, BMW, 98, 120
Thyssen, 193, 194
Trommsdorf, Wolf, 269
Turbine, T.1, 216
  T.2, 214
  T.3, 214
Turbine blades, ceramic, 215
Turbojet, ATAR 101 (French), 140
  BMW 109–003, 103 et seq.
  BMW 109–018, 131
  BMW P.3302, 97, 99, 102
  BMW P.3303, 97
  BMW P.3304 (109–002), 97
  BMW P.3306, 137

BMW P.3307, 139
Daimler-Benz P.100 (109–016), 150
ducted fan, Daimler-Benz 109–007, 144
  HeS 50, 37
Heinkel-Hirth HeS 011 (109–011) 40
  HeS 1, 18
  HeS 3, 21
  HeS 6, 21
  HeS 8 (109–001), 25
  HeS 9, 26
  HeS 10 (109–010), 26
  HeS 40, 33
INI-11, Spanish version of H-H 109–011, 55
Japanese, 130
Junkers 109–004, 59, et seq.
Junkers 109–012, 89
Lutz swing-piston, compound, 228
Milo (Swedish), 11, 18
Porsche 109–005, 153
RD-10 (Soviet), 93
RD-20 (Soviet), 140
TGAR-1008, (French), 93
Turbojet-rocket, BMW 109–003R, 128
Turboprop, AEG, 221
  BBC, 221
  BMW 109–028,
  Daimler-Benz 109–021, 55, 150
  Heinkel-Hirth 109–021, 55, 150
  Junkers 109–022, 91
Tuttlingen engine, 52

Udet, Ernst, 11, 16, 32

Velox steam generator, 196
Verkehrsministerium (predecessor of RLM), 11
Versuchsanstalt Heerte, 286
Volland, Dr, 246

Wachtel, Max, 249
Waffen SS, 172, 174, 252
Wagner, Herbert, 57
Waldmann, Emil, 17, 152
Walter, Hellmuth, 264
Warsitz, Erich, 21
Weinig, F. 172, 210
Weinrich, Helmut, 97, 174, 179, 181, 183, 191, 212
Wertheim, Joseph, 10
Whittle, Frank, 11, 56
Wiener Neustadter Flugzeugwerke (WNF), 83, 191, 212
Wiener Neustadter Flugzeugwerke (WNF), 231
WMF of Geislingen, 112, 113
WNF 342 jet helicopter, 231
Wolff, Harold, 40, 52

Zadnik, Otto, 156
Zahnradfabrik, 162
Zborovski, Helmut Philip von, 127
Zöbel, Dr, 246